Recent Advances in Ready-to-Eat Food Technology

Ready-to-Eat (RTE) describes foods that need not be cooked, reheated, or otherwise prepared before consuming them. ***Recent Advances in Ready-to-Eat Food Technology*** covers all the aspects of RTE from statistics, method of production, mechanization, thermal and non-thermal processing, gluten-free, consumer behavior, control of foodborne illness and hygiene, packaging requirements, and improved functionalization to application of nanotechnology.

Key Features:

- Covers the development of ready-to-eat products from meat, cereal, fruits, vegetables, dairy, and pulses
- Provides a global review of labeling and packaging for ready-to-eat products
- Discusses hygienic design and safety in the production and consumption, with an emphasis on pathogenicity issues

Written by a team of well-recognized researchers who present the latest advances in RTE food product development, this book is of interest to industry professionals and academicians as well as to undergraduate students and postgraduate researchers.

Recent Advances in Ready-to-Eat Food Technology

Edited by
Hari Niwas Mishra, Pradyuman Kumar, and Ajay Singh

CRC Press
Taylor & Francis Group
Boca Raton London New York

CRC Press is an imprint of the
Taylor & Francis Group, an **informa** business

Cover image: Mai.Chayakorn / Shutterstock.com AND Arcady / Shutterstock.com

First edition published 2025
by CRC Press
2385 NW Executive Center Drive, Suite 320, Boca Raton FL 33431

and by CRC Press
4 Park Square, Milton Park, Abingdon, Oxon, OX14 4RN

CRC Press is an imprint of Taylor & Francis Group, LLC

ISBN: 978-1-032-60778-8 (hbk)
ISBN: 978-1-032-62238-5 (pbk)
ISBN: 978-1-032-62240-8 (ebk)

DOI: 10.1201/9781032622408

Typeset in Times
by Newgen Publishing UK

Contents

Preface

Food products which are precooked, packaged, and ready for consumption without prior preparation or cooking comes under the category of ready-to-eat (RTE) foods. This category of food must be in an edible form without required additional preparation phase to accomplish food safety. Such food products are very popular among children, youth, age old, and working family due to their convenience and short or nil preparation time. This book deals with various aspects of RTE food processing such as mechanization, thermal processing, extrusion cooking, and application of non-thermal processing methods, viz. ultrasound and high-pressure processing. Consumer behavior to RTE has very important aspects linked to functionality which further increased awareness among consumer toward plant and animal-based RTE foods. Gluten-free RTE is gaining popularity among consumers with celiac disease. Therapeutic and nutritional information of RTE also creates awareness among consumers to select a particular RTE.

The RTE foods must be cooked during its production which allows to reach least temperature with minimum holding time that enables microbial destruction of public health concern. During its preparation ensuring hygiene and controlling cross-contamination with potential foodborne pathogens, through application of appropriate food safety practices, is crucial. Cooking step is generally attained by thermal processing which ensures a minimum lethality (F_0) for the target microorganism (usually a foodborne pathogen). A properly processed and packaged RTE food should be free from the target foodborne pathogen and is ready for consumption. The use of nanomaterials for enhanced nutrient delivery, improved sensory attributes, and extended shelf life through innovative packaging solutions are encouraged. Careful consideration of regulatory requirements and consumer acceptance is crucial to ensure the successful integration of nanotechnology in the RTE food industry. RTE formulation is considered as the nutritious and healthy requirement by different age groups, which are varying. Use of composite raw materials, supplementation, and fortification with certain nutrients can meet the individual requirements.

The book ***Recent Advances in Ready-to-Eat Food Technology*** provides a critical reference for scientists and professionals working and an insight into formulation, processing, packaging, and quality and safety attributes of RTE foods.

Dr. Hari Niwas Mishra
Dr. Pradyuman Kumar
Dr. Ajay Singh

Editors' Bio

Hari Niwas Mishra, having over 36 years of professional experience in teaching, research, and administration, has many laurels and awards to his credit. He has published 620 research papers including 258 in peer-reviewed international journals and 362 in conference proceedings. He has written 4 books, 4 e-books, 35 book chapters, 7 lecture compendium and laboratory manuals, 4 technology manuals, and several popular articles in newspapers and magazines and has 14 Indian patents to his credit. Besides, he is on the editorial boards of several reputed journals. He has supervised more than 290 student research projects including 12 PDF and 50 Ph.D. research scholars including 2 PMRF and 7 GYTI awardees. Professor Mishra has handled several international and national sponsored research and industrial consultancy projects. He has created state-of-the art research laboratory and pilot plants for food processing and preservation. Dr. Mishra has received several awards and honors such as AIFPA Presidents' Award, GYTI Award, NRDC Award, Best Teacher Award, Dr. J.S. Pruthi Award, etc. for his outstanding contribution in the field of food processing in the country. Professor Mishra has worked in different capacities on various academic and administrative committees of IIT Kharagpur and many other institutions in the country. He has visited several countries abroad including Taiwan, Singapore, Thailand, Malaysia, Australia, Turkey, UAE, France, Switzerland, Denmark, Germany, Sweden, Norway, Ireland, UK, Belgium, USA, and Canada.

Pradyuman Kumar has over 25 years of experience in teaching and research and has to his credit more than 190 papers in International and National Referred Journals and conferences/workshops. He has guided 9 Ph.D. and 27 M. Tech academicians and guiding 4 Ph.D. His area of research and teaching interests includes food processing, formulation, product development, characterization, and optimization. He received Best Paper award in various international/national conferences, Young Scientists Award for 2005 and Dr. J.S. Pruthi Award for 2019 from AFST, Mysore. He has published 16 chapters in various books published by CRC, Elsevier, Springer, 3 edited books by CRC press, and 1 book by Brillion Publishing Delhi. Serving as Guest Editor for two journals, namely Foods, an MDPI publisher, and Frontiers in Bioinformatics, a Frontiers publisher. Prof. Kumar has worked as member of various academic bodies for syllabus development for UG and PG courses in Food Technology.

Ajay Singh has about 10+ years of teaching and research experience. He has published around 40 papers and 30 book chapters in publishing houses of international repute, edited 6 books (4 CRC Press, 1 NIPA, and 1 Objective Food Science with Brillion Publishers). Dr. Singh presented his discourses, and demonstrated workshops at various national and international venues. He is the potential reviewer for many of international journals (Springer, Elsevier, Wiley, and Taylor & Francis) and serving as editorial board member for various SCI and SCOPUS listed Journals (*Journal of Food Processing and Preservation* (Wiley), and *Journal of Food Biochemistry* (Wiley)). Dr. Singh worked as guest issue editor for *Frontiers in Bioinformatics* (PUBMED & SCOPUS Indexed) and *Journal of Food Processing and Preservation* (Wiley). He worked as mentor for an international funded project from RONDO, Switzerland, on "Snacks-Technology-Process and Automation". He has one design patent to his credit. His area of research and teaching is composite blended gluten-free baked optimization and functional approaches among new product development.

Contributors

Shaikh Adil
Department of Dairy Technology
P.I.T., Parul University
Vadodara, Gujrat, India

Bindvi Arora
Division of Food Science & PHT
ICAR-IARI
Pusa, New Delhi, India

Prajya Arya
Department of Food Processing Technology and Food Technology
Sardar Vallabhbhai Patel University of Agriculture and Technology
Modipuram
Meerut, Uttar Pradesh, India

Subrata Kumar Bag
Department of Dairy Engineering
West Bengal University of Animal & Fishery Sciences
Mohanpur Campus
Nadia, West Bengal, India

Bindu Bazaria
Department of Food and Nutrition and Food Technology
Lady Irwin College
New Delhi, India

Ashok Bhosale
Department of Veterinary Microbiology
COVAS, MAFSU
Udgir, Maharashtra, India

Ankit Bihola
Dairy Technology Division
ICAR-National Dairy Research Institute
Karnal, Haryana, India

Mahima Bora
Agricultural and Food Engineering Department
Indian Institute of Technology Kharagpur
West Bengal, India

Chandrakant Genu Dalbhagat
Department of Food Process Engineering
National Institute of Technology
Rourkela, Odisha, India
&
Agricultural and Food Engineering Department
Indian Institute of Technology Kharagpur
West Bengal, India

Joyati Das
Agricultural and Food Engineering Department
Indian Institute of Technology Kharagpur
West Bengal, India

Chinju Davis
Quality Assurance and Management Division
ICAR-CIFT
Kochi, Kerala, India

Lakshmi E. Jayachandran
Faculty of Fisheries Engineering, Department of Food Technology
Kerala University of Fisheries and Ocean Studies
Panangad, Kochi, Kerala, India

Alka Joshi
Division of Food Science & PHT
ICAR-IARI
Pusa, New Delhi, India

Payal Karmakar
Food Science and Nutrition Department
University of Minnesota
Minneapolis, Minnesota, USA

Gurpreet Kaur
Department of Zoology
Akal University
Talwandi Sabo, Punjab, India

Simrat Kaur
Department of Chemistry
Mata Gujri College
Fatehgarh Sahib, Punjab, India

Vijay Kele
Department of Livestock Products Technology
COVAS, MAFSU
Udgir, Maharashtra, India

Murod Kurbanov
Department of Food Technology
Bukhara Engineering Technological Institute
Bukhara, Uzbekistan

Sivashankari M.
ICAR-National Rice Research Institute
Cuttack, Odisha, India

Shubham Mandliya
Agricultural and Food Engineering Department
Indian Institute of Technology Kharagpur
West Bengal, India

Neeraj
Agriculture Department
Jharkhand Rai University
Namkum, Ranchi, India

A. Nithya
Agricultural and Food Engineering Department
Indian Institute of Technology Kharagpur
Kharagpur, India

Ranjana Pande
Life Sciences Department, School of Basic Sciences and Research
Sharda University
Greater Noida, Uttar Pradesh, India

Indira Dey Paul
Department of Meat and Marine Sciences
CSIR-Central Food Technological Research Institute (CFTRI)
Mysuru, Karnataka, India

Rekha Rani
Department of Dairy Chemistry
College of Dairy and Food Technology, AU
Jodhpur, Rajasthan, India

Swati Ray
Department of Food Technology
Haldia Institute of Technology
Haldia, West Bengal, India

Mohammad Raziuddin
Department of Livestock Products Technology
COVAS, MAFSU
Udgir, Maharashtra, India

Sadvatha R. H,
ICAR-Central Institute of Agricultural Engineering
Regional Station
Coimbatore, Tamil Nadu, India

Pardeep Kaur Sandhu
Department of Zoology
Mata Gujri College
Fatehgarh Sahib, Punjab, India

Dipali Saxena
Shri Vaishnav Institute of Home Science
Shri Vaishnav Vidyapeeth Vishwavidyalaya
Indore, Madhya Pradesh, India

Shruti Sethi
Division of Food Science & PHT
ICAR-IARI
Pusa, New Delhi, India

Sourav Sharma
Department of Management Studies
Mata Gujri College
Fatehgarh Sahib, Punjab, India

Nitin Shinde
College of Dairy Technology
Pusad, MAFSU
Nagpur, Maharashtra, India

Rakesh Mohan Shukla
Processing and Food Engineering Division
College of Agricultural Engineering and Technology
SKUAST-Kashmir
Shalimar Campus, Srinagar, India

Amritpal Singh
Department of Food Technology
Mata Gujri College
Fatehgarh Sahib, Punjab, India

Anamika Thakur
Division of Food Science & PHT
ICAR-IARI
Pusa, New Delhi, India

Ira Tripathi
Department of Home Science
Kashi Naresh Government Post Graduate
College Gyanpur
Bhadohi, Uttar Pradesh, India

Siddharth Vishwakarma
Agricultural and Food Engineering
Department
Indian Institute of Technology Kharagpur
West Bengal, India

Latika Yadav
Department of Home Science
Vijay Singh Pathik Government PG Degree
College
Kairana, Shamli, Uttar Pradesh, India

1 Ready-to-Eat Foods and Mechanization Novelties

Mechanization Upgradation in RTE Equipment

Chandrakant Genu Dalbhagat[1*] *and Rakesh Mohan Shukla*[2]
[1]Department of Food Process Engineering, National Institute of Technology Rourkela, Rourkela, Odisha, India
[2]Processing and Food Engineering Division, College of Agricultural Engineering and Technology (CoAE&T), Sher-e-Kashmir University of Agricultural Sciences and Technology of Kashmir (SKUAST-K), Srinagar, Jammu & Kashmir, India
*Corresponding Author

1.1 INTRODUCTION

With rapid urbanization and busy lifestyles, there is a growing reliance on fast and convenience foods which are readily offered by quick-service restaurants and fast-food chains. In urban areas, to save time in a hectic busy life, ready-to-eat (RTE) foods emerged as a culinary time-saving product and have become popular, offering meals that save time for individuals with hectic schedules (Temgire et al., 2021). Traditional home-cooked meals gradually give way to increased online food orders or going out to dine-in restaurants. The availability of different tempting cuisine options, as well as pre-cooked meals, instant foods such as noodles, pasta, rice, and packaged snacks like chips, cookies, and biscuits, etc., shifted the trend to use RTE products as an alternative to the home cooked foods (Hawa et al., 2014). Apart from this, with the daily busy schedules and increased workload and responsibilities, people often like to do on-the-go snacking, increasing the demand for easily consumable RTE snacks like energy bars, granola bars, and other quick bites. The food habits of Indian people are evolving due to urbanization, globalization, technological advancements, and changing lifestyles (Patel & Rathod, 2017). Although there is an inclination of consumers toward fast and convenient foods, there is a large segment of consumers who seek healthier options and demand health-boosting foods.

RTE foods are those foods that are processed and packaged and are ready to eat immediately, without requiring further cooking or preparation (Temgire et al., 2021). The RTE foods frequently require little or less extra work, making them more convenient for people with hectic schedules and busy lifestyles. These RTE foods or meals are designed in such a way as to fit into the schedules of individuals who do not have the time or desire to prepare large, complex meals (Hawa et al., 2014). They can easily be consumed at home, in the office, or while traveling, advocating convenience as its special feature. The typical processing operations in manufacturing RTE foods involve cleaning, grading, sorting, handling, mixing, cooking, freezing, shaping, etc., while the packaging is done to prolong their shelf life and preserve freshness. Packaging methods include form fill and sealing (FFS), vacuum-sealing, nitrogen flushing, or other methods to prevent spoilage and maintain food

DOI: 10.1201/9781032622408-1

quality (Patel & Rathod, 2017). RTE foods are available in a wide range of variety products, from full meals to single products, including fresh-cut fruits and vegetables, instant foods, dairy and bakery products, etc.

1.1.1 Factors Driving the Growth of the RTE Food Sector

The RTE food market has seen significant growth in the last few years and is expected to grow in the coming years. As the research and development on the different raw ingredients, processing technologies, and packaging innovations are going on, the development of RTE food products with varying nutritional contents is possible. Due to their convenience, they are globally accepted as an alternative food preference to cooked meals. Many food industries are trying to develop RTE foods with balanced nutritional content with low amounts of preservatives and additives and natural ingredients to meet consumer preferences (Contini et al., 2016; Patel & Rathod, 2017). The growth of The RTE food market is driven by many factors that cater to consumers changing lifestyles, preferences, and demands. The key factors contributing to the growth of the RTE food sector are discussed below.

1.1.1.1 Convenience and Rising Urbanization

Convenience is the primary driving force for increasing the demand of the RTE food market. Urbanization is associated with rapid growth, increased workload and hours, a faster pace of life, and a higher number of sources of dual income. Due to busy lifestyles and hectic time schedules, there is a need for quick and easy healthy meal options for the working class. Moreover, due to the increase in urbanization, demographic changes are also to be seen, including the rising number of working professionals, single-person households, and smaller families. Small families may find it more convenient to buy processed and packaged RTE foods.

1.1.1.2 Increased Disposable Income and Busy Lifestyle

The rise in disposable income in many countries has given the liberty to an individual to spend more on convenience foods, irrespective of the higher price than the raw ingredients to save time. The modern lifestyle comprising long working hours, daily travelling, commitments, and multitasking work has paved the way for quick and easy-to-prepare food options. RTE foods fit well into this lifestyle by eliminating the need for extensive cooking.

1.1.1.3 Globalization, Culinary Diversity, Health, and Wellness Trends

Consumers are now exposed to various cuisines and flavours globally and are keen to have novel health-boosting foods that have organic ingredients, low chemical preservatives, and high nutrient content. This has increased the acceptance and demands of various RTE foods of different compositions as well as culinary traditions (Bursey et al., 2021). Moreover, the availability of a wide range of healthy RTE foods through e-commerce websites made it more accessible to the consumer.

1.1.1.4 Technological Advancements and Packaging Innovations

Technology advancement and innovation in packaging have played a pivotal role in the extension of the shelf life of RTE foods. Novel and emerging technologies are being used to preserve food products for longer shelf life by maintaining nutritional integrity and microbiological safety.

1.2 RTE FOODS AND EQUIPMENT USED IN PRODUCTION

Proper selection of the equipment is crucial in manufacturing the RTE food products of the specific category because their quality is largely dependent on the raw materials, processing techniques, and machinery used in preparation.

1.2.1 RTE Food Products

As consumers seek a wide variety of RTE foods, there are a lot of options available in the market. The categorization of RTE foods may be based on the use of raw ingredients, dietary requirements, government regulations, sensory perception, different health benefits, and personal requirements. RTE food products are fruits, vegetables, cereals and pulses, dairy, and meat-based.

1.2.1.1 Fruit-Based RTE Products

Fruit-based RTE products have emerged as healthy alternatives for those who seek quick, nutritious, and raw products with hassle-free preparation to enjoy the natural goodness of fruit (Santeramo et al., 2018). These products can be made in different ways, viz. fruit cups, fruit bars, dried fruits, etc.

1.2.1.1.1 Fruit Cups

Fruit cups are pre-packaged sealed cups containing fresh sliced/diced fruits to maintain the freshness of fruits. These cups are convenient to carry and visually appealing to enjoy the fruits which do not require washing, peeling, and chopping. They can be often taken for breakfast or lunches, in travel or picnic, and even in offices. They include the sole or combinations of papaya cubes, pineapple chunks, apple slices, orange slices, grapes, and berries, providing a diverse and flavourful combination. Canned fruit cups are also available in the market; however, it is filled with sugar syrup, juice, or water.

1.2.1.1.2 Dried Fruit Snacks

Dried or dehydrated fruit snacks like banana chips, orange slices, fig slices, mango slices, or fruit cubes or chunks are another category of RTE fruit-based snacks. Drying removes the moisture content from the fruit slices, thereby extending the shelf life of the fruits. Due to lower moisture content and water activity, the microorganisms cannot grow, and spoilage of the product gets reduced, making it a self-stable food product.

1.2.1.1.3 Fruit Pouches

The fruit juice or puree can be packaged in suitably designed packaging pouches after suitable thermal treatment like sterilization and pasteurization to confirm the microbiological safety. The fruit juices may be with or without added sugar and preservatives. These pouches are more popular among children, providing a mess-free option for parents to choose them as healthy on-the-go snacks for their kids. Fruit pouches are also convenient for athletes or anyone needing a quick energy boost.

1.2.1.1.4 Fruit Bars

Different dried fruits as per formulation can be combined with nuts, cereal grains, coarse grain granules or powder, sugar, or jaggery to make a compacted bar having chewy or dense texture. They are available in various flavours, shapes, and sizes, and taste preferences and packaged efficiently for easy carrying in bags or pockets.

1.2.1.2 Vegetable-Based RTE Products

Similar to fruits, vegetable-based RTE products are also available providing a nutritious and convenient way to add vegetables into the daily diet, especially in salads. These products provide hassle-free solutions for a busy lifestyle, to experience the goodness of vegetables. Some vegetable-based RTE products are explained below.

1.2.1.2.1 Veggie Snack Packs

Veggie snack packs include washed pre-cut vegetables in the appropriate portions in the packs. These packs often include carrots, cherry tomatoes, bell peppers, lettuce, onion springs, etc. They can be consumed with the sauce or chutney.

1.2.1.2.2 Vegetable Chips

Similar to dried fruit chips, the dried/dehydrated vegetable chips are made to provide a crunchy and flavourful alternative to traditional potato chips. The vegetable chips include sweet potato chips, carrot chips, beet chips, kale chips, etc. These snacks retain the nutritional benefits of vegetables while offering a satisfying texture and a burst of natural flavours.

1.2.1.2.3 Vegetable Dips

Pre-cut vegetables and dips are another choice of RTE food that can be enjoyed as a snack. These are the balanced combination of cut vegetables as a source of vitamins, minerals, and fibres and tasty and flavourful dips, perfect for social gatherings, family dining, or solo snacking.

1.2.1.2.4 Pickled Vegetables

Pickling is the traditional process of Indian cuisine to preserve vegetables. The vegetables are pickled in jars or pre-packaged containers to provide a tangy and crisp option. It preserves the vegetables while improving their flavour profile. Commonly pickled vegetables include cucumbers, carrots, and cauliflower. Pickled vegetables can be enjoyed as a snack or a flavourful addition to salads and sandwiches.

1.2.1.3 Cereal and Pulse-Based RTE Products

Cereal and pulse-based RTE products consist of different roasted cereals or pulse grains or powders in the preparation of quick and nutritious options. It has starch and protein combinations to decide the formulation of RTE food products. These products offer a wide range of flavours and textures to suit various preferences.

1.2.1.3.1 Breakfast Cereals

It is made from processed cereal grains and has become a ubiquitous and convenient choice in breakfast worldwide. It is often served with milk or yogurt and offers a quick and satisfying start to the day. Cornflakes, oatmeal, and corn grit are widely used cereals, catering to diverse tastes and dietary preferences.

1.2.1.3.2 Granola Bars

Granola is the combined product of rolled oats, nuts, puffed rice, and sweeteners such as sugar, jaggery, or honey. Granola bars are delicious, chewy, convenient, and energy-boosting snacks which are good sources of fibre and essential nutrients. The bars can be made in different shapes and sizes with various flavours and provide a good source of fibre and essential nutrients. They come in various flavours, catering to both sweet and savoury preferences.

1.2.1.3.3 Instant Oatmeal Packets

Instant oatmeal is prepared from oatmeal which is cooked, dried, rolled, and pressed. It is also a popular nutritious breakfast choice, providing fibre and energy to the individual. These packets often include a variety of toppings such as nuts, dried fruits, or spices.

1.2.1.3.4 Pulse Snacks

Pulse snacks include the sole or combination of different roasted or seasoned pulses such as Bengal gram, lentils such as red lentils, green gram, pigeon pea, etc. These are protein and fibre-rich and low-fat crunchy snacks, offering an alternative to traditional snacks such as chips and fried snacks. They come in various flavours, from savoury to spicy, providing a customizable and guilt-free snack.

1.2.1.3.5 Cereal Bars

Like granola bars, cereal bars are made mainly from cereal grains which include rice, wheat, or corn as base material. The toppings may contain fruits, nuts, and chocolate as flavour ingredients.

1.2.1.3.6 Extruded Snacks

Extrusion technology is used to produce puffed or expanded extruded snacks with crispy textures, diverse shapes and sizes, and desired flavours. The extruder machine is the heart of the extrusion processing which transforms the starch or protein-based materials into expanded or textured products. The raw ingredients are subjected to heat, shear, and pressure as they traverse through the extruder barrel and die. At the extruder die, when the material exits, it is shaped, puffed, and cut which gives a crunchy texture to the product. The snack further is generally coated with a spice mix to enhance the flavour and savoury experience. The innovative shapes and flavour combinations make these products more acceptable and popular (Dalbhagat et al., 2019; Suri et al., 2020).

1.2.1.4 Dairy-Based RTE Products

As a source of calcium and protein, dairy products are another category of RTE foods providing essential nutrients without much need for preparation.

1.2.1.4.1 Yogurt Cups

Yogurt is a fermented dairy-based product with its characteristics texture and flavour. Yogurt cups may include other ingredients such as nuts, fruits, or granola to enhance the texture and flavour of the product. They are also convenient for on-the-go consumption, a popular choice for the busy individual.

1.2.1.4.2 Pudding Cups

Pudding cups are the dessert option, offering a sweet and delicious option served after the main meal. It may come in different flavours such as chocolate, vanilla, and butterscotch.

1.2.2 Equipment Used for Ingredient Handling

For manufacturing RTE products, the ingredient handling equipment is very important to ensure efficiency, hygiene, and product quality. The handling equipment connects the processing operations to the packaging of RTE products which involves the conveyors, elevators, and robotic systems to streamline the movement of raw and finished products in the production facility and storage tanks for storing the raw, intermediate, and finished products.

1.2.2.1 Conveyors

Conveyors are the material handling devices, most widely used to transport the raw ingredients, processed and finished products within or from the processing area to packaging storage or distribution facilities. These conveyors are made up of food-grade materials, viz. food-grade stainless steel (SS304/SS314), Teflon, or plastic. The design of conveyors is such that it prevents product damage and maintains its integrity during transportation. Conveyors are not only used for conveying materials but also to accumulate products during operations, like singulating conveyors make single rows of products, sorting conveyors separate materials according to their shape and size provided with automation control, elevating conveyors lift the materials and stagger belt conveyor splits the product line into two streams (Fellow, 2022). The different types of conveyors used in RTE food manufacturing include belt, screw, bucket, pneumatic, drag and chain conveyors, etc.

1.2.2.1.1 Belt Conveyor

Belt conveyor generally transports material in horizontal and inclined (restricted) directions, from one location to another. The construction elements include belts, pulleys, idlers, drive, structure, belt trippers, belt cleaners, belt training devices, take-ups, etc. The belt provides a supporting surface to the material to be conveyed; the pulley supports the belt and controls belt tension; idlers support the belt for troughing run and flat run-in return, and the drive provides power to the pully for the movement of the belt, the structure provides support, the belt tripper dislodges materials, and take-ups maintain the tension. The design of belt conveyors depends upon the horizontal or inclined distance between which the material has to be transported, the type and nature of the material, the capacity required, the temperature of the material, the bulk density of the material, the number of loading and discharge points, and choice of power source (Rashid et al., 2023; Sahay & Singh, 2001).

1.2.2.1.2 Screw Conveyor

These conveyors consist of a tubular pipe or U-shaped trough in which the horizontal rotating screw conveys the material. The screw is supported at both ends with the bearings. As the screw rotates the screw spiral pushes the material forward in the trough or pipe. The capacity of the screw conveyors varies with the diameter, pitch, and length of the screw and the speed of operation. The material is loaded at one end and discharged at the other end by the screw. The screw conveyors are generally used to transport the material horizontally; however, as per requirement they may be used in a certain degree of inclination to vertically up to 90° from horizontal. The screw generally consists of a shaft and a helical blade or flight. The flight is a continuous spiral mounted on the horizontal shaft. The loading and unloading of the material can be done at several places in the screw conveyors. The discharge of material can be done by opening the bottom of the trough at specified locations (Rashid et al., 2023; Sahay & Singh, 2001).

1.2.2.1.3 Bucket Elevator

It is a vertical conveying system used for bulk materials such as grains, powders, and granular materials. It consists of small buckets attached to the belt or chain which revolve around the driven (top) and return (bottom) pulleys. The buckets scoop the material at the bottom and discharge it at the top of the conveyor. The different components of a bucket elevator consist of a flat, endless belt or chain, buckets, drive pulley, take-up pulley, head section, boot section, and casing. Centrifugal discharge, positive discharge, and continuous discharge elevators are the different types of bucket elevators used in the food industry. The chains and buckets are housed in single or double vertical legs. The buckets may be made up of stainless steel or plastic while the legs are made up of stainless steel. The top curved hood is designed for the discharge of material while the lower boot is for loading the material. The advantages of this conveyor include limited horizontal space requirement, enclosed dust-free, and fairly quiet operation (Rashid et al., 2023; Sahay & Singh, 2001).

1.2.2.1.4 Pneumatic Conveyor

Pneumatic conveyors transport granular and powdered materials in a closed pipe or duct using compressed air under pressure or suck the material under vacuum. It is a continuous and flexible transporting method from the inlet to the discharge end of a pipeline (Rashid et al., 2023). The material moves in the enclosed pipeline due to the air pressure, minimizing dust generation and airborne particles. The material can be transported horizontally, vertically, or at inclination as per requirements and navigating around equipment, obstacles, and structural constraints. Multiple inlets and outlets can be integrated at different stages of production. The positive air supply system consists of a fan or blower and an air lock rotary feeder for feeding the material into the pipe or duct. The capacity of this conveyor depends upon the size and length of the pipe of the duct, volumetric flow rate of air, feed rate, and material properties (Sahay & Singh, 2001).

1.2.2.2 Storage

Storage vessels are used to store the raw material, intermediate processed and finished products at various stages of production process. For continuous production of RTE products, the stock levels are maintained in the storage tanks. These vessels ensure the proper product quality, safety, and efficiency in production of RTE foods. Batch processing is mostly used in many RTE food manufacturing where the raw ingredients are cleaned, washed, cooked, and processed in large batches. Large tanks are used to hold the ingredients in these operations, enables efficient production and minimizes the downtime between batches. Some RTE foods require the preservation of freshness, flavour or development of texture based on temperature control. For this purpose, the storage vessels are equipped with the jacket around it in which the hot water or steam or chilled water circulates to maintain the uniform temperature in the vessel. Moreover, the storage vessels protect the materials from the external agent that affect its quality including deteriorating, i.e., contaminants, moisture, light, and air exposure. It may also provide a checkpoint for quality evaluation and glass window to inspect the product stored in the vessel for visual defects or other quality issues. Any product that fails to meet the standard quality requirements can be discarded from the storage vessel. Storage tanks can be categorized as fixed and transportable. The fixed tank may be horizontal or vertical, while most transportable tanks are horizontal. Tanks are mainly used to store liquid foods such as fruit juices, dairy products, fruits and vegetable concentrates, etc. Food grade stainless steel is a widely used construction material for the food industry storage tanks. The food industry tanks have different components such as manhole, viewport, sampling valve, air vent, liquid level indicator, agitator, ladder and top platform (Baeghbali et al., 2023).

1.2.3 Processing Technology and Machinery

Processing technology and machinery play vital role in the transforming the raw ingredients into the finished products ensuring quality, safety, and consistency. The most common machinery used in the manufacturing RTE food products include sorting and grading machinery, mixing and blending equipment, thermal processing equipment, cooling and freezing systems, forming, shaping, and packaging equipment.

1.2.3.1 Sorting and Grading Machinery

Sorting and grading are terms that are frequently used interchangeably but denote two different operations in food industry. These operations are used for mostly the classification, inspection, and categorization of raw or processed products according to the predefined criteria. The criteria are predefined as per the requirement like quality and safety aspect or physical parameters. Sorting process is generally referred to the separation of materials based on individual properties such as size, shape, weight, density, colour or any defect in the material (Sahay & Singh, 2001). It is the primary operation done prior of production process to remove any undesired material which can potentially alter the finished product quality. It is done by manually or using automated machinery like air classifiers, vibratory conveyors, or optical sorters, etc. Grading is another classification-based operation generally carried after sorting. In this method, the product is graded as per predetermined quality criteria to fetch the better price. It helps in getting the consistent and uniform product quality. The different graders available in market include screen grader, roller grader, weight-based grader, and image-based graders (Kheto et al., 2023).

Screen graders involve the perforated screens or sieves through which the material passes. The material is graded according to the sieve size. The desired product may be larger or smaller than the sieve size opening, and according to the requirement, the shape and size of the screen may be considered (Saravacos & Kostaropoulos, 2016). The screening devices may be classified as oscillating sieve graders, inclined screen graders, and rotary screen graders. Belt graders are widely used to grade round-shaped fruits and vegetables. It includes V-type belt graders, laterally tilted belt graders, wire-type belt graders, and expanding pitch graders. V-type belt graders consist of several belts and pulleys with a progressive increase in the gap between the belts along their length. The laterally tilted belt graders have a lateral inclination of 10–20° (Kheto et al., 2023). A weight grader works on the principle of weight measurement. It has a conveyor belt with movable cups and variable weight sensors. The cup discharges the fruit and vegetables based on weight at a certain discharge point. Roller graders have a pair of rollers inclined downwards with progressively increasing gaps between them. The produce (fruit or vegetable) slide along the roller and fall through the gap between the roller in the collection trough. The grading efficiency and percent damage to produce are affected by the roller speed, feed rate, and angle of inclination. The digital image-based sorting and grading technique is also efficient method for grading the fruits and vegetables based on colour and size. It enables detection of damaged, bruised produce and determine the maturity level of fruits and vegetable (Davies, 2013; Kheto et al., 2023).

1.2.3.2 Mixing and Blending Equipment

For making RTE foods having different ingredients as a part of composition, the mixing and blending equipment are very important to get uniform blending of ingredients. The formulation of nutritional contents is largely dependent on the type and design, operation of blender or mixer, and handling of the ingredients. Moreover, the mixing process defines the physical, rheological, textural, and organoleptic properties of the developed RTE food product.

1.2.3.2.1 Batch Mixer

For low-capacity mixing of ingredients, batch mixers are mostly suitable. The common batch mixers include paddle blender, ribbon blender, planetary mixer, turbine mixers, anchor mixer, etc. It consists of the mixing vessel and blades mounted on the shaft. The raw ingredients are loaded into the mixing vessel where the blades mix the material at appropriate operational speed. The process parameters such as mixing speed, paddle angle, mixing time are decided as per properties of ingredients being handled. Paddle and anchor mixers are operated relatively at low speed. Anchor mixers are often jacketed to heat the ingredients while processing (Berk, 2018). High-speed mixers such as turbine mixers blend the ingredients rapidly exerting considerable shear and useful in mass transfer or phase dispersion applications. High-speed mixers are commonly used in the production of RTE sauces, dressings, and emulsions. Propeller mixers are used to blend the low-viscosity liquids, while helical mixers for the high-viscosity fluids (Berk, 2018).

1.2.3.2.2 Continuous Mixers

Continuous mixers involve the long screw conveyor and mixing chamber assembly in which the ingredients are fed at the inlet, mixed while traveling the chamber path and removed at the outlet. These mixers are useful in the large-scale continuous production of batters, slurries, or dough, etc., required to produce RTE foods.

1.2.3.3 Thermal Processing Equipment

Thermal processing equipment are used to heat or cook the raw ingredients to achieve the desirable cooked texture, flavour, and appearance to the product as well as to attain the microbiological safety. The mode of applications defines the designs of the equipment that can accommodate the different kind of RTE food products. The thermal processing equipment includes ovens, fryers, blanchers, pasteurizers, and sterilizers, etc. Oven is the versatile equipment which can be used for drying, baking, and roasting the raw ingredients using radiant heat energy. It has the insulated working chamber or cavity to accommodate a number of trays in which raw ingredients are loaded. The temperature controller regulates the optimum temperature required for any process by controlling the heaters, i.e., heating elements. The heat is uniformly distributed to the raw ingredients or processed product for evenly cooking so that the desired quality of the end product can be achieved. Oven types include batch oven, vacuum oven, rotary oven, and convection oven, etc. Another mostly used thermal equipment is fryer, in which the food products are deep fried in oil to make it ready to eat. The deep frying makes the texture of RTE food crispy and enhanced the flavour. Fryers may be batch or continuous type. They consist of the fry basket or conveyor belt which passes through the hot oil wherein the products are fried. It is also equipped with the temperature control system, oil filtration and circulation system to maintain to frying consistency and quality of fried products (Berk, 2018).

Some fruits and vegetables required the pre-heating process to deactivate the enzymes before further processing, for which, the blanchers are used. It involves exposing the raw materials to high temperature for short time followed by rapid cooling to stop the cooking process. Blanchers are operated using hot water and steam in the operations solely or in combination. The products are immersed in the hot water or exposed to the hot steam for short duration depending upon the size of product or capacity of blancher. The heat penetrates into the product, soften the tissues, remove the contaminants and deactivate the enzymes (Berk, 2018).

For inactivating the harmful pathogens, bacteria, and spoilage microorganisms, the pasteurizers and sterilizers are used (Amit et al., 2017; Holdsworth & Simpson, 2007). It does include the balance tank, primary filters, heat exchangers, temperature control system as well as the flow diversion systems. The RTE products, mostly juices, are pasteurized or sterilized to ensure the microbiological safety and extend the shelf life (Amit et al., 2017). High-moisture fruits and vegetables are dried to lower the moisture content and increase its storability and shelf life. Drying is the techniques through which the moisture can be reduced to desired level where the product would be self-stable (Berk, 2018). The dryers used for such process include the hot air, vacuum, microwave, spray, fluidized and freeze dryers. While using the appropriate process parameters, the desired quality, i.e., texture, flavour, and nutritional value of the dried RTE product, can be maintained. Thus, the dried products can be used to manufacture of RTE snack bars, cereal bars or mixes, etc. (Berk, 2018; Singh & Heldman, 2014).

1.2.3.4 Freezing Systems

RTE food products available in frozen forms such as dairy products, or cut fruits, etc., for which the freezing systems are employed for rapid chilling and freezing process to preserve its freshness, texture and extend shelf life (Amit et al., 2017). The RTE food products are kept in the freezing chambers and using specific freezing medium the products are frozen. The different types of freezers include air blast, tunnel, fluidized bed and plate freezers. The air blast freezer involves blowing the extremely cold air at higher velocity for rapid freezing of the RTE food products using powerful blower fan. The turbulence in the cold air increases the heat transfer, ensuring rapid heat removal and uniformity in freezing of food. The RTE foods such as baked goods, dairy products, vegetables, cut fruits, and meats are exposed to the stream of cold air in air blast freezer. Plate freezer is another freezer in which the food product is kept between two refrigerated plates, which rapidly remove the heat from the product to freeze it. Common products frozen in plate freezers include meat patties, seafood, prepared meals, etc. (Singh & Heldman, 2014). In tunnel freezers, the food products are passed through the insulated long tunnel in which cold air is circulated around the food. The tunnel may have multiple sections to control the temperature and residence time, ensuring proper freezing of the RTE food product (Saravacos & Kostaropoulos, 2016). The irregular shaped fruits and vegetables are frozen in fluidized bed freezer. It has the perforated conveyor belt or vibrating bed on which the product is kept, and the cold air or cryogenic fluid is allowed to pass through it. The cold air or cryogenic fluid surrounds the product uniformly and removes the heat quickly. Individual quick freezing (IQF) is one of the types of fluidized bed freezing (Amit et al., 2017; Saravacos & Kostaropoulos, 2016).

1.2.3.5 Forming and Shaping Machinery

The forming and shaping machinery used to provide the definite shape and size to the RTE food product such as energy bars, extruded snacks, bakery and dairy products, confectionery, etc. Extruders are the widely used equipment for manufacture of varied range of RTE foods such as expanded snacks, breakfast cereals, etc. The starchy materials are extruded through a small die and expanded under high temperature and pressure. According to the die dimensions, the size, shape, and texture of the final product can be managed. It is a versatile equipment in which numerous operations, such as conveying, mixing, shearing, heating, cooling, melting, etc., can be done using appropriate design of barrel and screw (Berk, 2018; Dalbhagat et al., 2019). For shaping the processed ingredients, moulding is the other technique which is used to produce RTE foods in specific sizes and shapes, sometime, to achieve definite volume. Moulding machines are available in automatic, semi-automatic, or manual mode depending upon the capacity of material to be handled. The examples of moulded RTE foods products include chocolate, candies, protein bars, and confectionery items.

The products like cookies, cakes, muffins, and filled pastries are manufactured by using depositors which deposit the processed ingredients on to the trays, moulds, or conveyor belts with predefined production rate. The raw ingredients used in depositors include batters, creams, and sauces. The depositors are equipped with the different nozzles or attachment to accommodate various types of shapes and sizes. Sheeting machine is used for manufacturing products from the sheet of the dough which is cut into desired shapes. These machines can be standalone having high production capacity with precision in making uniform sheet thickness and width. The RTE food products such as pizza crusts, pie crusts, crackers, and cookies are made using sheeting machines. For products like bread rolls, buns, pizza dough balls, the dough dividers and rounders are used. These machines divide the dough into equal proportion and round them in consistent size and shape.

1.2.3.6 Packaging Equipment

Packaging equipment is used to pack a wide variety of RTE food products such as snacks, chips, powders, roasted grains, and frozen foods. The packaging equipment handles various packaging materials in different sizes and able to deliver the material in varying quantities based on volume or mass. The machines available in market are vertical form fill seal (VFFS), horizontal form fill seal (HFFS), tray sealers, shrink wrapping machines, etc. The VFFS machine packs the material vertically in flexible packaging films, while the HFFS machine operates horizontally, to pack material, in trays, cups, or containers. VFFS machines can handle different kinds of packaging films such as polyethylene, polypropylene, laminates with options of printing the necessary information such as batch code, packaging date, expiry date on it. The HFFS forms the desired shape of packaging film, and fill it with the product such as RTE meals, sandwiches, dairy and confectionery products, and seal it with the heat or pressure (Tucker, 2016).

Tray sealers are used to seal the pre-formed trays with the RTE food products using a film lid under heat or pressure creating leak-proof seal. The fresh produce such as cut fruits and vegetables, RTE meals, and seafoods are packed using tray sealers. Shrink wrapping machines are used to enclose RTE food products in a tightly sealed film by applying heat to shrink the film around the product. These machines can package individual items or bundles of products, providing tamper-evident and attractive packaging for display purposes. Shrink wrapping machines are commonly used for RTE products such as beverage bottles, canned goods, and multi-packs of snacks (Saravacos & Kostaropoulos, 2016). Thermoforming machines are used to create rigid or semi-rigid packaging from roll-fed or pre-cut plastic sheets. These machines heat the plastic material until it becomes pliable, then form it into a mould to create the desired shape. Thermoforming machines are commonly used for packaging products such as meats, cheeses, baked goods, and fresh produce (Tucker, 2016).

1.3 MECHANIZATION NOVELTIES AND UPGRADATION IN RTE FOOD PRODUCTION

Mechanization novelties refer to the latest technological advancement including automation, instrumentation, and process control to enhance the efficiency of process maintaining the quality and safety of the product throughout the production process including processing, packaging, quality control, and traceability.

1.3.1 Continuous Production Line

Nowadays, highly automated continuous production lines (Figure 1.1) are used for manufacturing the RTE foods on commercial level to produce large quantities in continuous and uninterrupted

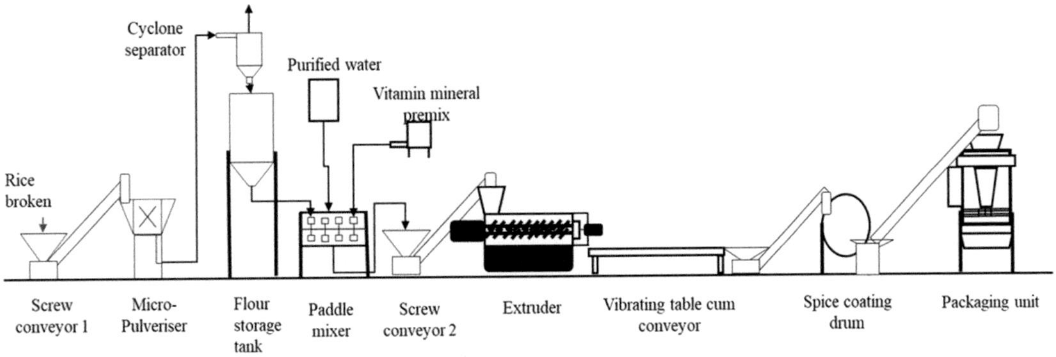

FIGURE 1.1 Schematic of continuous production line for RTE snacks.

manner. Minimal manual intervention, continuous production, and maximum efficiency are the key features of the continuous lines. The raw ingredients are automatically weighed in batches from the storage silos and conveyed to processing machinery according to the recipe requirements. The ingredients may undergo initial primary processing such as washing, peeling, cutting, or milling, before further processing. The automated blending and mixing units are used to mix the different pre-weighed ingredients to form the base mixture of RTE food. The installed sensors in the equipment control and monitor the mixing process to ensure the consistency and uniformity of the mix. Continuous cooking equipment such as oven, fryers, steamers, or pasteurizers are used to cook the material independently on a conveyor belt or in cooking chamber where the temperature is controlled and monitored to achieve the desired cooking characteristics in the final product. The shaping and forming equipment such as extruder, former, moulding machines shape the cooked product into the desired shape and size continuously as per set online programme or recipe. High-speed cutters and portioning equipment precisely weigh and cut the product in desired dimensions. Continuous rapid chilling and freezing systems are employed to lower down the temperature of RTE products to preserve its freshness, flavour, and texture. Packaging machinery packed the final product to make them marketable form into different packaging materials.

1.3.2 Automation and Process Control

Automation and process control (Figure 1.2) ensures the quality and safety of the finished product in the continuous production line. The programmes are designed in control panel in such a way that the automated system can precisely select, measure, and dispense the ingredients into the formulation ensuring consistency. The handling equipment such as conveyors, feeders, and batching system efficiently transfer the pre-weighed ingredients to the processing machinery improving efficiency and reducing the manual labour (Gray & Davis, 2013). The continuous process line is controlled by installing different sensors such as temperature, pressure, pH, humidity, moisture content, level, and flow sensors, at various stages of production. The control system collects the real-time data from the different sensors, analyses it, and controls or adjusts the process parameters to maintain the optimum operational conditions in the production unit. Programmable logic controllers (PLCs) or distributed control systems (DCS) also regulate the operations of the equipment, different valves and sensors installed in the production line to ensure the precise process control. Automated cooking systems such as ovens, fryers, and steamers, involve the programmable control to maintain the consistent cooking temperature and time to obtain desired properties in the product. Thermal processing systems such as pasteurizers and sterilizers are controlled in such a way that if the temperature of pasteurized or sterilized product does not reach the set temperature, the whole product is again

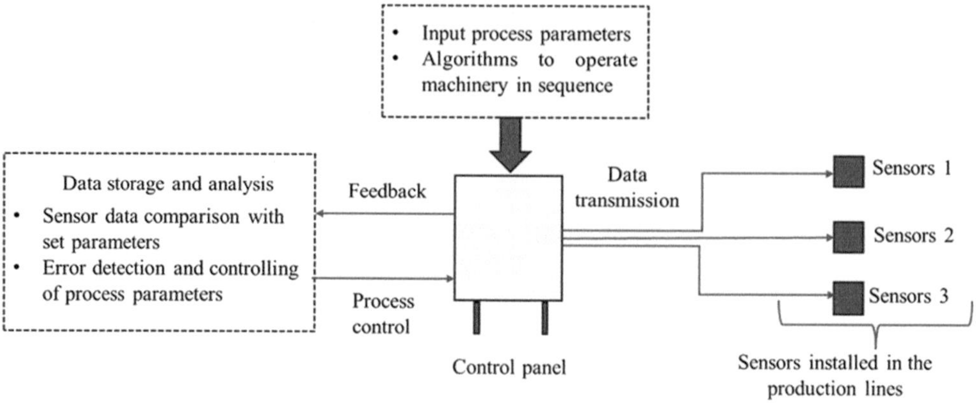

FIGURE 1.2 Automation and process control.

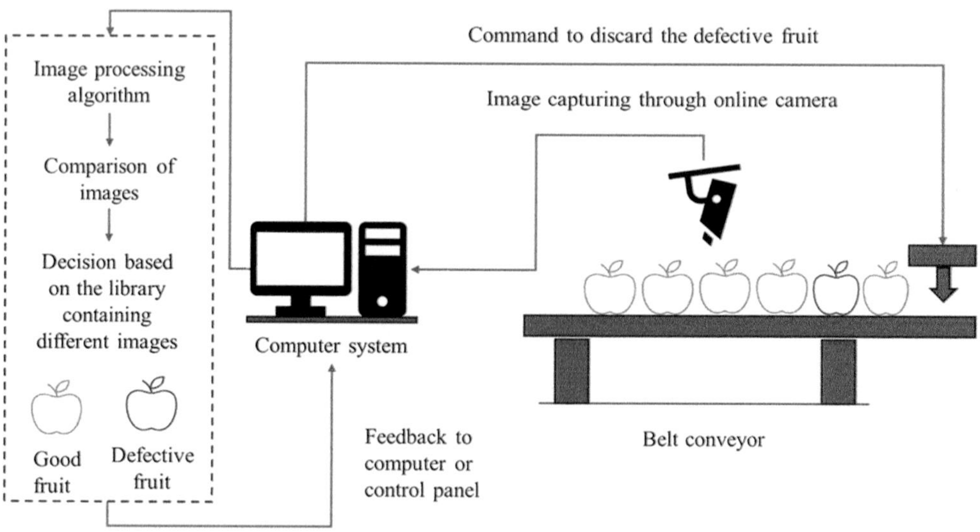

FIGURE 1.3 Use of digital image processing in quality inspection.

recirculated in the heating section, unless until it satisfies the set quality criteria. Similar kind of automation and process control is involved in the extruder, former, high-speed cutting, and forming units. Robotics and pick-and-place systems handle the raw, intermediate, or processing products at loading, filling, labelling or palletizing sections with high speed, precision, and accuracy. Automated quality control systems inspect the packaging integrity, label accuracy, and product appearance, ensuring compliance with regulatory requirements.

1.3.3 Digital Image Processing

Digital image processing is a valuable novel technique explored for the inspection and process optimization, manufacturing, product sorting and grading, colour control, and automated quality inspection. It involves the analysis of the images captured in various stages of the production process (Davies, 2013). The computer algorithms are used to analyse the image of products to detect the defects, irregularities, or any foreign objects (Figure 1.3). It is linked with the computer vision system (CVS) wherein the camera captures the image of the product in continuous production line which is analysed by the software and compared with the standard image in the library to find any deviation from the quality standards (Addanki et al., 2022; Ahmad et al., 2017). The defective or non-complying product can be rejected or diverted automatically from the production line ensuring the consistent quality of the product. Automatic sorting and grading are the best examples of CVS. The captured images of fruits and vegetables are mapped for the consistent colour, shape, and appearance, and decision is made for acceptance and rejection of the material (Addanki et al., 2022; Davies, 2013). The CVS can also be used for inspection of packaging materials to detect the defects as they move along the production line. Moreover, the package information also be confirmed and corrected if misprinted in real time ensuring compliance with regulatory requirements and brand standards. (Ahmad et al., 2017; Davies, 2013). By analysing images of RTE food products, the production process can be optimized and rectified to reduce the waste, minimize the downtime, and increase the production capacity. Real-time data logging, processing, and monitoring enable proactive decision-making and continuous improvement in the production process (Davies, 2013).

1.4 FOOD SAFETY AND QUALITY THROUGH MMECHANIZATION

The proper selection of the processing machinery, hygienic design, materials of construction plays a crucial role in preserving the quality of RTE food and ensuring the physical and microbial safety of the food product (Kotzekidou, 2016). Controlled, automated and integrated systems should be employed for the cleaning and sanitation of the equipment used for the manufacture of RTE products (Baeghbali et al., 2023).

1.4.1 Continuous Monitoring of Critical Parameters

Real-time monitoring of process parameters such as temperature, pressure, pH, moisture content, operational speeds, etc. are crucial for getting good-quality products. The strategic use of high-tech sensors and instrumentation in the continuous production line provides the insight of the process conditions during the production (Kotzekidou, 2016). Automation systems are employed to regulate the process parameters maintaining the optimal conditions required for the production of RTE products. The deviations in the process parameters may lead to compromise in the product quality. Timely detection and monitoring variation in process parameters would save the product, ensuring the improved efficiency of process and increasing economical profit. The prompt feedback loops and algorithms in control system can swiftly adjust the processing conditions to mitigate the variation in product and elevate both the product consistency and quality. Moreover, the data acquisition software collects the information of real-time process parameters and analyses the data, thereby providing suitable information to the operators with insights into the process performance and product quality. The identification of trends, anomalies, and opportunities for optimization of process can be done from the analysed data.

1.4.2 In-Line Inspection and Detection Systems

The instruments or systems, viz. metal detectors, X-ray scanners, and cameras are integrated into the production line to detect the defects in the product, foreign matter (organic and inorganic), contaminants, etc. in the RTE food products (Ahmad et al., 2017; Gray & Davis, 2013). The scanning of the product automatically inspects the product as it moves under different processing operations and takes decision for immediate acceptance or rejection of the material based upon the set criteria to ensure the consistent product quality and safety.

1.4.3 Automated Sampling and Analysis

The product samples are automatically collected in specific intervals for the physical, chemical, and microbiological analysis. Analytical tools that measure product qualities quickly and accurately include spectrophotometers, chromatographs, and microbiological testing apparatus. Online analytical data is utilized to confirm the nutritional value, microbiological safety, and composition of the product, enabling fast modifications to the production process when necessary.

1.4.4 Traceability and Documentation

Traceability of the raw ingredients and finished products is another important task of online quality control system. Database systems, barcoding, and radio frequency identification (RFID) tagging monitor batch history, product lineage, and pertinent process variables. Automation generates the electronic record of the process and equipment, i.e., production logs, inspection reports, which are used to measure the quality of product and indicator for regulatory compliance and audit readiness. Automated record-keeping ensures documentation of critical control points, sanitation procedures, and quality assurance measures for audit and regulatory purposes.

1.5 EMERGING TRENDS AND TECHNOLOGIES IN RTE FOOD MECHANIZATION

The food production environment is constantly changing due to emerging trends and technology in food mechanization, which improve sustainability, quality, and efficiency.

1.5.1 Robotics

The robotic arm is typically used to handle the raw or processed ingredients automatically. According to the set programme for certain recipes, it can measure, weigh and dispense the ingredients (Wang et al., 2022). They are trained with arm tools to make the formulations of food materials and ingredients in RTE food processing. It ensures proper and thorough mixing, time of operation, and preserves the product quality in the production line. It can be used to provide shape and size to the food products. They can lift and handle the products and place into packaging containers. They make sure to seal the packets and then place them automatically in the stocking place (Gray & Davis, 2013).

The Automated Guided Vehicles (AGVs) are automated vehicles with the navigation system that are used to transport the food materials and products within the area of food production facilities. They can make the material easy to handle and processes by carrying the ingredients, intermediate products, and finished products through various stage of production line. AGVs are able to move automatically through the production facilities by following predetermined or selected paths and changing the production layout configurations or plans. These automated vehicles enable the flexibility, agility, and effective material transport. The effective management of material flow is critical to ensure the smooth and safe operation of a food manufacturing plant but can be done and ensure by using them. AGVs are combined with conveyors and plant system to make the smooth transportation of material and workflow coordination. They can connect automatically the loading and unloading stations, providing uniformity and regular material flow and minimizing the wasting of time. AGVs are automated with safety features to avoid the trouble shooting such as collision avoidance, sensors, etc., and are equipped with emergency stop buttons and obstacle detection system to make sure smooth operation in the plant site. This automation features mitigate the risk of accidents in the workplace.

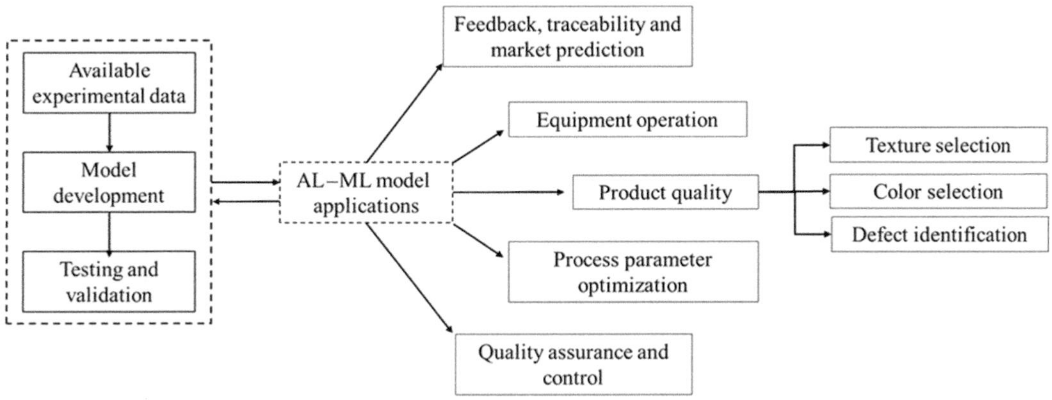

FIGURE 1.4 Integration of AI–ML for quality evaluation.

1.5.2 Integration of Artificial Intelligence and Machine Learning in RTE Food Processing and Quality Control

At present, artificial intelligence (AI) and machine learning (ML) technologies are offering advanced abilities to optimize processes, ensure quality control, and predict the maintenance and able to guarantee the efficiency and transformation in RTE food production. AI and ML algorithms are mostly used to analyse data in production processes to identify patterns, correlations, and in competence (Figure 1.4). This analysis helps manufacturers in optimizing the processing parameters mainly temperature, pressure, and mixing times, to maximize output by reducing energy consumption and maintain the product quality (Manthou et al., 2020). AI and ML algorithms are used for real-time monitoring, and in identifying any areas that need improvement and ensure that product meets the desired standards. Vision inspection systems are automated and equipped with AI algorithms to detect the foreign matter and inconvenience in the RTE (Davies, 2013). These models are analysing the data from the equipped sensors to forecast maintenance need and prevent unexpected problems. The wastage of time can be minimized by optimizing equipment lifespan by monitoring the equipment performance and by identifying the malfunctioning of equipment (Addanki et al., 2022; Davies, 2013).

AI-powered predicting models are used to analyse historical sales data, market trends, and other external factors are typically used to predict demand for RTE food products. ML algorithms promote inventory management, distribution processes, and securing the adequate amount of stock levels by lowering excess inventory and interference in the flow of goods or services from their initial production to their final delivery. In today's scenario, AI and ML are playing crucial roles to understand consumer preferences, feedback, and market trends and provide valuable insights for what products customers are looking for and what features they value most. By analysing these trends and values, AI and ML algorithms can make us to design and develop the product according to the customer needs and demands. These technologies provide facility to create new RTE food products by incorporation of dietary aspects, such as nutritional requirements, colour, and flavour profiles progressing the innovative and setting oneself apart from competitors in the market. To keep the ingredients, materials and finished product on track and trace them which are present in the plant as well as to trace and monitor their origin and movement throughout the supply chain, the AI-powered analytics platforms are utilized to make sure for transparency and accountability. ML algorithms are able to identify the potential of food safety and risks, which are mainly microbial contamination or allergen cross-contact, insisting manufacturers to implement preventive measures and maintain product safety for human consumption. AI and ML reduce resource inputs and minimize the waste and decrease the environmental impact in RTE food production. Using these technologies, it is easy to optimize packing designs, production schedules, and distribution routes and time to reduce the carbon footprint and increase sustainability across the plant and supply chain.

1.5.3 Smart Sensors and Internet of Things (IoT) Devices for Real-Time Monitoring and Data Analysis

Smart sensors and IoT devices play acute roles in the product development and preservation of RTE products by delivering the continuous updated data from the systems to secure the products by increasing the production and efficiency during storage.

1.5.3.1 Manufacturing Process

Smart sensors collect and analyse the data from RTE food production facility such as temperature, relative humidity, and air quality index (AQI). These sensors can generate the data of the food product being processed such as product temperature, pH, moisture content, colour, etc., and track the efficiency of the equipment used in production of RTE foods (Manthou et al., 2020). IoT devices can facilitate the real-time data transmission to the control system, empowering the operator to control the process parameters to update the optimum conditions for better quality and food safety. Moreover, the predictive maintenance of the equipment can be determined from the analysed data of smart and IoT sensors to avoid the downtime of equipment in production process (Ammar et al., 2021).

1.5.3.2 Storage and Distribution

Similar to manufacturing process, the smart sensors can observe, monitor, and control the environmental conditions continuously in storage areas and vehicles equipped with the refrigeration techniques for transportation of raw and heat-sensitive products. Not only the observation and monitoring of environmental conditions, these IoT and smart sensors can provide the global positioning system (GPS) information of movement of products. Additionally, these sensors constantly record the data of temperature, humidity, and light exposure throughout the storage process to ensure the product quality (Escobedo et al., 2021). Data collection and analysis are required for maintaining the product freshness and safety. The information is also used to ensure the regulatory compliance and food safety regulations (Kodan et al., 2020).

1.6 CHALLENGES AND OPPORTUNITIES

1.6.1 Initial Capital Investment and Equipment Maintenance

Automation in the RTE food production needs startup funds for procuring and installing the machinery and equipment, including smart and IoT sensors, system control units, high-end data processing machines, software programming and algorithms, etc. The automation not only requires the initial investments but also needs the maintenance costs, operating costs, repairing, and replacing of damaged parts in the machines. Despite high initial investment, this automation can reduce the labour costs, less wastage of materials for processing/developing the RTE product, and also increase the production efficiency which can boost the profits.

1.6.2 Skilled Personnel and Training

Skilled persons with deep knowledge of automation technologies are required for the maintenance and operation of automated and robotic systems. It somewhat challenging for hiring a skilled person in the remote areas, as well as to train the available manpower with limited access to technical programmes. The technical skill development with certification programmes can be arranged at the education institutes or high-end research laboratories. Collaboration with industries can increase the skills and workforce for operation and maintenance of the systems.

1.6.3 Leveraging Automation, AI, and ML for Process Optimization, Predictive Maintenance, and Quality Control

Incorporation of AI and ML technologies in the RTE food production needs a precise planning, customization, and integration of the systems which are in operation. Automation of AI and ML technologies provide excellent opportunities for process enhancement, proactive maintenance, and quality control of the RTE food production. Product developers can also improve the production capacity and quality of the product by leveraging data analytics, predictive modelling, and real-time monitoring capabilities.

1.7 CONCLUSION

Advanced technologies and novelties in mechanization continue to boost the growth of the RTE food sector. The upgradation of equipment and instrument enhances the quality and safety which attracts the consumers towards the RTE food products. RTE food products can be manufactured in various categories for which the production efficiency and food quality and safety can be improved by use of smart sensors and IoT devices. The online adjustments of process parameters can be done by automation to optimize the process parameters. Mechanization and automation coupled with AI–ML are revolutionizing industry, enabling real-time data collection, transmission, storage, analysis, and process optimization. Though these processes provide certain advantages in the production process, it comes with challenges including initial capital investment, the need for skilled personnel, and the integration of existing systems. By using these techniques, RTE food producers can harvest the benefits such as improved process efficiency, less wastages, online quality management, and high-quality RTE food products.

REFERENCES

Addanki, M., Patra, P., & Kandra, P. (2022). Recent advances and applications of artificial intelligence and related technologies in the food industry. *Applied Food Research*, *2*(2), 100126. https://doi.org/10.1016/j.afres.2022.100126

Ahmad, M. H., Sahar, A., & Hitzmann, B. (2017). Measurement, Modeling and Automation in Advanced Food Processing. In B. Hitzmann (Ed.), *Advances in Biochemical Engineering/Biotechnology* (Vol. 161). Springer International Publishing. https://doi.org/10.1007/978-3-319-60111-3

Amit, S. K., Uddin, M. M., Rahman, R., Islam, S. M. R., & Khan, M. S. (2017). A review on mechanisms and commercial aspects of food preservation and processing. *Agriculture and Food Security*, *6*(1), 1–22. https://doi.org/10.1186/s40066-017-0130-8

Ammar, M., Haleem, A., Javaid, M., Walia, R., & Bahl, S. (2021). Improving material quality management and manufacturing organizations system through Industry 4.0 technologies. *Materials Today: Proceedings*, *45*, 5089–5096. https://doi.org/10.1016/j.matpr.2021.01.585

Baeghbali, V., Hedayati, S., & Jafari, S. M. (2023). Storage vats, vessels, and tanks. In Transporting Operations of Food Materials Within Food Factories*: Unit Operations and Processing Equipment in the Food Industry* (pp. 15–30). Woodhead Publishing (Elsevier), Cambridge. https://doi.org/10.1016/B978-0-12-818585-8.00012-X

Berk, Z. (2018). *Food Process Engineering and Technology*. Elsevier. https://doi.org/10.1016/C2016-0-03186-8

Bursey, A. S., Wiles, N. L., & Biggs, C. (2021). The nutrient quality and labelling of ready-to-eat snack foods with health and/or nutrition claims. *South African Journal of Clinical Nutrition*, *34*(2), 65–71. https://doi.org/10.1080/16070658.2019.1682242

Contini, C., Romano, C., Scozzafava, G., & Casini, L. (2016). Food habits and the increase in ready-to-eat and easy-to-prepare products. In *Food Hygiene and Toxicology in Ready-to-Eat Foods*. Academic Press, London. https://doi.org/10.1016/B978-0-12-801916-0.00001-7

Dalbhagat, C. G., Mahato, D. K., & Mishra, H. N. (2019). Effect of extrusion processing on physicochemical, functional and nutritional characteristics of rice and rice-based products: A review. *Trends in Food Science and Technology*, *85*, 226–240. https://doi.org/10.1016/j.tifs.2019.01.001

Davies, E. R. (2013). Machine vision in the food industry. In *Robotics and Automation in the Food Industry: Current and Future Technologies* (pp. 75–110). Woodhead Publishing Limited, Cambridge. https://doi.org/10.1533/9780857095763.1.75

Escobedo, P., Bhattacharjee, M., Nikbakhtnasrabadi, F., & Dahiya, R. (2021). Flexible strain and temperature sensing NFC tag for smart food packaging applications. *IEEE Sensors Journal*, *21*(23), 26406–26414. https://doi.org/10.1109/JSEN.2021.3100876

Fellow, P. J. (2022). *Food Processing Technology*: *Principle and Practice* (Fifth). Woodhead Publishing (Elsevier). https://doi.org/10.1016/C2019-0-04416-0

Gray, J. O., & Davis, S. T. (2013). Robotics in the food industry: an introduction. In *Robotics and Automation in the Food Industry* (pp. 21–35). Woodhead Publishing Limited, Cambridge. https://doi.org/10.1533/9780857095763.1.21

Hawa, A., Kanani, H., Patel, M., Taneja, N., Maru, P., Kaliwala, S., Gopani, S., Sharma, S., Sharm, S., & Patel, S. (2014). A study on consumer purchase intention towards ready-to-eat food in Ahmedabad. *Asian Journal of Management Research*, *202*(2), 202–209.

Holdsworth, D., & Simpson, R. (2007). Sterilization, Pasteurisation and Cooking Criteria. In *Thermal processing of packaged foods* (pp. 125–148). Springer, New York. https://doi.org/10.1007/978-0-387-72250-4

Kheto, A., Nema, P. K., Sharanagat, V. S., & Kumar, P. (2023). *Technological Development in the Grading of Fruits and Vegetables*, *25*(4), 246–265.

Kodan, R., Parmar, P., & Pathania, S. (2020). Internet of Things for food sector: Status quo and projected potential. *Food Reviews International*, *36*(6), 584–600. https://doi.org/10.1080/87559129.2019.1657442

Kotzekidou, P. (2016). Factors influencing microbial safety of ready-to-eat foods. In *Food Hygiene and Toxicology in Ready-to-Eat Foods*. Academic Press (Elsevier Inc), London. https://doi.org/10.1016/B978-0-12-801916-0.00003-0

Manthou, E., Lago, S. L., Dagres, E., Lianou, A., Tsakanikas, P., Panagou, E. Z., Anastasiadi, M., Mohareb, F., & Nychas, G. J. E. (2020). Application of spectroscopic and multispectral imaging technologies on the assessment of ready-to-eat pineapple quality: A performance evaluation study of machine learning models generated from two commercial data analytics tools. *Computers and Electronics in Agriculture*, *175*, 105529. https://doi.org/10.1016/j.compag.2020.105529

Patel, D., & Rathod, R. (2017). Ready to eat food perception, food preferences and food choice: A theoretical discussion. *World Wide Journal of Multidiscipl Inary Research and Development WWJMRD*, *3*(8), 198–205.

Rashid, S., Bashir, O., & Majid, I. (2023). Different mechanical conveyors in food processing. In *Transporting Operations of Food Materials within Food Factories: Unit Operations and Processing Equipment in the Food Industry* (pp. 253–263). Woodhead Publishing (Elsevier), Cambridge. https://doi.org/10.1016/B978-0-12-818585-8.00007-6

Sahay, K. M., & Singh, K. K. (2001). *Unit Operations of Agricultural Processing* (p. 388). Vikas Publishing House Pvt. Ltd., New Delhi.

Santeramo, F. G., Carlucci, D., De Devitiis, B., Seccia, A., Stasi, A., Viscecchia, R., & Nardone, G. (2018). Emerging trends in European food, diets and food industry. *Food Research International*, *104*, 39–47. https://doi.org/10.1016/j.foodres.2017.10.039

Saravacos, G., & Kostaropoulos, A. E. (2016). Handbook of food processing equipment. Springer Cham(Springer International Publishing), Switzerland. https://doi.org/10.1007/978-3-319-25020-5

Singh, R. P., & Heldman, D. R. (2014). Introduction to food engineering: Fourth edition. In *Introduction to Food Engineering: Fifth Edition*. Academic Press (Elsevier), London. https://doi.org/10.1016/C2011-0-06101-X

Suri, S., Dutta, A., Chandra Shahi, N., Raghuvanshi, R. S., Singh, A., & Chopra, C. S. (2020). Numerical optimization of process parameters of ready-to-eat (RTE) iron rich extruded snacks for anemic population. *LWT*, *134*(June), 110164. https://doi.org/10.1016/j.lwt.2020.110164

Temgire, S., Borah, D. A., Kumthekar, S., & Idate, A. (2021). Recent trends in ready to eat/cook food products: A review. *The Pharma Innovation*, *10*(5), 211–217. https://doi.org/10.22271/tpi.2021.v10.i5c.6207

Tucker, G. S. (2016). Modified atmosphere packaging. In *Food Preservation and Biodeterioration* (pp. 167–192). Wiley. https://doi.org/10.1002/9781118904657.ch7

Wang, Z., Hirai, S., & Kawamura, S. (2022). Challenges and opportunities in robotic food handling: A review. *Frontiers in Robotics and AI*, *8*, 1–12. https://doi.org/10.3389/frobt.2021.789107

2 Thermal Processing of RTE Foods
Process Design and Quality Aspect

Shaikh Adil,[1*] *Ankit Bihola,*[2] *Ajay Singh,*[3]
Mohammad Raziuddin,[4] *and Nitin Shinde*[5]
[1]Department of Dairy Technology, P.I.T., Parul University, Vadodara, Gujarat, India
[2]National Dairy Research Institute, Karnal, Haryana
[3]Department of Food Technology, Mata Gujri College, Fatehgarh, Punjab, India
[4]Department of Livestock Products Technology, MAFSU, Nagpur, Maharashtra, India
[5]College of Dairy Technology, Pusad, MAFSU, Nagpur, Maharashtra, India
*Corresponding Author

2.1 INTRODUCTION

Food consumption and selection are largely influenced by the lifestyles of individuals today. As it relates to food, convenience can be defined as the reduction of time and effort required for food preparation, storage, and consumption. The main category of convenient foods at this time, ready-to-eat (RTE) food, serves as the answer to these issues (Buckley *et al.*, 2007). RTE meals have become a vital part of modern food consumption due to the fast-paced lifestyles we lead today. This category of foods comprises of pre-cooked or prepared meals that are packaged with appropriate materials, providing consumers with quick meal options (Spencer, 2005). These dishes are exactly what their name implies: they are ready to eat right away, without further need for additional cooking or lengthy preparation.

RTE meals include a wide range of products, from canned soups and frozen dinners to supermarket-available pre-packaged salads and sandwiches and fast-food items such as burgers and pizzas. RTE foods have become a popular choice for people and families alike because of their convenience and accessibility, transforming our dietary patterns and gourmet preferences. As food standardization becomes more prevalent and our lives become busier, there is a growing need for convenient meal options. However, the conventional methods used to process food often lead to faster decline in quality, and the presence of microorganisms poses a safety risk to RTE foods (Chen *et al.*, 2020). RTE foods also contribute to minimizing food wastage by providing portion-controlled servings and longer shelf lives. Moreover, they play a vital role in emergency situations and disaster relief efforts, ensuring a stable and reliable source of sustenance when conventional cooking facilities are unavailable (Bajzelj *et al.*, 2020).

Food processing has several primary objectives including making foods edible, protecting them from microbial spoilage and hazards, improving their shelf life, and increasing consumer convenience. Thermal and non-thermal processing are two innovative methods commonly used in the production of RTE foods. Specific processing methods for RTE foods include "cook-chill processing, pasteurization, sterilization, sous vide, high-pressure processing, irradiation, ohmic heating, microwave processing, smoking, and curing" (Lajnaf, 2023).

 DOI: 10.1201/9781032622408-2

In the food industry, traditional thermal processes are widely used for microbiologically stabilizing foods such as low-acid foods (Spilimbergo *et al.*, 2002). In recent years, there have been new methods developed to make sure that food is safely processed for thermal purposes while also minimizing any negative impact on its quality. The current high temperatures and holding periods required have been shown to negatively affect the nutritional value and flavor of the food (Ramos *et al.*, 2018).

One of the key factors behind the success and safety of RTE foods is thermal processing. Thermal processing refers to the application of heat to food products, either through cooking, canning, or pasteurization, to kill or inactivate harmful microorganisms, enzymes, and spoilage agents while preserving the food's flavor, texture, and nutritional value. This preservation technique has been a game changer in the food industry, enabling the production of safe, shelf-stable RTE foods that can be stored for extended periods without refrigeration. The application of heat also enhances the organoleptic qualities of many food items, making them more appealing to consumers (Fellows, 2009).

The chapter is designed to provide a comprehensive understanding of RTE foods and thermal processing. Additionally, it also focuses on the significance of thermal processing in food preservation, discussing the principles, methods, and equipment involved. This will be followed by an examination of the safety considerations and quality control measures associated with thermal processing. Finally, we will explore emerging trends and innovations in the RTE food industry, considering their implications for the future of food consumption.

2.2 PRINCIPLES OF THERMAL PROCESSING

The transfer of heat can be accomplished in three modes: conduction, convection, and radiation. Conduction occurs when molecules in solid materials change direction and transmit heat. Convection involves the movement of fluid, which carries heat due to changes in buoyancy and density. Finally, radiation involves the transfer of electromagnetic energy between objects with different temperatures (Farid, 2018).

2.3 MODES OF HEAT TRANSFER

2.3.1 Conduction

Conduction is the process by which heat energy is transferred from a higher energy state to a lower energy state when an object is exposed to different conditions (Wang & Sun, 2006; Shahidian *et al.*, 2020).

2.3.2 Convection

When temperature changes cause density variations in liquid, free or natural convection can occur, which can transfer heat inside containers (Kachalov *et al.*, 2021). Alternatively, heat can be deliberately transferred through forced convection by rotating the liquid axially or end-over-end inside the container.

When heat is transferred through conduction to the outer layers of fluid near a heated surface, natural convection occurs. This causes the density of the fluid to decrease, making the heated layer rise. Once it reaches the liquid's top, the fluid motion causes it to descend at the center, pushing colder fluid from the core to the wall and displacing the heated fluid. As the fluid's temperature becomes more uniform, the driving force for convection decreases, and the fluid velocity slows down. Eventually, when the fluid is evenly heated, convection stops.

2.3.3 Radiation

Thermal radiation encompasses wavelengths between 0.8 and 400 m and can transfer energy without requiring a medium. This process involves converting thermal energy from a hot source into electromagnetic waves, propagating those waves from the hot source to a colder receiver, and then absorbing and transforming the electromagnetic waves into heat energy. Overall, these three processes work together to transfer energy through radiation (Wang & Sun, 2006; Bhattacharjee *et al.*, 2019).

2.3.4 F_0 Concept

To assess the success of a saturated steam sterilization method on microorganisms with a *Z*-value of 10, one must calculate the time in minutes required at a temperature of 121°C. This time frame is an indication of the level of lethal impact that the process imparts on the final containerized product. To derive the total F_0 of the process, one must analyze the fatality rates over a given time period at specific temperature ranges, accounting for the heating and cooling phases of the cycle as well (Koszucka & Nowak, 2019).

When selecting a steam sterilization cycle based on the F_0 concept, it is important to exercise caution to ensure that a sufficient level of sterility is always achieved. To demonstrate that microbiological parameters are within the specified limits and provide a Sterility Assurance Level (SAL) of 10-6 or greater, it may be necessary to carry out continuous, intensive microbiological monitoring during routine manufacturing as well as validating the process.

The *D*-value, also referred to as the decimal reduction value, is the quantity of a sterilization parameter (such as sterilization time or absorbed dosage) that must be applied to achieve a 90% reduction in the number of viable organisms. This value is only significant under very specific experimental conditions.

The following mathematical relationships apply:

$$F_0 = D_{121}\left(\log_{10} N_0 - \log_{10} N\right) = D_{121} \log \mathrm{IF}$$

D_{121} = *D*-value of the reference spores at 121°C
N_0 = initial number of microbes
N = final number of microbes
IF = inactivation factor

$$Z = \frac{T_2 - T_1}{\log_{10} D_1 - \log_{10} D_2}$$

D_1 = *D*-value of microorganism at temperature T_1
D_2 = *D*-value of microorganism at temperature T_2

2.4 THERMAL PROCESSING METHODS

2.4.1 Pasteurization

Pasteurization is a process that involves moderate heating to eliminate harmful bacteria in food and beverages. It was named after the French scientist Louis Pasteur who pioneered the technology in the 1860s (Tucker & Featherstone, 2021).

The amount of time and temperature required for pasteurization is determined by the food or beverage being treated. However, the goal is always to eliminate as many harmful germs as possible without compromising the quality or nutritional value of the product. One common method of pasteurization is to heat food or drinks to a high temperature for a short period of time. For example, milk is typically pasteurized by being heated to 72°C for 15 seconds. Another technique is to cook the food or beverage for a longer period of time at a lower temperature. Yogurt, for instance, is typically pasteurized by being heated to 63°C for 30 minutes. Pasteurization is an effective way to reduce the risk of foodborne illnesses. It is also a commonly used practice in the food industry because it is cost-effective (Chiozzi *et al.*, 2022).

2.4.2 Blanching

When preserving vegetables through drying, freezing, or canning, it is recommended to use a technique called blanching. Blanching involves quickly boiling or steaming the vegetables to prevent the activity of enzymes that cause deterioration. This technique not only enhances the taste, color, texture, and mineral content of vegetables but also makes them more enjoyable to consume. In addition to these benefits, blanching eliminates dirt, kills surface germs, and makes certain vegetables, such as broccoli and spinach, more compact and easier to store and package (Hong *et al.*, 2017).

To blanch vegetables, simply bring a large pot of water to a rolling boil and add the vegetables for the appropriate cooking time depending on the type of vegetable. Once the vegetables are blanched, quickly submerge them in a bowl of icy water to halt the cooking process. After the vegetables are cold, pat them dry before storing them in containers. Blanching vegetables can extend their shelf life to up to a week in the fridge or several months in the freezer.

Blanching is often used as a pre-treatment for freezing or canning and helps to inactivate enzymes in fruits and vegetables. Although there are various blanching techniques, hot water and steam blanching are the most commonly used methods. While hot water blanching is simple, it can lead to a loss of nutrients and energy. Steam blanching can minimize nutritional loss, but it may not completely inactivate all enzymes (Phopin *et al.*, 2022). A newer technique called high-pressure blanching aims to address the drawbacks of hot water and steam blanching. However, it has not yet been proven to completely inactivate all enzymes. Peroxidase (PO) is used as a measure of blanching effectiveness, as it is the most heat-resistant oxidative enzyme. Moist hot gas and ohmic heating are rarely used for blanching vegetables.

2.4.2.1 Benefits of Blanching

1. Inactivates enzymes that can cause off-flavors, loss of texture, and color change.
2. Improves the color and brightness of vegetables
3. Reduces the microbial load on vegetables
4. Removes pesticide residues
5. Extends the shelf life of vegetables

2.4.2.2 Drawbacks of Blanching

1. Can leach out water-soluble nutrients
2. Can be energy intensive
3. May not be effective at inactivating all enzymes.

2.4.3 Sterilization

When it comes to ensuring food safety, sterilization is a key process that involves using heat to kill harmful bacteria and enzymes. Canned goods, for example, undergo commercial sterilization in a large pressure cooker called a retort, which eliminates all biological organisms and gives them a shelf life of over two years (López *et al.*, 2019). However, foods with a pH higher than 4.6, like meat and most vegetables, require high-temperature cooking to eliminate any potential pathogens. This is typically done under pressure at 121°C for varying periods of time, with extra measures taken to ensure that *Clostridium botulinum* spores are completely eradicated. These spores can produce a deadly "botulinum toxin" in oxygen-free environments but can be killed by heat or inhibited at pH levels below 4.6. As a result, acidic foods with a pH below 4.6 that are packed without oxygen, such as spaghetti sauce, do not need to undergo such intensive heat treatment. The ultimate goal of commercially sterilized foods is to eliminate all vegetative and spore-forming organisms and prevent illness.

2.5 THERMAL PRESERVATION PROCESSES FOR RTE FOOD PRODUCTS

These are processes that employ heat to make food safe for consumption and/or extend shelf-life by reducing or eliminating microbiological contamination (pathogens or spoilage) and enzymatic activity of the food. These processes could also affect the texture and flavor of food products. Table 2.1 is a detailed description of non-thermal processes for RTE food products and their preservation techniques.

2.5.1 Non-Thermal Processes for RTE Food Products and Their Preservation

"Processes that make the food safe for consumption and extend the shelf-life of a food using alternative technologies without heat to reduce/eliminate/inactivate microbial contamination in the food are non-thermal processes."

Food Process	Purpose	Food Use Examples	Benefits	Drawbacks
High-pressure processing	A process where a food product in a sealed container is subjected to high iso-static pressure, usually via water	Fruit and vegetable juice, sauces, dips, dressings, hummus, and deli meats	Food safety; enables taste closer to freshly made	High cost associated with equipment and packaging; usually requires refrigerated distribution; limited product application due to high moisture requirement of the food product
Aseptic filling	Filling and sealing a food or beverage product in a sterile environment after the product has been sterilized, usually using ultra-high heat treatment	Dairy beverage products, juices, fruits, vegetables, soups, and RTE meals shipped and sold at room temperature	Allows perishable products to be packaged, shipped, sold, and stored at room temperature; could improve flavor; reduces loss of vitamin(s) due to a shorter heat treatment	Very complex and expensive technology to install and operate; requires specialized expertise to ensure food safety requirements are met
Membrane filtration (microfiltration)	A liquid product is passed through multiple membranes with controlled porosity to remove large pathogenic or spoilage microorganisms that cannot pass through the membranes	Beverages including juices, milk, wine, water, and beer	Food safety; avoids or minimizes use of heat treatment to extend product shelf-life; less burnt flavor	Requires significant redundancy in filtration system to ensure product safety; expensive technology, applicable to liquid products only with particles smaller than the microorganisms

Food Process	Purpose	Food Use Examples	Benefits	Drawbacks
Acidification	Organic and/or inorganic acid(s) is added to a food product to increase acidity/reduce the pH below a threshold to prevent growth of pathogenic and spoilage causing microorganisms	Beverages (e.g., fruit and vegetable juice, soda, sports and energy drinks), salad dressings, mayonnaise, salsa, canned fruits and vegetables, dips, and some types of soft cheese	Food safety; requires only low-moderate heat treatment; reduces cooked or burnt taste; and maintains food structure (e.g., salsa)	Impacts taste and appearance of the food product; high acid content may not be tolerated by some people
Fermentation	A process where non-harmful, live microorganisms (e.g., lactic acid bacteria) use sugars in the food product and generate organic acids which increases acidity/reduces pH below a threshold to prevent growth of pathogenic microorganisms	Dairy products (e.g., cheese, kefir, buttermilk, yogurt, and sour cream), sausages (e.g., salami), sauerkraut, pickled vegetables, tofu, kombucha, soy sauce, wine, vinegar, and beer	Food safety; shelf life; requires less heat treatment	Changes protein structure; could impact taste, texture, and appearance; high acid content may not be tolerated by some people
Refrigeration	Growth of spoilage and pathogenic microorganisms is prevented or reduced by decreasing the temperature of the food product and then storing the product at a lower temperature for future use	Meat, cheese, butter and spreads, yogurt, sour cream, dips, sauces, dressings, milk, juice, eggs, fruits and vegetables, salad mixes, and prepared meals (ready to eat)	Food safety; shelf life; prevents degradation of some vitamins; preserves flavor and texture of the food product	Causes thickening of some liquid or soft products; some spoilage microorganism can still grow, although slowly, thereby limiting the shelf life
Brining/marinating	Salt and/or sugar is added to a food with high water content (e.g., meat, fruit, and vegetables)	Meat, seafood, fruits and vegetables, feta cheese, jams and jellies, and pie fillings	Food safety; shelf life	Significantly changes taste and texture of the food product; increases sodium and/or sugar content of the food product

(continued)

Food Process	Purpose	Food Use Examples	Benefits	Drawbacks
Freezing	Food temperature is lowered to below freezing. Subsequent storage at or below freezing temperature prevents spoilage by slowing or stopping microbial growth	Frozen meat, vegetables, fruits and fruit concentrates, sauces, prepared meals (e.g., pizza), desserts (cakes & ice cream), frozen potatoes, bagels, and breads	Food safety; shelf life; preserves vitamins; maintains much of the original taste	Generally, will not destroy microorganisms in the food; changes food texture and sometimes taste
Modified atmosphere packaging/ storage	Food product is packaged in a sealed container with minimal or no oxygen, via flushing with an inert gas (e.g., nitrogen), followed by another preservation method such as refrigeration for some food products, to inhibit growth of microorganisms.	Salad mixes, fresh dips and spreads, meat, cheese and dairy products, fresh fruits and vegetables, and some breads	Shelf life; reduces deterioration of flavor, color, and vitamins caused due to exposure to oxygen	Does not inhibit growth of some pathogenic and spoilage causing microorganisms that require minimal or no oxygen; requires additional preservation method to ensure food safety
Irradiation	Application of high energy (gamma rays, X-rays, electron beams) particles or waves to a food product	Spices, meat, nuts, fruits and vegetables, wheat and wheat flour, salad mixes, potatoes, and food packaging	Food safety; inhibits sprouting	Potential loss of vitamin(s); could impact the color of some food products

Adopted from Institute of Food Technologists (2023).

2.6 PROCESS DESIGN

Process design in the context of thermal processing for RTE foods is a critical aspect that directly influences the safety, quality, and shelf life of the final product. It involves making decisions regarding the selection of equipment, process parameters, and control systems to achieve the desired product characteristics. Successful process design ensures that the RTE food product is safe for consumption and meets sensory and nutritional requirements.

Thermal processing is an effective method to eliminate microbes, and its success depends on the processing time and temperature. Safety is the top priority when processing most products, although certain food products, including canned dairy and food, must attain commercial sterility.

TABLE 2.1
Various Thermal Preservation Process for RTE Foods

Processing Methods	Process Description	Food Use Examples	Benefits	Drawbacks
Cooking/heating	Increasing the temperature of a food and beverage product to destroy microorganisms, inactivate enzymes, preserve the food, and improve the edibility of some foods	Eggs, meat, vegetables, potatoes, cereal grains (rice, oats, wheat), and pasta	Food safety; shelf life; impact flavor and texture of some food products	Significant damage of heat-sensitive vitamin(s); changes color; influences taste in a number of food products
Pasteurization	Heating food and beverage products at a specific temperature for a particular length of time	Milk, juices, some egg products, jams, jellies, and some sauces	Food safety; shelf life	Significant loss of heat-sensitive vitamin(s); modifies protein structures
Drying	Using heated air or light (infrared, solar) to substantially reduce water content	Meat, milk, cheese, and other dairy products, egg products, grains, and fruits and vegetables	Shelf life	Significant loss of heat-sensitive vitamin(s); modifies product texture and flavor; increases calorie content due to water loss
Direct steam injection	Heating a liquid food product by injecting steam under pressure to achieve a particular temperature (e.g., pasteurization)	Milk, processed cheese, and other dairy products, nutrition shakes, juices and other beverages, and soups	Food safety; shelf life	Increases moisture content of the food product; potential loss of heat-sensitive vitamin(s); modifies protein and fiber structures from the original state; imparts burnt flavor
Ultra-high heat treatment (indirect steam heating)	Rapid heating of a product above the boiling point of water for a very short time	Shelf-stable liquid dairy products (e.g., milk and nutrition shakes), juices, and other beverages, pastes, sauces, and purees	Food safety; shelf life	Significant loss of heat-sensitive vitamin(s); modifies protein structures; imparts burnt flavor
Frying	Rapid heating or cooking of a food product in hot oil (300–500°F)	Fried meat, potatoes (e.g., French fries or potato chips), fried vegetables, and eggs	Food safety; shelf life; changes texture and color and flavor	Adds fat to the food, thereby increasing calorie content of the food; potential loss of heat-sensitive vitamin(s); modifies protein structures; could impart burnt flavor; could generate components that are potentially carcinogenic (e.g., acrylamide)
Baking	Heating or cooking a food product using either a batch or continuous oven at a specified temperature and time to achieve a desired temperature within the food product	Meat, breads, cakes, potatoes, vegetables, mixed dishes (e.g., casserole), and fruits	Food safety; shelf life; imparts brown color on the surface and flavor; modifies texture	Potential loss of heat-sensitive vitamin(s); modifies protein structures; increases calorie content due to water loss

(*continued*)

TABLE 2.1 (Continued)
Various Thermal Preservation Process for RTE Foods

Processing Methods	Process Description	Food Use Examples	Benefits	Drawbacks
Grilling	Cooking a food product over open flame (usually 350–600°F) for a specified time to achieve a desired temperature within the food product	Meat, potatoes, and vegetables	Food safety; shelf life; imparts favorable taste, texture, and appearance	Potential loss of heat-sensitive vitamin(s); modifies protein structures; imparts burnt flavor; generates compounds that are potentially carcinogenic (e.g., acrylamide)
Smoking	A method used to preserve meat by using salts to dehydrate the meat and cooking the meat at a lower temperature ranging between 190 and 250 °F in the presence of smoke, generated by burning organic material(s)	Various types of meat and some cheeses	Food safety; shelf life; imparts unique flavor; modifies texture	Modifies protein structures in meat; could impart unacceptable flavor due to the variation in the absorption of smoky flavor
Radio frequency heating (microwave, pulsed electric field, ohmic)	Heating food via radio waves (e.g., microwaves and pulsed energy) causing rapid molecular vibration of water or via electrical resistance (ohmic) to achieve pasteurization or sterilization	Meat, potatoes, grains, and vegetables	Food safety; shelf life; causes less structural changes compared to other thermal processes	Potential loss of heat-sensitive vitamin(s); modifies food structure
Infrared heating	Heating the surface of the food by absorption of infrared radio waves followed by heat conduction to achieve a desired internal temperature in the food	Meat, potatoes, grains, and vegetables	Food safety; used in food service to maintain the required temperature of cooked foods	Slow heating process makes it inefficient for high-throughput production systems; leads to more moisture loss compared to other thermal processes and modifies texture
Canning and pressure cooking	Cooking food in a can or jar over 250°F for a set time period to potentially destroy microorganisms, including spores (e.g., *Clostridium botulinum*), to achieve food sterility	Meat, vegetables, potatoes, jams, jellies, and some dairy products	Food safety; shelf life	Substantial loss of heat-sensitive vitamin(s); container seal integrity is a critical factor in canning process to ensure correct treatment and that the package remains commercially sterile post processing
Extrusion cooking	Blended food mixture is heated to a specific temperature and simultaneously driven by the extrusion screw toward the head of the press and forced through a die to potentially destroy microorganisms	Grain-based pastas, ready-to-eat cereals, and proteins derived from plant sources	Food safety; makes the food convenient to use, for example pasta can be easily rehydrated and cooked	Potential loss of heat-sensitive vitamin(s); changes texture
Pressure assisted thermal sterilization	Combining high-pressure processing with heat to achieve sterilization temperature	RTE meals, vegetables, fruits, meats	Food safety; less nutrient, flavor, texture, and color loss than traditional pressure cooking	Limited scaling and availability; high cost associated with equipment and packaging

Source: Adopted from Institute of Food Technologists (2023); Holdsworth *et al.* (2008).

The temperature range of 10–63°C is a breeding ground for microorganisms that cause spoilage and sickness. It is crucial to understand that fresh meat and the ingredients used in preparation are breeding grounds for bacteria that can thrive in them. Therefore, it is imperative to develop efficient heat methods that can remove foodborne microorganisms and reduce spoilage (Tadini & Gut, 2022).

There are some key considerations in the process design of thermal processing for RTE foods:

2.6.1 Spoilage Microorganism Destruction

Heat treatment enhances organoleptic characteristics and adds value to meat products. Additionally, heating extends product shelf life by eliminating spoiling microorganisms. The absence of competing microflora makes recontamination of thermally processed items crucial, as infecting organisms can multiply uncontrolled (Li *et al.*, 2021) as is the case of *Listeria monocytogenes* (a pathogenic bacteria) that can grow at refrigeration temperatures. About 25% of foodborne outbreaks can be traced to recontamination of processed items (WHO, 2021). Some causes contributing to product contamination include poor hygiene (1.6%), cross-contamination (3.6%), poor sanitation in processing or storage facilities (4.2%), contaminated equipment (5.7%), and staff handling (9.2%) (Augustin *et al.,* 2020). Environmental and personnel recontamination are major causes. RTE foods can develop off-colors, odors, and flavors due to multiplication of spoilage microorganisms. Microorganisms such as *Pseudomonas* spp. produce green pigments on meat products kept in refrigerator in aerobic conditions which ultimately affect color of these products. These microbes also create lipopolysaccharides, which slime product surfaces (NACMCF Report, 2022).

The cooking methods commonly used for thermally processed meats aim to reach internal temperatures of 68°C. This temperature is effective in eliminating the majority of non-spore-forming pathogens such as *Salmonella* spp., *Escherichia coli* O157:H7, *S. aureus*, and *Listeria* spp. among others (Holloywood, 1991). Such observed microorganisms would be indicative of recontamination of the products.

2.6.2 Safety Measures: Destruction of Pathogens

Foodborne diseases can be categorized into two main types: infections, in which the consumer ingests food that contains live organisms, typically in significant quantities, and intoxications, in which the consumer ingests food that contains inherent toxins (Priyanka *et al.*, 2016). The occurrence of foot-and-mouth disease and bovine spongiform encephalopathy in cattle serves as evidence that new safety concerns will definitely emerge in future times. This section will mostly center on the bacterial infections that are of importance in RTE products and their eradication by thermal processing in conjunction with other approaches. The meat products are associated with many pathogens that are of particular concern, namely *E. coli* O157:H7, *Salmonella* spp., *L. monocytogenes, S. aureus*, and *Clostridium* spp. (specifically *C. perfringens* and *C. botulinum*).

2.7 RTE FOODS: QUALITY ASPECTS

The conditions that foods are subjected to and the hazards that can cause them to deteriorate are important factors in determining a product's shelf life, as well as the best materials and packaging techniques. Here, the extrinsic and intrinsic aspects affecting each type of RTE food's shelf life will be investigated. First off, ready meals are precooked foods that primarily suffer from microbiological growth and oxygen, which should not make up more than 1% of the product in the packaging. Additionally, prepared meals must be kept at a minimum temperature of 4°C due to the effectiveness of modified atmosphere packaging (MAP) application at lower temperatures, which

TABLE 2.2
Some Destructive Reactions for Ready Foods

Food Product	Deterioration Reaction	Effects on Product
Frozen ready meals containing fruit and vegetable derivatives	Enzymatic reactions	Color, flavor, texture changes, nutrient and bioactive compound loss
	Oxidative reactions	Off-flavor formation, color changes, nutrient and bioactive compound loss toxic compound formation
Chilled ready meals	Bacterial and mold growth	Off-flavor formation, visible colony formation, texture, and color changes

Source: Calligaris and Manzocco (2012).

increases CO_2 solubility and preventative activity (Subramaniam, 1993). Table 2.2 displays the primary degradation reactions together with the ready meals.

The respiration rate of fruits and vegetables can be influenced by various factors, including the cutting procedure, environmental temperature, and MAP (Kandasamy, 2022). An elevation in storage temperature leads to a corresponding rise in respiration rates, resulting in a decrease in the shelf life of the product. Furthermore, it is worth noting that certain fruits, such as lemons, apples, and oranges, exhibit high levels of acidity. Conversely, most vegetables tend to possess higher pH values. Consequently, to mitigate the potential risk of *C. botulinum* contamination, it is necessary to store these vegetables in anaerobic containers at temperatures below 3°C. Furthermore, the utilization of modified environment packing and cooling temperature is employed to inhibit the growth and spread of spoilage and pathogenic bacteria in fruits and vegetables with high water activity (a_w). Moreover, the impact of ethylene activity on shelf life, namely its role in accelerating ripening, is mitigated using MAP. In addition to temperature, relative humidity (RH) is a significant extrinsic element that influences the shelf life of a product. When the RH of the environment falls below 80–95%, it results in the dehydration of fruits and vegetables. A water loss of 3–6% is considered sufficient to induce a decline in quality (Table 2.3) (Day, 1993).

Both raw and cooked cured meats can be influenced by several elements during the storage process. The unstable color of these products and the emergence of a gray hue can be attributed to the interaction between the air and the substance. Additionally, the process of fading can be achieved through exposure to light in relation to the product. Furthermore, it has been shown that when cured meats are stored in environments with high temperatures and low humidity, the flesh pigments undergo drying and change, resulting in a discoloration known as browning (Shakil *et al.*, 2022). The primary forms of deterioration observed in cured meats include discoloration, rancidity resulting from oxidation, and microbial development (Robertson, 1993).

Due to moisture, temperature fluctuations, and endogenous lipases, frozen RTE pizza can undergo oxidative rancidity and hydrolytic rancidity during storage. In pizza ingredients, lipase and lipoxygenase enzymes help break down polyunsaturated fatty acids (PUFAs), producing peroxide and hydroperoxide radicals. These are responsible for undesirable odor. The process of oxidative rancidity in pizza results in the formation of "aldehydes, ketones, and short-chain free fatty acids", leading to a reduction in the overall acceptability as well as quality of the food product (Smith, 1993). To ensure the microbiological safety of frozen foods, it is imperative to prevent temperature fluctuations. Furthermore, Calligaris and Manzocco (2012) have reported that the potential deterioration processes for frozen salami pizza include dehydration, lipid oxidation, and pigment discoloration in the tomato sauce.

Fresh pasta is an RTE dish that necessitates a distinct technique for extending its shelf life when compared to dry pasta. This is mostly because fresh pasta contains a higher moisture content than its dried counterpart. Fresh pasta typically contains a moisture content ranging from 24% to 30%, rendering it a perishable food item. Consequently, it is advisable to store fresh pasta at a temperature no higher than 4°C. The primary cause of deterioration in fresh pasta is microbiological activity, namely the activity of bacteria, mold, and yeast. This activity has a significant impact on the shelf stability and taste qualities of the product (Lucera *et al.,* 2014). RTE foods, such as sandwiches, are typically comprised of multiple ingredients. Therefore, it is imperative to consider the destructive characteristics of both the rolls and other components while analyzing this sort of food. Galic *et al.* (2009) identified several elements that influence the shelf life of fresh bakery products. These factors primarily include oxygen, temperature, and moisture, which can lead to microbiological decomposition and moisture loss, resulting in hardening, staling, and oxidative rancidity of the product. In addition to the primary constituent, it is imperative to have clear definitions for all influential elements pertaining to each component of sandwiches. Additionally, the degradation of quality can occur due to moisture transfer between components, which is influenced by the RH of the environment and the moisture gradient of the components (Roudaut and Debeaufort, 2011). Another example of an RTE dish with multiple components is a delicatessen salad. This salad typically consists of a combination of cooked or uncooked fruits and vegetables, as well as meat particles. Additionally, it occasionally comes with dressing choices like mayonnaise or vinaigrette. Various types of delicatessen salads include potato salad, macaroni salad, ham salad, seafood salads, and other specialty salads.

According to Hwang (2010), certain regions exhibit this phenomenon. The storage of delicatessen salads necessitates chilling at temperatures ranging from 0 to 5°C due to the perishable nature of the ingredients involved. The observation of the interaction between dressing and other components and its impact on stability over shelf life is evident. In general, dressings are known to possess a significant acidic content, which serves to enhance stability against microbial growth and contribute to the desirable taste profile of salads (Brocklehurst, 1994). The primary cause of spoilage in these salads is cross-contamination, a phenomenon that can be effectively mitigated by the implementation of good manufacturing practices (GMP), proper hygiene, and appropriate handling protocols throughout the stages of production, distribution, and storage (Hwang, 2010). Furthermore, it is imperative to examine the potential for deteriorative reactions among all ingredients during the packaging and shelf-life processes.

RTE desserts, sometimes referred to as dairy desserts, mostly consist of custards, jellies, trifles, mousses, and cheesecake-like Foods. The elements that influence the shelf life of RTE chilled desserts include the quality of raw materials, the formulation of the product, operational parameters,

TABLE 2.3
Factors Affecting Shelf Life of RTE Fruits and Vegetables

Intrinsic Factors	Extrinsic Factors
Respiration rate	Harvesting
Acidity	Handling
Water activity	Hygiene
Biological structure	Temperature
Ethylene production and sensitivity	Water loss and RH
	Packaging material
	Gas/product ratio

Source: Day (1993).

adherence to GMP, storage conditions, distribution processes, and consumer handling practices. During the storage period, it is imperative to maintain certain features of RTE desserts to ensure their quality. These properties include the preservation of a crunchy structure, steady aeration of the dessert, emulsion stability, and color retention. To illustrate the preservation of product stability, it is imperative to keep milk-based desserts, which serve as a conducive environment for microorganisms, at low temperatures to inhibit microbial growth. Furthermore, the breakdown of lipase enzymes can lead to the occurrence of rancidity, a common phenomenon observed in dessert creams that possess a high fat content. The regulation of temperature is a crucial factor in the preparation of starch-based sweets such as custard, mostly due to the inherent gelling properties of starch. The primary external elements affecting the quality of RTE desserts are oxygen and light, as they contribute to the oxidation of milk fat and the discoloration of tastes (Clarke *et al.*, 2021).

2.8 CONCLUSION

The importance of ensuring the quality and safety of RTE foods cannot be underestimated. Thermal processing is of utmost importance in ensuring the safety and sustainability of these items. The process involves the utilization of heat to eliminate detrimental microbes and enzymes, while concurrently maintaining the sensory attributes, such as flavor, texture, and nutritional composition of the food. Conventional thermal processing techniques, although successful in eliminating microorganisms, may occasionally have detrimental effects on the sensory and nutritional attributes of the end product. To tackle these concerns, novel and emerging technologies are being devised with the aim of enhancing the microbiological safety of thermal processing, while simultaneously mitigating any adverse effects on the quality of food products. These procedures encompass several techniques, including pasteurization, sterilization, and sous vide, among other approaches. Each methodology possesses distinct advantages and limitations, and their choice is contingent upon the specific demands of the RTE food product.

The efficacy of thermal processing in RTE foods is contingent upon various elements, encompassing the eradication of spoilage bacteria and pathogens, implementation of safety protocols, and maintenance of product quality. The proper design of processing processes is of utmost importance to efficiently eradicate foodborne pathogens and minimize the occurrence of spoiling types. In combination with thermal processing, the shelf life of RTE foods is influenced by many intrinsic and extrinsic factors. These factors encompass respiration rates, acidity levels, water activities, and ambient conditions. The aforementioned characteristics exhibit variability contingent upon the specific type of RTE food, hence exerting an impact on its overall quality and safety.

REFERENCES

Augustin, J. C., Kooh, P., Bayeux, T., Guillier, L., Meyer, T., Jourdan-Da Silva, N., ... & Anses Working Group on Consumer Information on Foodborne Biological Risks. (2020). Contribution of foods and poor food-handling practices to the burden of foodborne infectious diseases in France. *Foods*, 9(11), 1644S, 01-11.

Bajzelj, B., Quested, T. E., Roos, E., & Swannell, R. P. (2020). The role of reducing food waste for resilient food systems. *Ecosystem Services*, 45, 101140.

Bhattacharjee, C., Saxena, V. K., & Dutta, S. (2019). Novel thermal and non-thermal processing of watermelon juice. *Trends in Food Science & Technology*, 93, 234–243.

Brocklehurst, T. F. (1994). Delicatessen salads and chilled prepared fruit and vegetable products. In: Man, C. M. D., & Jones, A. A. (Eds), *Shelf Life Evaluation of Foods* (pp. 87–123). Chapman & Hall.

Buckley, M., Cowan, C., & McCarthy, M. (2007). The convenience food market in Great Britain: Convenience food lifestyle (CFL) segments. *Appetite*, 49, 600–617.

Calligaris, S., & Manzocco, L. (2012). Critical Indicators in Shelf-Life assessment. In: Nicoli, M. C. (Ed), *Shelf Life Assessment of Food* (Chapter 4, pp. 61–75). CRC Press.

Chen, F., Zhang, M., & Yang, C. H. (2020). Application of ultrasound technology in processing of ready-to-eat fresh food: A review. *Ultrasonics Sonochemistry*, 63, 104953.

Chiozzi, V., Agriopoulou, S., & Varzakas, T. (2022). Advances, applications, and comparison of thermal (pasteurization, sterilization, and aseptic packaging) against non-thermal (ultrasounds, UV radiation, ozonation, high hydrostatic pressure) technologies in food processing. *Applied Sciences*, 12(4), 2202S, 01-10.

Clarke, H. J., McCarthy, W. P., O'Sullivan, M. G., Kerry, J. P., & Kilcawley, K. N. (2021). Oxidative quality of dairy powders: Influencing factors and analysis. *Foods*, 10(10), 2315.

Day, B. P. F. (1993). Fruits and vegetables. In: Parry, R. T. (Ed), *Principles and Applications of Modified Atmosphere Packaging of Food* (pp. 114–132). Chapman & Hall.

Farid, M. (2018). Heat and mass transfer in food processing. In: *Handbook of Farm, Dairy and Food Machinery Engineering* (3rd edn., pp. 439–460). Academic Press. https://doi.org/10.1016/B978-0-12-814803-7.00017-8

Fellows, P. J. (2009). Heat processing. In: Fellows, P. J. (Ed), *Woodhead Publishing Series in Food Science, Technology and Nutrition, Food Processing Technology* (3rd edn., pp. 339–366). Woodhead Publishing.

Galic, K., Curic, D., & Gabric, D. (2009). Shelf Life of Packaged Bakery Goods—A Review. *Critical Reviews in Food Science and Nutrition*, 49(5), 405–426.

Holdsworth, S. D., Simpson, R., & Barbosa-Cánovas, G. V. (2008). In: *Thermal Processing of Packaged Foods* (Vol. 284). Springer.

Holloywood, N. H., Varabioff, Y., & Mitchell, G. E. (1991). The effect of microwave and conventional cooking on the temperature profiles on microflora of minced beef. *International Journal of Food Microbiology*, 14, 67–76.

Hong, W. X., Zhongli, P., Li-Zhen, D., Hamed, M., Xu-Hai, Y., Arun, S. M., ... & Qian, Z. (2017). Recent developments and trends in thermal blanching—A comprehensive review. *Information Processing in Agriculture*, 4(2), 101–127.

Hwang, C. A. (2010). Delicatessen salads. In: Hwang, A., & Huang, L. (Eds), *Ready-to-Eat Foods Microbial Concerns and Control Measures* (Chapter 2, pp. 63–66). CRC Press.

Institute of Food Technologists (2023). Non-Thermal Preservation Processes. Retrieved from www.ift.org/policy-and-advocacy/advocacy-toolkits/food-processing/non-thermal-preservation

Kachalov, A. B., Sánchez, P. S., Porter, J., & Ezquerro, J. M. (2021). The combined effect of natural and thermocapillary convection on the melting of phase change materials in rectangular containers. *International Journal of Heat and Mass Transfer,* 168, 120864.

Kandasamy, P. (2022). Respiration rate of fruits and vegetables for modified atmosphere packaging: A mathematical approach. *Journal of Postharvest Technology*, 10(1), 88–102.

Koszucka, A., & Nowak, A. (2019). Thermal processing food-related toxicants: A review. *Critical Reviews in Food Science and Nutrition,* 59(22), 3579–3596.

Lajnaf, R. (2023). Introductory Chapter: Novel thermal and non-thermal technologies for food processing. In: *Food Processing and Preservation.* IntechOpen.

Li, R., Wang, C., Zhou, G., Li, C., & Ye, K. (2021). The effects of thermal treatment on the bacterial community and quality characteristics of meatballs during storage. *Food Science & Nutrition*, *9*(1), 564–573.

López, M., Calvo, T., Prieto, M., Múgica-Vidal, R., Muro-Fraguas, I., Alba-Elías, F., & Alvarez-Ordóñez, A. (2019). A review on non-thermal atmospheric plasma for food preservation: Mode of action, determinants of effectiveness, and applications. F*rontiers in Microbiology*, 10, 622.

Lucera, A., Costa, C., Padalino, L., Conte, A., Lacivita, V., Saccotelli, M. A., … & Nobile, M. A. D. (2014). Combination of process technology and packaging conditions to improve the Shelf Life of fresh pasta. *Journal of Food Processing & Technology*, 5(12), 1–5.

NACMCF Report 2018–2020. (2022). National Advisory Committee on Microbiological Criteria for Foods. (2022). Microbiological testing by industry of ready-to-eat foods under FDA's jurisdiction for pathogens (or appropriate indicator organisms): Verification of preventive controls. *Journal of Food Protection*, 85(11), 1646–1666.

Phopin, K., Wanwimolruk, S., Norkaew, C., Buddhaprom, J., & Isarankura-Na-Ayudhya, C. (2022). Boiling, blanching, and stir-frying markedly reduce pesticide residues in vegetables. *Foods*, 11(10), 1463.

Priyanka, B., Patil, R. K., & Dwarakanath, S. (2016). A review on detection methods used for foodborne pathogens. *The Indian Journal of Medical Research*, 144(3), 327.

Ramos, S. J., Millán, D., Ortiz, L., Alonso, D., & García de la Torre, S. (2018). Effect of high-pressure thermal sterilization on the inactivation of *Geobacillus stearo thermophilus* spores in ready to eat meals. *The International Journal of Engineering and Science*, 7(8), 65–74.

Robertson, G. L. (1993). Packaging of flesh foods. In: Hughes, H.A. (Ed), *Food Packaging: Principles and Practice* (Chapter 15, pp.431–470). Marcel Dekker.

Roudaut, G., & Debeaufort, F. (2011). Moisture loss, gain and migration in foods. In: Kilcast, D., & Subramaniam, P. (Eds), *Food and Beverage Stability and Shelf Life* (pp. 63–100). Woodhead Publishing.

Shahidian, A., Ghassemi, M., Mohammadi, J., & Hashemi, M. (2020). *Bio-Engineering Approaches to Cancer Diagnosis and Treatment*. Academic Press.

Shakil, M. H., Trisha, A. T., Rahman, M., Talukdar, S., Kobun, R., Huda, N., & Zzaman, W. (2022). Nitrites in cured meats, health risk issues, alternatives to nitrites: A review. *Foods*, 11(21), 3355.

Smith, J. P. (1993). Bakery products. In: Parry, R. T. (Ed), *Principles and Applications of Modified Atmosphere Packaging of Food* (pp. 134–166). Chapman & Hall.

Spencer, K. C. (2005). Modified atmosphere packaging of ready-to-eat foods. In: *Innovations in Food Packaging* (pp. 185–203). Academic Press.

Spilimbergo, S., Elvassore, N., & Bertucco, A. (2002). Microbial inactivation by high-pressure. *The Journal of Supercritical Fluids,* 22(1), 55–63.

Subramaniam, P. J. (1993). Miscellaneous applications. In: Parry, R. T. (Ed), *Principles and Applications of Modified Atmosphere Packaging of Food* (pp. 170–187). Chapman & Hall.

Tadini, C. C., & Gut, J. A. (2022). The importance of heating unit operations in the food industry to obtain safe and high-quality products. *Frontiers in Nutrition*, 9, 853638.

Tucker, G., & Featherstone, S. (2021). *Essentials of Thermal Processing*. John Wiley & Sons.

Wang, L., & Sun, D. (2006). Heat and mass transfer in thermal food processing. *Food Science and Technology*, 150, 35–50.

WHO. (2021). Food Safety. Retrieved from www.who.int/news-room/fact-sheets/detail/food-safety. Accessed on 19 May, 2022.

3 Ready-to-Eat Foods Development through Composite Extrusion Technological Development

Shubham Mandliya[1*], *Siddharth Vishwakarma*[1], *Mahima Bora*[1], *Subrata Kumar Bag*[2], *Chandrakant Genu Dalbhagat*[1,3], *and Hari Niwas Mishra*[1]
[1]Agricultural and Food Engineering Department, Indian Institute of Technology Kharagpur, Kharagpur, West Bengal, India
[2]Department of Dairy Engineering, West Bengal University of Animal & Fishery Sciences, Mohanpur Campus, Nadia, West Bengal, India
[3]Department of Food Process Engineering, National Institute of Technology, Rourkela, Odisha, India
*Corresponding author

3.1 INTRODUCTION

Ready-to-eat (RTE) meals are gaining popularity as customer demand has grown because of how simple they are to prepare, consume, and store. Modern living has long placed a high value on snack foods, which are a unique, continually expanding, and changing set of food items. Globally, the market for snack food is anticipated to reach USD 762 billion by 2024, growing at a compound annual growth rate (CAGR) of 5.34% from 2019 to 2024. Sales of snacks increased by around six times between the years 2004 and 2014, from 8000 crores to 47,000 crores (Nielsen 2014). Consumers are being targeted by several food processing firms, including Parle Agro, ITC Foods, PepsiCo, and Nestle.

Due to its simplicity of usage and capacity to generate a range of textures and forms that appeal to customers, extrusion technology has been widely employed in the manufacturing of cereal RTE snacks. According to Singh and Singh (2002) and Xu et al. (2016), extrusion is a flexible process that includes mixing, kneading, shearing, cooking, shaping, and forming. It is a multifaceted technology in which moistened protein and starch materials are transformed under various combinations of processing conditions, offering a special chance to create an innovative product (Liu et al. 2011). It causes changes in the product's macroscopic shape, chemical properties, and microstructure by gelatinizing the starch, degrading the protein, and forming complexes between lipids, starch, and proteins (Hagenimana et al. 2006). For products like breakfast cereals, savory snacks, crispy flatbread, precooked flours, cereal-based infant food, and textured proteins, it becomes a crucial production step. Based on the mode of operation and the kind of structure, such as either a single- or twin-screw extruder, extrusion can be categorized as cold, warm, or hot (Dalbhagat et al. 2019).

DOI: 10.1201/9781032622408-3

Most of the extruded RTE food products, which are currently available in the market, are rich in sugar and salt contents but poor in nutritional value. The majority of the snacks available in the market are cereal-based, high in calories, and low in nutrients and micronutrients. Consumers have been much more concerned about health since the last decade and are demanding healthier and more nutritious snacks. From the perspective of the customer, the snack food business is going through unprecedented changes. Snacks should not just smell nice, but also feel and look good to consumers. Designing snack foods nowadays may be a challenging process as consumers' tastes and expectations are evolving. For example, consumers are looking for products that are "good for health," "rich in protein," "offer a unique flavor," and the illusive "something unique that also appeals to a wide variety of people." Due to its popularity as pleasant and healthful snacks, cereal-based snacks have become a new trend in the market.

Efforts have been made to study and incorporate different ingredients, blend composition, and process parameters to develop nutritious snacks. The incorporation of different fruits and vegetables was used to develop a delicious and nutritious snack. Composite extrusion is a feasible way to design a novel extruded snack with functional and nutritional benefits of different ingredients used by combining them into several layers, filling, or coatings.

3.2 CLASSIFICATION OF EXTRUDERS

One of the most flexible processes for transforming the components into partially cooked or completed foods is extrusion. Regarding crude fiber, extrusion has proven to be the most effective method for incorporating fiber into an RTE food item (Artz et al. 1990). Extrusion cooking has become more popular, especially in the snack food business. Because of its ability to handle a variety of raw materials, energy efficiency, persistent nature, lack of process effluents, and ease and adaptability in the form and texture of finished products, it has largely replaced other working systems in the snack food industry (Abd El-Hady and Habiba 2003; Meng et al. 2010). The fundamental input factors for extruded snack foods are the raw material composition, feed level, feed moisture content, barrel temperature profile, and screw speed. These factors have an impact on the material's structural, chemical, and functional properties when combined with other factors (Guy 2001).

Depending on the working procedures and moisture level of the end product, an extruder may be referred to as a wet or dry extruder. While heating is accomplished in dry extruders by mechanical resistance during extrusion, it is necessary to pump steam or water into the barrel in wet extruders (Rokey 2000). Single-screw wet extruders, single-screw dry extruders, single-screw interrupted-flight extruders, and double-screw extruders are the most frequently used extruder types.

A wet extruder typically runs at a low speed in a smooth barrel with a deep sliding screw, producing little to no contact. Wet extruders are used to make wieners, pasta, baked goods, and a specific type of confectionery. Extrusion cooking (or hot extrusion) refers to the process of heating food to temperatures above 100°C. The temperature quickly increased because of frictional heat and any extra heating employed in this process. Along with the internal heating caused by the contact between the rotating screw and the (notched) barrel, hot extrusion also uses electrical or steam heating via the exterior walls to quickly raise the temperature over 100°C. Hot extrusion is a common process for producing morning cereals, crisp breads, and snack foods like corn chips, pretzels, and Cheetos. Extruders with interrupted flights were used to make pet goods.

Extruders are divided into single-screw and twin-screw varieties based on the screw configuration. Twin-screw extruders, which offer higher versatility and better control, have been installed in response to rising demands for novel objects with bundled shapes and microscopic dimensions that exceed the capabilities of a single-screw system (Dalbhagat et al. 2019). Later, they are further divided into counter-turning and co-pivoting twin-screw extruders, which can be either intermeshed or non-intermeshed (Figure 3.1). Co-rotating self-cleaning types are most commonly used in the food industry where there is a need for frequent product change, uniform measurement and shape,

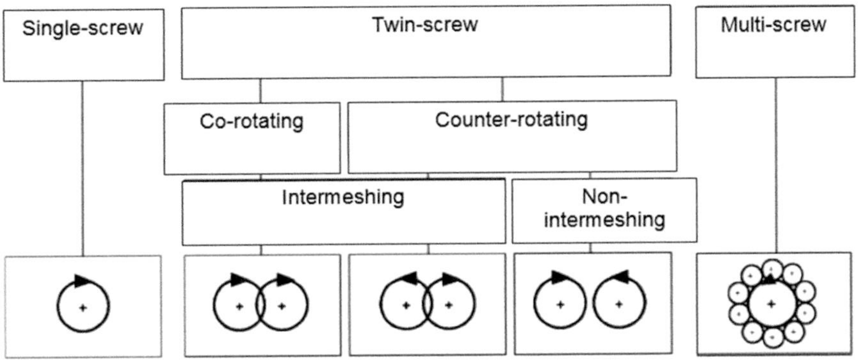

FIGURE 3.1 Types of screws used in extruder. (Adapted with permission from Schuchmann (2008).)

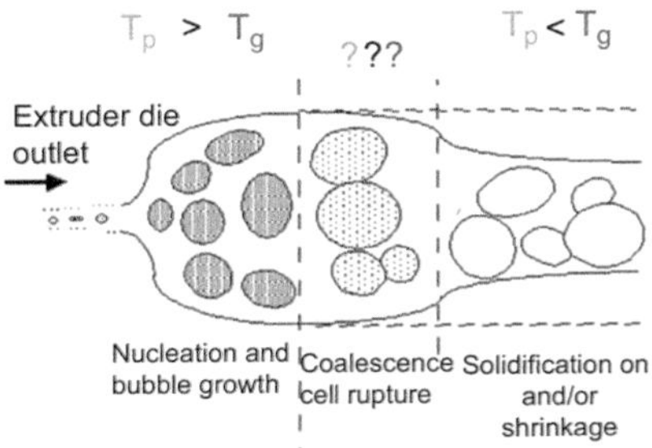

FIGURE 3.2 Mechanism of direct expansion of snack at die outlet. (Adapted with permission from Kristiawan et al. (2020).)

items made with low-density powders, and items ranging from noodles, pet treats, grains and corn pieces, high-fat aquatic feeds, corn chips tortillas, chocolate-filled snacks, and so on (Riaz 2000a).

Additionally, extruders can be classified as having low, medium, or high shear pressure. Low-shear extruders are used to densify and shape highly moist materials such as pasta, treats, and some confections. These extruders have a smooth barrel, high length/diameter ratio, and generally non-intermeshing type screws are used. Products that are low in moisture and demand a lot of mechanical energy can be handled by medium-shear extruders with a furrowed barrel and a high-pressure screw. They are used to produce textured vegetable proteins (TVP). Second-generation extruded snacks require a high-shear extruder with scored barrel and screws with a pressure expansion, low length/diameter, and shallow flights (Riaz 2000b).

3.3 MECHANISM FOR THE DEVELOPMENT OF EXTRUDED SNACKS

3.3.1 Expansion Mechanism

The result of the pressure differential that forms between the extruder's internal pressure and the outside air pressure is expansion. When a food product is forced into a barrel by way of screw limitations, pressure and heat are created. When the product is released into the outer environment (out of the die), there is an abrupt pressure drop, which causes the product to expand (Figure 3.2). Excessive moisture evaporation causes the rupture of starch cells. The degree of expansion is affected by several variables, including the product's starch concentration, inlet–outlet temperature, pressure, and the moisture content of the feed.

3.3.2 Bubble Growth and Bubble Collapse

The biaxial extension of individual bubbles serves as the foundation for extrudate growth. The pressure difference between the two sides of the barrel is the driving factor for bubble development and growth. The relationship between the bubble formation in a viscous fluid and the extrudate's precise volume depends on the ratio of vapor pressure to melt viscosity. Due to shrinking that took place in the high moisture content extrudates, the correlation was insufficient at higher moisture contents. The final texture of the expanded product is directly impacted by the cessation of bubble expansion. The expansion of the extrudate, which achieves a glassy condition after traversing the rubbery zone (retro gradation), ends when the melt exits the die orifice and water vapors flash out. This happens because of the temperature dropping caused by evaporative cooling.

The bubble wall collapses when it can no longer resist the pressure inside the bubble, which usually happens above 20% moisture content. The temperature of the extrusion and the rheological characteristics of the material control the bubble collapse process. Low extrusion temperatures are highly conducive to the collapse. Low melt viscosities, in terms of rheological characteristics, assist the bubble collapse.

3.4 MAJOR COMPONENTS REQUIRED FOR EXTRUDED RTE FOOD PRODUCT DEVELOPMENT

The quality of extruded products is significantly influenced by various material characteristics, including starch content, protein composition, fiber presence, moisture levels, and fat content. These components play crucial roles in determining the texture, nutritional value, and overall sensory attributes of the final extrudate. Understanding how each of these materials' characteristics impact the extrusion process and the properties of the end product is essential for optimizing the production of high-quality extruded foods and materials (Singha et al. 2018).

3.4.1 Moisture

Moisture is a fundamental component in the extrusion process, serving as the primary plasticizer that enables the transformation of raw materials into a cohesive, processable mass. It plays a crucial role in several key stages of the extrusion process, particularly in facilitating starch gelatinization and ensuring the even dispersion of ingredients throughout the mixture. The presence of water is indispensable in initiating the gelatinization process, which is essential for achieving the desired texture and structural properties in the final extrudate (Shruthi et al. 2017; Van Hoan et al. 2010).

During the initial stages of extrusion, the addition of water aids in converting the dry, free-flowing mixture of ingredients into a viscous, homogeneous fluid that can be easily conveyed and processed. Water acts as a lubricant, reducing friction between particles and assists in maintaining the appropriate temperature for the gelatinization of starch and the denaturation of proteins. This process is pivotal for creating the desired texture, taste, and digestibility in the finished product.

In the extrusion barrel, the control of moisture content significantly impacts the viscosity of the material being processed. Lower moisture content in the feed leads to an increase in the viscosity of the mixture. This elevated viscosity affects the flow characteristics, leading to changes in the expansion behavior and texture of the extruded RTE product. Moreover, the moisture content of the final product directly influences its physical properties. Controlling the moisture content is thus crucial for achieving the desired product characteristics, ensuring optimal texture, appearance, and overall consumer acceptability (Shruthi et al. 2017; Van Hoan et al. 2010).

3.4.2 Starch

Starch serves as the primary structural component, forming the fundamental polymer in the majority of cereal-based systems and playing a pivotal role in the expansion process. For achieving optimal expansion, it is crucial to maintain a starch content of 60%–70% as a lower limit. Comprising amylose and amylopectin, starch composition influences the texture of the final product. Higher levels of amylopectin facilitate the development of light, elastic, and uniformly expanded textures, while increased amylose content results in denser and less expanded extrudates.

Research indicates that a 50% amylose content yields the maximum expansion of starch. During expansion, the linear amylose chains align within the shear fields, rendering them more resistant to separation. In contrast, amylopectin-rich starches, with the same moisture content, tend to expand more easily, producing softer textures. Through the gelatinization and dispersion of starch granules during extrusion, a continuous melt phase forms within the extruder, leading to reduced average molecular weight. This phenomenon promotes the optimal formation and stability of air cells at the die exit, ultimately enhancing the overall expansion characteristics. The combined presence of both amylose and amylopectin is vital for achieving the best possible expansion outcomes (Huang et al. 2022).

3.4.3 Protein

Protein significantly impacts the quality characteristics of extruded products, particularly in terms of maintaining the structural integrity of the protein matrix. Various protein types and enrichments, including those from meat, dairy, and legumes, can be incorporated into extruded snacks. However, the extrusion process subjects proteins to high pressure, shear forces, and temperature, leading to the disruption of their native structures and a decrease in solubility due to cross-linking reactions and covalent bond formation.

The choice of protein type and additives in the formulation significantly influences the texture of the final product. Moreover, during extrusion, the formation of a separate phase of polysaccharides enhances protein aggregation along the extrusion direction while reducing it perpendicular to the extrusion direction. Manipulating the temperature and macromolecular alignment can enhance protein–protein interactions, resulting in the formation of crystalline aggregations that manifest as parallel fibers with varying thickness and length. These interactions involve a diverse array of amino acids, including hydrophobic, cation-mediated electrostatic interactions, and covalent bonds, contributing to the stabilization of the extruded product network.

Recent research suggests that the formation of new peptide bonds significantly affects the extrudate structure, while the impact of disulfide bonds appears relatively insignificant. Structurally, extruded soy meal exhibits a coextensive fibrous strand embedded within a porous expanded matrix. Microscopic examination using differential staining and light microscopy reveals carbohydrate inclusions and steam-generated voids enclosed within a protein-rich matrix. Notably, protein content and the second-order protein effect contribute to approximately 77% of the variation observed in product shear values (Singh et al. 2007).

3.4.4 Fiber

Fibrous materials like bran constitute the dispersed phase within extrudates, embedded in the continuous starchy phase. Importantly, the extrusion process leaves the fiber chemically unchanged, influencing the final product's expansion dynamics. The impact of fibers on extrudate expansion appears to be contingent on their concentration. For instance, the radial and axial expansion of rice flour extrudates demonstrated an increase with the addition of 10% rice bran, but this expansion diminished at higher levels (20% and 30%).

Studies indicate that augmenting oat or wheat bran content in cornmeal up to 20% and 30%, respectively, elevates bulk density and longitudinal expansion but reduces radial expansion. At lower concentrations, the elongated and rigid fiber molecules align with the flow direction in the extruder, reinforcing the feed material's expansion and enhancing its mechanical resilience longitudinally. However, the structural anisotropy of fibers proves detrimental to the biaxial extensional properties of extrudates, consequently limiting their radial expansion. Beyond a critical concentration, fiber molecules disrupt the melt's continuous structure, impeding its elastic deformation during the expansion process.

Furthermore, fibers tend to bind with the moisture present in the matrix, thereby reducing its availability for expansion. It is generally recommended that the body requires soluble dietary fibers (SDFs) and insoluble dietary fibers (IDFs) in a ratio of 1:3 for optimal physiological benefits. Maintaining the appropriate balance of these dietary fibers is crucial for promoting digestive health and overall well-being (Camire et al. 1990).

3.5 RTE PRODUCTS USING EXTRUSION TECHNOLOGY

Categories of snacks have expanded to include options such as granola bars, protein bars, and baked chips. Additionally, there is a rising need for snacks that meet particular dietary requirements, such as low-sugar, gluten-free, and vegan choices. Manufacturers of snacks are always coming up with new ideas to cater to these various tastes and design goods that not only satiate cravings but also fit consumers' lifestyles and health preferences.

3.5.1 Breakfast Cereals

Breakfast cereal refers to a type of processed food product that is typically consumed as a breakfast meal. They are traditionally dominated by extruded snack products, such as breakfast cereals, extruded cereal shapes, and cereal biscuits/bars. It is commonly made from grains, such as wheat, corn, oats, rice, or barley, which are processed and often fortified with vitamins and minerals. These grains are usually cooked, flattened, and shaped into various forms, such as flakes, loops, puffs, or clusters. Breakfast cereals are often consumed with milk, yogurt, or other dairy/nondairy milk alternatives, and sometimes with fruits, nuts, or sweeteners to enhance their flavor and nutritional value.

Breakfast cereals can be categorized into two main types:

1. Hot cereals: These are cooked or heated before consumption. Examples include oatmeal, cream of wheat, and rice porridge.
2. Cold cereals: These are RTE without the need for cooking. Cold cereals are often packaged in boxes and can be found in a wide variety of flavors and shapes, including traditional flakes, granola, extruded shapes, and more.

Breakfast cereals are basically RTE that are processed grain formulations suitable for human consumption without further cooking in the home. The concept was first established in USA to meet the need of preparation of time in the shortest time at home. They are prepared by first cooking the grains (optional—addition of fortifying agents before cooking) with flavoring material and sweeteners and has higher stability. There are 11 types of breakfast cereals available in market that are flaked cereals (corn flakes, wheat flakes, and rice flakes), including extruded cereals, gun-puffed whole grains, extruded gun-puffed cereals, shredded whole grains, extruded and other shredded cereals, oven-puffed cereals, granola cereals, extruded expanded cereals, baked cereals, compressed flake biscuits, muesli-type products, and many more. A basic flow diagram for the manufacturing of breakfast RTE snacks from extrusion is given in Figure 3.3. Several other ingredients like sweeteners, flavoring or

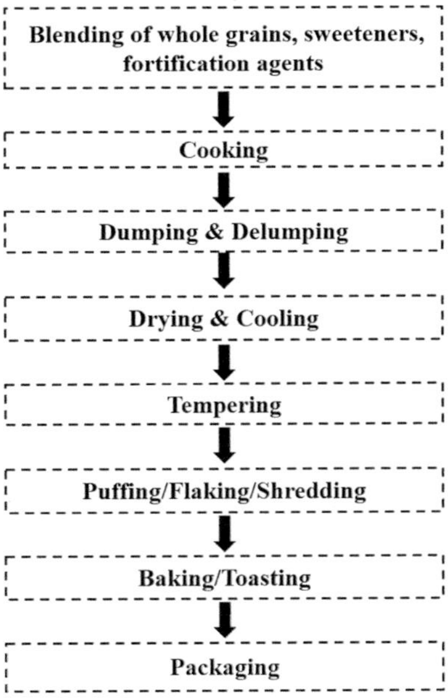

FIGURE 3.3 Process flowchart for manufacture of breakfast RTE snacks.

TABLE 3.1
Ingredients and Their Functions in RTE Cereals

Ingredients	Functions
Sugar	Imparts sweetness and used as a coating material (frosting)
Buffering salts (mono, di-, or tri-sodium phosphates)	Adjust the pH of cooked dough for browning and gelatinization
Salt	Flavoring
Emulsifiers (distilled mono glycerides and distilled di-glycerides)	Reduces the stickiness of the product
Fruits, dried fruits, milk products	Improve the nutritional value and sensory characteristics
Modified starches and stabilizers	Texturizing agent
Oil and leavening agents	Contributes to shape, texture and eating quality
Antioxidants	Improves shelf life
Flavorings and colorings	Improve the sensory and appearance of product

texturizing agents, nutritional fortification, and shelf-life preservation agents are added in the snacks for different functions (Table 3.1).

The unit operations performed during the preparation of RTE breakfast cereal snack are described as follows:

- Blending: All the ingredients can be blended in three different forms—namely, dry blending for dry ingredients, liquid blending for liquid ingredients, and combining liquid and dry ingredients.
- Cooking: It is usually done by two methods, viz., batch cooking and extrusion cooking, to provide desirable flavor.

- Dumping: To empty a batch of cooked grains/grits from the cooker, it takes about 7 minutes. But no dryers in industry can dry them to flakable moisture in 7 minutes.
- Delumping: After dumping, cooked grains/grits pass through a delumping equipment so as to obtain particles small enough for good circulation of heated air around each particle for uniform drying.
- Drying: The temperature used for drying should be <120°C and under controlled humidity conditions. Controlled humidity prevents case-hardening of the grit surface and greatly decreases the time needed for drying to the desired end moisture, usually 10%–14%. Drying is done in conveyor dryer either by single-pass or multi-pass conveyer dryer (most preferred).
  - Flake products are dried to 16%–22% moisture from 30% to 33% moisture.
  - Gun-puffed products are dried to 10%–12% moisture from 30% to 32% moisture.
- Cooling: After drying, cooked grains are cooled (present in the drier) to bring them down to the ambient temperature.
- Tempering: This allows the moisture content to equilibrate among the grains/grits as well as from the center to the surface of the individual particles. The retrogradation of starch (firmness of grain due to starch crystallization) allows moisture equilibration.
- Flaking: Flaked products are produced by passing tempered grits or pellets through two large counter rotating metal rolls (one adjustable, one fixed) so that the distance between them or roll gap can be set to produce a flake of the desired thickness. Optimum temperature of the roll surface required for flaking is 43–46°C. If the temperature is greater than 50°C, there may be a chance of excessive roll wear and sticking off the product to roll surface, whereas at temperature lower than 43°C, it lacks the grabbing ability. During flaking, excessive amount of chilled water is required.
- Toasting: Flakes are usually toasted by keeping them suspended in a hot air stream rather than by laying them out on a flat baking surface like those used for cookies and crackers. Oven time–temperature should be from 275°C to 330°C for about 90 s and the final moisture content of the flakes should be in between 1.5% and 3%.
- Puffing: Puffing is a thermal process in which rapid heat transfer takes place to phase-shift the water to vapor. Two things are important for grain to puff:
  1. The grain must be steeped or cooked.
  2. Large and sudden pressure drop must occur in atmosphere surrounding the grain.

3.5.2 RTE Snacks

Snacks are basically divided into three categories based on the product developed and technology used. Extruded snacks can be classified as natural snacks (first generation), direct expanded products (second generation), and half-cooked or pellets or coextruded snacks (third generation) that are made from a combination of multiple ingredients (Figure 3.4).

The directly expanded snacks likewise called "second-generation" snacks are delivered by direct steam expansion of the extrudate where water in the melt is superheated which flashes as steam when pressure falls in the die end. Moisture loss and cooling make the material solid in the expanded state (Singh and Smith 2007). These snacks are frequently promoted as being high in fiber, low in calories, high in protein, and nutritious (Lazou and Krokida 2011). They typically have low bulk densities. The most noteworthy level of expansion relies upon the starch content and the amylose and amylopectin content. Considering by and large course, the ingredients detailing 60% or less of starch level might appear in somewhat extruded hard and crispy items after frying and popping (e.g., corn chip snack). As per Riaz (2000a), the lowest content of starch required for development is 60%–70%; he additionally said that ingredients' composition having starch of 40%–50% might appear in an ordinary expansion of a few folds. The process flowchart for the manufacture of second- and third-generation RTE snacks using extrusion technology is shown in Figure 3.5.

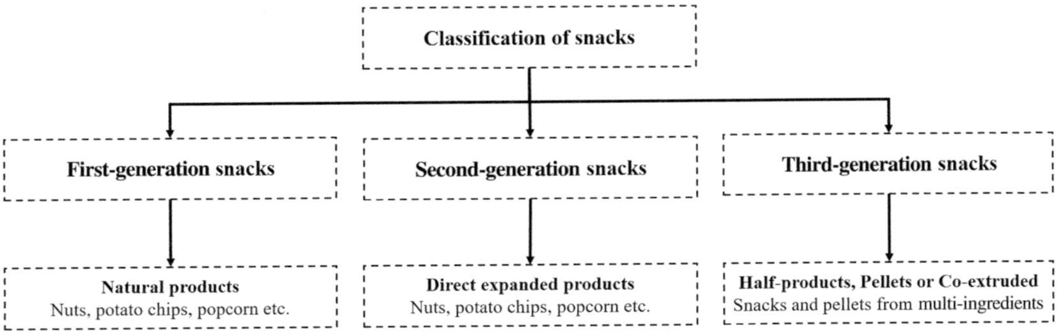

FIGURE 3.4 Classification of snacks.

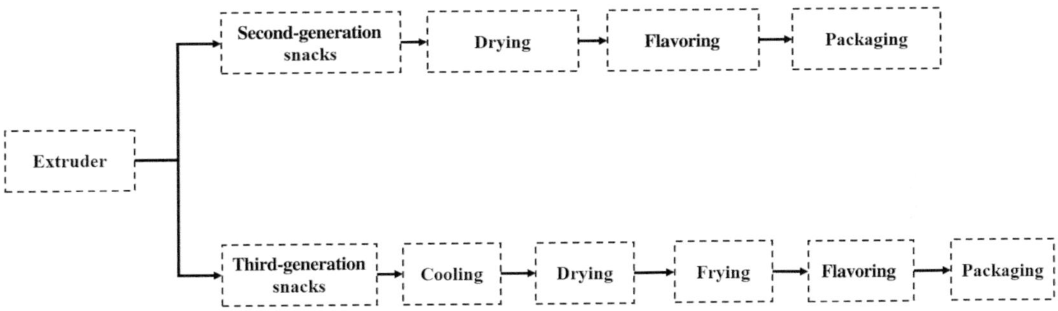

FIGURE 3.5 Process flowchart for the manufacture of second and third generation extruded RTE snacks.

Third-generation snacks or pellets are sometimes referred to as "half products" or "pellets." These snacks are made essentially identically to second-generation snacks, except as the result comes out of the extruder, it takes on the shape of the die and is dried in this state rather than expanding. This developed extruded pellet has a shelf life of a year or more under ambient conditions. Afterward, the product expands by frying, hot air heating, or microwave heating (Riaz 2000a; Carvalho et al. 2009). According to Carvalho et al. (2009), this type of product has a low moisture content (between 7% and 10%), high density, and stability to be stored for an extended period of time without microbiological harm. They are prepared in homes or restaurants using different techniques such as frying, puffing, or baking as per the requirement mentioned by the manufacturer.

3.5.3 Confectionary

The adoption of extrusion technology in confectionery products has allowed the development of complex gummy shapes, multi-layered filled candies, textured gummy bars, medicated confectionery, etc., by forcing a mixture of ingredients (including sugar, cereal flour, fat, starch, milk powder, cocoa, flavors, and colors) through an extruder using dies and nozzles of specific characteristics. Licorices, toffees, flour-based or cocoa-based confectionaries, creams and pastes, gums and jellies, aerated confectionaries, etc. (see Figure 3.7) can be easily manufactured using extrusion technology (Maskan and Altan 2011) with a variety of raw materials in precooked, raw, paste, wet, dry, or paste form. Extruder has a major advantage over other technologies due to its better control during polymorphism of fats, sugar crystallization, starch gelatinization, Maillard reactions, and changes in enzymes occurring during processing and can easily be scaled up for multiple purposes.

Sweets and candies: Twin-screw extrusion cookers with a plasticization unit are employed for the production of sweets and candies (commonly called licorice) with an elastic and spongy nature.

Licorice extract, sugar, starch, gum Arabic, or gelatin, and water are fed, mixed, and cooked. Heat treatment is applied to the mixture, and gelatin is pumped into the plasticization unit at the end of the extruder. Flavorings and other additives are applied at the end section of the barrel. For tough rubber licorice, 43% wheat flour, 44.5% sucrose, and water are fed with double kneading, compaction, and degassing. Commonly, concurrent twin-screw extruders and single-screw extruders are employed for the production of toffee to produce stable and consistent quality toffees along with a 35% saving in electrical energy. The application of these extruders extends to co-extrusion techniques, which are utilized in the form of filled sweets and crystal tubes with different fillings (Mościcki 2011).

Gums and jellies: Chewing gums are produced in specially prepared food extruders that allow lower energy consumption, better production stability, and extruded gums, which have higher flavor retention and are more durable. In contrast to the conventional method, chewing gum with mint flavor produced from extrusion-cooked gum has a fresh taste and 50% longer chewing. At the beginning of barothermal treatment, additives are applied to the gums.

Jellies are produced in a simpler way than gums. Glucose, agar, syrup, pectin, and gelatin are used as ingredients and fed to the single-screw extrusion cooker, followed by degassing and subsequent cooling. A higher degree of gelatinization is observed in starch jellies. Higher pressure can be observed in low-humidity mixtures for jellies. Thus, steam can be injected directly into the barrel, causing an increase in efficiency and no significant rise in moisture.

Creams and pastes: Extrusion cookers, in the case of creams and pastes, allow direct feeding of ingredients (fat, sugar, sorbitol, gelatin, ground nuts, dry fruits) and assist by maintaining consistency throughout and reducing the production time. In comparison to the conventional method of cream production, the risk of contamination is reduced. Additives are fed at the end section of the barrel. Co-extrusion equipment can be employed for the production of multi-layer or filled products with a coating of chocolate or sugar icing.

Advancement and application of extrusion-cooking have led to the development of new confectionery products and the recognition of consumers. This field has great potential and is developing rapidly to overcome the limitations of the conventional procedures followed in the production of confectionery products.

3.5.4 Pet Foods

Pet food is another category where extrusion is widely used to develop products, about 95% of the commercially available pet food is extruded (Tran et al. 2008). The predominant use of extrusion in the pet food industry is due to the versatile nature of the technology and its ability to increase nutrient digestibility, improve functionality, and texturize, pasteurize, and sterilize a large variety of food ingredients. Extrusion processing in this involves mixing ingredients, steam conditioning, compressing, and forcing out.

- Raw materials commonly selected for pet food include cereals with their milling waste, fish, animal fats, premixes, vegetable oils, meat, etc.
- Preconditioning plays an important role during which water, steam, or a few percent fats are added causing starch gelatinization in a few seconds, which ultimately enhances extruders' productivity.
- Grinding and mixing for uniform moisture uptake and extrusion cooking achieved as a result of uniform particle size distribution enhances the palatability and appearance of the pet food.
- It is fed through the extruder and processed with controlled parameters as desired by the end product.
- Extrudates are dried to achieve the required consistency, shelf life, and storage properties. In a few products, coating (fats and oil-based suspensions) with the help of spray is done to add an aroma or flavor to the product. Fat is coated to the dried and hot extrudate before cooling such

that it is absorbed rapidly. In the extruded material, vitamins are lost due to thermal treatment till they reached the exit end, thus at the end section of the barrel, vitamins are added during the coating process.

The adoption of extrusion in the pet food industry has not only become economical and useful but has also led to the development of new products with increased functionality and palatability that were earlier time-intensive, underutilized less accepted, and costly to produce.

3.5.5 Texturized Vegetable Protein

Texturized vegetable protein is a new category of snack products developed using extrusion. The term "textured vegetable proteins" refers to "texturates" formed from water- and plant-based protein sources that have undergone a process to change them from a powder-like substance to a structured material (Baune et al. 2022). Extrusion is a major process that provides the desired texture to these vegetable proteins (Mandliya et al. 2022a). These snacks can have different characteristics such as fibrous structure, particle size, and protein content. These TVPs are designed in dry form for snacks and flavors are added externally afterward. They may appear as strips, shreds, chunks, balls, bits, granules, slices, or other forms.

TVPs can be prepared from various protein-rich ingredients such as legume material, cereal material, lentil flour, and soy protein using low-moisture extrusion technology. Some novel protein ingredients have been explored recently to develop a variety of snack products such as mycelium (Mandliya and Mishra 2021; Mandliya et al. 2022a) that needs to be dried beforehand for application in extrusion (Mandliya et al. 2022b), spirulina (Palanisamy et al. 2019). Hong et al. (2022) investigated 28 commercial TVP samples from soy protein, pea protein, wheat gluten, chickpea protein, pea/chickpea mixed protein, and pea/navy bean mixed protein for analysis and prepared them into patties (Figure 3.6). Meat chunks is a common product made from plant protein and has spongy texture when hydrated. There are three different types of meat extenders: Chunks (15–20 mm), minced (2–2 mm), and flaked (2–2 mm). Compared to minced goods or chunk-style meat extenders, extender flakes rehydrate more quickly (Riaz 2000a).

The application of fruits and vegetables by-products in TVP can provide the desirable fiber for the snack production (Dey et al. 2021). The cost of producing texturized protein snacks can vary depending on the choice of ingredients and the production process. However, the cost of these products is generally higher than starch and cereal-based RTE snack products and thus available in market in a quantity of 200 g or more.

3.6 RECENT ADVANCES IN RTE FOOD PRODUCTS FROM EXTRUSION

With the advancement in technology, RTE food products produced from extrusion technology have gained widespread attention and development in terms of energy saving, cost cutting, nutritional enhancement, healthier food alternatives, wide variety of taste and texture, enhancement of shelf life, waste reduction, etc. Among the technologies developed, hot-melt extrusion, supercritical CO_2 extrusion, and extrusion-based three-dimensional (3D) printing hold great potential for the production of a wide variety of RTE food products. Apart from technological advancements, process modification and utilization of novel and sustainable ingredients can create various functional and nutritional RTE food products.

3.6.1 Valorization of Fruits and Vegetable Waste for Development to RTE Snacks

RTE snacks developed from fruits and vegetable waste are an innovative way to contribute toward a circular economy, and sustainability, and reduce the environmental impact of food industries.

FIGURE 3.6 Commercial TVPs available (1–14: soy protein; 15–21: pea protein; 22–25: wheat gluten; 26: chickpea protein; 27: pea/chickpea mixed protein; 28: pea/navy bean mixed protein). (Adapted from Hong et al. 2022.)

During industrial processing, high amounts of fruit and vegetable wastes are generated which are rich in phytochemicals and nutrients that can further be utilized to contribute to RTE-based food products. Valorization can be done using different extraction methods to extract vitamins, minerals, or antioxidants from waste generated. Further, it can be used in the form of dry powder and can be used as an ingredient for nutritional, functional, or flavor enhancement. The fiber content can be increased, or micronutrient and bioactive enrichment. Different studies conducted on the development of RTE food products utilizing biowastes are presented in Table 3.2.

As consumers are focused on consuming dietary supplements, the utilization of by-products in the fruit and vegetable industries has not gained the potential popularity and the bioactive, and natural sources-rich products are discarded within the food industries. Thus, the valorization of fruits and vegetables through extrusion technology to produce RTE-based snacks can offer great advantages to consumers as well as contribute toward circular economy.

3.6.2 Supercritical CO_2-Based Extrusion

In this method, first, the liquid CO_2 is cooled, its pressure is increased above the supercritical pressure, and the supercritical CO_2 is heated. Then, the supercritical CO_2 is injected into a dough or melt fed into the extruder and replacing the use of steam to control puffing. Water acts as a plasticizer when mixed with the ingredients. Once the supercritical CO_2 is injected, it breaks down due to the shear action of mixing the screw into a nucleated gas bubble, distributes it uniformly, and produces products with a puffed empty core. This method causes a change in the rheological properties of the material and develops a product with a uniform regular shape, unconnected internal pores, uniform internal cell structure, less intensive puffing, and a smoother outer surface (Prabha et al. 2021). For

TABLE 3.2
Utilization of Different By-Products of Fruits and Vegetables in RTE Snacks

By-Product Utilized	Results	References
Citrus pomace with corn extrudates	58.33% increase in fiber content and 2.12 times increase in dietary fiber content was observed with the incorporation of 15% pomace.	Asif et al. (2023)
Fig molasses by-product (FMP) and corn semolina	A proportion of FMP-7% was found to be optimum for snack production. The end product was found to have a fiber content of 2.04%. At optimum condition, no significant difference was observed between the actual values and response variables.	Boluk et al. (2023)
Beetroot with corn grits	10% beetroot and 25% water content were suggested to incorporate in corn mixture for value-added snacks	Igual et al. (2023)
Tender coconut mesocarp with RTE Kheer	10% mesocarp was found to have maximum acceptability in terms of texture, sweetness, flavor, and consistency	Naik et al. (2023)
Avocado seeds	A porous puffed snack was successfully developed with less crispiness, hard and more brittle in comparison to rice and barley extrudates. When oven drying replaced by extrusion antioxidant capacity increased from 58% to 69%	Permal et al. (2023)
Defatted sesame flour with corn	Protein content of 19.21%, water solubility index of 19.37%, water absorption index of 4.75 g/g, and hardness of 214.07 N was estimated in the extrudates	Gojiya et al. (2022)
Rosehip puree with corn extrudates	Extrudates rich in bio-active compounds were developed with higher hygroscopicity, swelling index, and water absorption index.	Igual et al. (2021)
Carrot pomace with corn starch extrudates	Carrot pomace content of 5 g/100 g was observed to be best in terms of extrudate expansion. With the increase in ample pomace concentration, surface of the extruded product also increased.	Kaisangsri et al. (2016)
Apple pomace and cheese whey into puffed extruded product	The extruded product was very light in weight with density of 0.21–0.35 g/cm^3 with dietary fiber 14 g/100 g, phenolics 93 mg/100 g, vitamin C 652 mg/100 g. Addition of cheese whey increase mineral content from 0.3% to 0.6% (w/w)	Paraman et al. (2015)
Partially defatted hazelnut flour (PDHF) and fruit waste with durum clear flour	Responses were majorly dependent on PDHF in comparison to fruit waste. Bulk density and water solubility index increased with PDHF content and porosity and water absorption index had a decreased with PDHF content. 95% confidence interval existed between the responses.	Yağci and Göğüş (2008)

starch–water mixture, supercritical CO_2 acts as an effective plasticizer. The rate of injection and rate of pressure drop are key parameters during supercritical CO_2-based extrusion (Figure 3.7).

Advantages of supercritical CO_2 extrusion:

- Extruded products with low-shear expansion and at lower temperature reduces
- Varying texture with a smooth external surface and porous internal surface
- Control over barrel pressure and temperature
- Reduced melt viscosity and pH
- Energy savings and preservation of nutritionally high-quality food, specifically heat-labile products

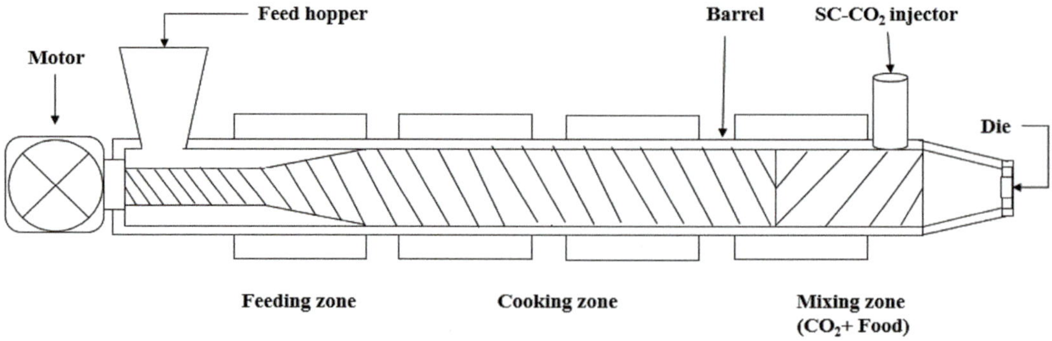

FIGURE 3.7 Schematic of supercritical CO_2-based extrusion. (Adapted with permission from Prabha et al. (2021).)

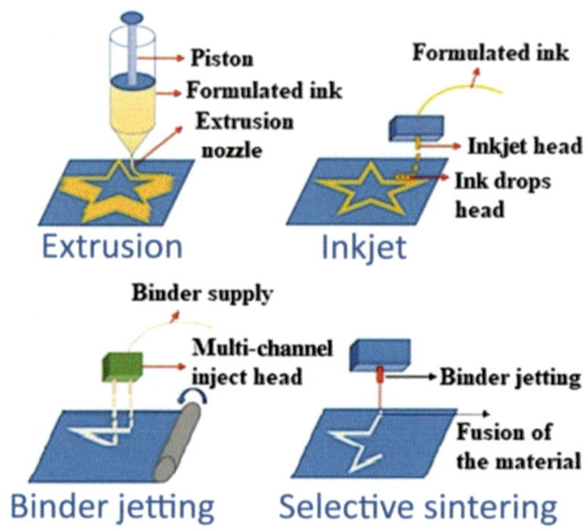

FIGURE 3.8 Various types of 3D printing technologies. (Adapted with permission from Sharma et al. 2023.)

This technology can be used for waste utilization of by-products obtained from fruit pomace and whey when combined with cereal formulations, which ultimately leads to the development of nutritionally enriched puffed snacks with increased palatability and economic feasibility.

3.6.3 Extrusion-Based 3D Printing

Extrusion-based 3D printing technology involves the utilization of a robotic arm that is attached to a syringe/cylinder, and extruded products are produced from the cylinder nozzle (Figure 3.8).

In this technology, extrusion is controlled digitally and can be useful to develop intricate food design, internal structure, geometries, controlled nutrient content, and higher surface-to-volume ratio. Further, the material can be extruded in a layer-by-layer fashion with the help of specific nozzles. Different parameters including ability to retain the structure, nutritional and rheological properties, thermal stability, and ease in deposition play a key role in the overall aspects of the end product developed.

Anukiruthika et al. (2020) conducted research on development of 3D printed cereal-based snacks and assessed the quality, printability, and nutritional aspect of the product. An increase of total essential amino acid from 32.5% to 38.2%, and protein content of 41.3 g/100 g was observed for

10%–20% of ground insect-enriched snacks. From 0% to 20% snack enrichment amino acid score also increased from 41.6 to 65.2.

For the application of this technology in a precise manner and to obtain high-quality products, certain parameters, including rheological properties, extrusion temperature, 3D printing time, and post-processing operations, are very important. This technology finds its application in products including pasta, cakes, meat and cheese products, cookies, chocolates, etc. It can further be extended to healthier food alternatives, targeting specific consumer groups (people suffering from dysphagia), fruits and vegetable matrices for printing, etc.

3.7 FUTURE SCOPE AND CONCLUSION

This chapter explained the mechanism of development of RTE snacks using extrusion technology, i.e., expansion as well as bubble growth and bubble collapse mechanisms. Additionally, a brief but detailed overview of different types of extruder designs were reviewed to explain the process of manufacture of RTE snacks. Other than the basic raw materials such as corn, rice, wheat, barley, soy, etc., the application and function of different key components, viz., starch, moisture, protein, carbohydrates, and vitamins and minerals were discussed to give the idea of how these key components affect the development of RTE snacks during extrusion. Different categories of extruded RTE products and their manufacture process were reviewed such as breakfast cereals, RTE snacks, confectionary, pet foods, and texturized vegetable protein. The use of protein to develop protein-rich products are gaining popularity with the increase in consumer awareness. Soy protein is mostly used for protein enrichment or texturization. However, soy has been linked with a lot of problems such as deforestation, allergens, etc. Thus, exploration of novel ingredients that can reduce the carbon footprint providing less allergens and is healthy and nutritious is required, which needs to be further explored in future. The utilization of by-products for the manufacture of antioxidant- and fiber-rich RTE snacks is a promising field with environmental situation. Novel technologies such as supercritical CO_2 along with extrusion needs to be explored in depth. There is limited information on the texturized RTE snacks made from novel ingredients. For their efficient use, commercial, environmental, and dietary factors should also be taken into consideration.

REFERENCES

Abd El-Hady, E. A., & Habiba, R. A. "Effect of soaking and extrusion conditions on antinutrients and protein digestibility of legume seeds." *LWT-Food Science and Technology* 36(3), (2003) 285–293.

Anukiruthika, T., Moses, J. A., & Anandharamakrishnan, C. "3D printing of egg yolk and white with rice flour blends." *Journal of Food Engineering* 265, (2020) 109691.

Artz, W. E., Warren, C., & Villota, R. "Twin-screw extrusion modification of a corn fiber and corn starch extruded blend." *Journal of Food Science* 55(3), (1990) 746–754.

Asif, M., Khan, M. K. I. Khan, M. I., Maan, A. A., Helmick, H., & Kokini, J. L. "Effects of citrus pomace on mechanical, sensory, phenolic, antioxidant, and gastrointestinal index properties of corn extrudates." *Food Bioscience* 55, (2023) 103012.

Baune, M. C., Terjung, N., Tülbek, M. Ç., & Boukid, F. "Textured vegetable proteins (TVP): Future foods standing on their merits as meat alternatives." *Future Foods* 6, (2022) 100181.

Boluk, I., Kumcuoglu, S., & Tavman, S. "Development, characterization and sensory evaluation of an extruded snack using fig molasses by-product and corn Semolina." *Foods* 12(5), (2023) 1029.

Camire, M. E., Camire, A., & Krumhar, K. "Chemical and nutritional changes in foods during extrusion." *Critical Reviews in Food Science & Nutrition* 29(1), (1990) 35–57.

Carvalho, A., Vasconcelos, M., Silva, P., & Aschieri, J. "Produção de snacks de terceira geração por extrusão de misturas de farinhas de pupunha e mandioca." *Brazilian Journal of Food and Technology* 12(4), (2009) 277–284.

Dalbhagat, C. G., Mahato, D. K., & Mishra, H. N. "Effect of extrusion processing on physicochemical, functional and nutritional characteristics of rice and rice-based products: A review." *Trends in Food Science & Technology* 85, (2019) 226–240.

Dey, D., Richter, J. K., Ek, P., Gu, B. J., & Ganjyal, G. M. "Utilization of food processing by-products in extrusion processing: A review." *Frontiers in Sustainable Food Systems* 4, (2021) 603751.

Gojiya, D., Davara, P., Gohil, V., & Dabhi, M. "Process standardization for formulating protein-augmented corn-based extrudates using defatted sesame flour (DSF): Sesame oil industry waste valorization." *Journal of Food Processing and Preservation* 46(12), (2022) e17203.

Guy, R. *Extrusion Cooking: Technologies and Applications* (pp. 206). Woodhead Publishing Ltd., Cambridge, UK (2001).

Hagenimana, A., Ding, X., & Fang, T. "Evaluation of rice flour modified by extrusion cooking." *Journal of Cereal Science* 43(1), (2006) 38–46.

Hong, S., Shen, Y., & Li, Y. "Physicochemical and functional properties of texturized vegetable proteins and cooked patty textures: Comprehensive characterization and correlation analysis." *Foods* 11(17), (2022) 2619.

Huang, X., Liu, H., Ma, Y., Mai, S., & Li, C. "Effects of extrusion on starch molecular degradation, order–disorder structural transition, and digestibility—a review." *Foods* 11(16), (2022) 2538.

Igual, M., Chiş, M. S., Păucean, A., Vodnar, D. C., Muste, S., Man, S., Martínez-Monzó, J., & García-Segovia, P. "Valorization of rose hip (*Rosa canina*) puree co-product in enriched corn extrudates." *Foods* 10(11), (2021) 2787.

Igual, M., Moreau, F., García-Segovia, P., & Martínez-Monzó, J. "Valorization of beetroot by-products for producing value-added third generation snacks." *Foods* 12(1), (2023) 176.

Kaisangsri, N., Kowalski, R. J., Wijesekara, I., Kerdchoechuen, O., Laohakunjit, N., & Ganjyal, G. M. "Carrot pomace enhances the expansion and nutritional quality of corn starch extrudates." *LWT-Food Science and Technology* 68, (2016) 391–399.

Kristiawan, M., Chaunier, L., Sandoval, A. J., & Della Valle, G. "Extrusion—cooking and expansion." In: *Breakfast Cereals and How They Are Made* (pp. 141–167). AACC International Press (2020).

Lazou, A., & Krokida, M. "Thermal characterization of corn-lentil extruded snacks." *Food Chemistry* 127(4), (2011) 1625–1633.

Liu, C., Zhang, Y., Liu, W., Wan, J., Wang, W., Wu, L., Zou, N., Zhou, Y., & Yin, Z. "Preparation, physicochemical and texture properties of texturized rice produce by improved extrusion cooking technology." *Journal of Cereal Science* 54(3), (2011) 473–480.

Mandliya, S., & Mishra, H. N. "Effect of different drying methods on quality attributes and microstructure of mycelium (*Pleurotus eryngii*)." In: *Biology and Life Sciences Forum* (vol. 6, no. 1, p. 10). MDPI (2021).

Mandliya, S., Pratap-Singh, A., Vishwakarma, S., Dalbhagat, C. G., & Mishra, H. N. "Incorporation of mycelium (*Pleurotus eryngii*) in pea protein based low moisture meat analogue: Effect on its physicochemical, rehydration and structural properties." *Foods* 11(16), (2022a) 2476.

Mandliya, S., Vishwakarma, S., & Mishra, H. N. "Modeling of vacuum drying of pressed mycelium (*Pleurotus eryngii*) and its microstructure and physicochemical properties." *Journal of Food Process Engineering* 45(10), (2022b) e14124.

Maskan, M., & Altan, A., eds. *Advances in Food Extrusion Technology.* CRC Press, Boca Raton, FL (2011).

Meng, X., Threinen, D., Hansen, M., & Driedger, D. "Effects of extrusion conditions on system parameters physical properties of a chickpea flour-based snack." *Food Research International* 43(2), (2010) 650–658.

Mościcki, L. "Snack pellets." In: *Extrusion-Cooking Techniques: Applications, Theory and Sustainability.* John Wiley and Sons, Inc., Hoboken, NJ (2011).

Naik, B., Kumar, V., & Gupta, A. K. "Valorization of tender coconut mesocarp for the formulation of ready-to-eat dairy-based dessert (Kheer): Utilization of industrial by-product." *Journal of Agriculture and Food Research* 12, (2023) 100572.

Nielsen. "Nielsen global snacking report September 2014, the Nielsen Company." (2014). www.nielsen.com

Palanisamy, M., Töpfl, S., Berger, R. G., & Hertel, C. "Physico-chemical and nutritional properties of meat analogues based on spirulina/lupin protein mixtures." *European Food Research and Technology* 245, (2019) 1889–1898.

Paraman, I., Sharif, M. K., Supriyadi, S., & Rizvi, S. S. H. "Agro-food industry byproducts into value-added extruded foods." *Food and Bioproducts Processing* 96, (2015) 78–85.

Permal, R., Chia, T., Arena, G., Fleming, C., Chen, J., Chen, T., Chang, W. L., Seale, B., Hamid, N., & Kam, R. "Converting avocado seeds into a ready to eat snack and analysing for persin and amygdalin." *Food Chemistry* 399, (2023) 134011.

Prabha, K., Ghosh, P., Abdullah, S., Joseph, R. M., Krishnan, R., Rana, S. S., & Pradhan, R. C. "Recent development, challenges, and prospects of extrusion technology." *Future Foods* 3, (2021) 100019.

Riaz, M. N. "Introduction to extruders and their principles." In: *Extruders in Food Applications*, M. N. Riaz *(Ed.)* (pp. 1–23). CRC Press, Boca Raton, FL (2000a).

Riaz, M. N. "Texturized vegetable proteins." In *Handbook of Food Proteins* (pp. 395–418). Woodhead Publishing, Cambridge, UK (2000b).

Rokey, G. J. "Single-screw extruders." In: *Extruders in Food Application*, M. N. Riaz (Ed.) (pp. 25–50). Technomic Publishing Co., Inc., Lancaster, Pennsylvania (2000).

Rokey, G. J., Plattner, B., & de Souza, E. M. "Feed extrusion process description." *Revista Brasileira de Zootecnia* 39, (2010) 510–518.

Schuchmann, H. P. "Extrusion zur Gestaltung von Lebensmittelstrukturen." *Chemie Ingenieur Technik* 80, (2008) 1097–1106.

Sharma, R., Nath, P. C., Hazarika, T. K., Ojha, A., Nayak, P. K., & Sridhar, K. "Recent advances in 3-D printing properties of natural food gels: Application of innovative food additives." *Food Chemistry* 432, (2023) 137196.

Shruthi, V. H., Sharanagouda, H., Udaykumar, N., Ramachandra, C. T., & Kisan, J. "Effect of temperature, feed moisture and feed composition on the physical properties on corn extrudates." *Environment and Ecology* 35(3), (2017) 1610–1617.

Singh, N., & Smith, A. C. "Application of extrusion cooking technology." In: *Food Science and Technology*, Khatkar, B.S., Ed. Daya Books Publishing House, Delhi (2007).

Singh, S., Gamlath, S., & Wakeling, L. "Nutritional aspects of food extrusion: A review." *International Journal of Food Science & Technology* 42(8), (2007) 916–929.

Singh Gujral, H., & Singh, N. "Extrusion behaviour and product characteristics of brown and milled rice grits." *International Journal of Food Properties* 5(2), (2002) 307–316.

Singha, P., Singh, S. K., Muthukumarappan, K., & Krishnan, P. "Physicochemical and nutritional properties of extrudates from food grade distiller's dried grains, garbanzo flour, and corn grits." *Food Science & Nutrition* 6(7), (2018) 1914–1926.

Tran, Q. D., Hendriks, W. H., & van der Poel, A. F. B. "Effects of extrusion processing on nutrients in dry pet food." *Journal of the Science of Food and Agriculture* 88(9), (2008) 1487–1493.

Van Hoan, N., Mouquet-Rivier, C. L. A. I. R. E., & Trèche, S. "Effects of starch, lipid, and moisture contents on extrusion behavior and extrudate characteristics of rice-based blends prepared with a very-low-cost extruder." *Journal of Food Process Engineering* 33(3), (2010) 519–539.

Xu, E., Pan, X., Wu, Z., Long, J., Li, J., Xu, X., Jin, Z., & Jiao, A. "Response surface methodology for evaluation and optimization of process parameter and antioxidant capacity of rice flour modified by enzymatic extrusion." *Food Chemistry* 212, (2016) 146–154.

Yağcı, S., & Göğüş, F. "Response surface methodology for evaluation of physical and functional properties of extruded snack foods developed from food-by-products." *Journal of Food Engineering* 86(1), (2008) 122–132.

4 Ultrasound and High-Pressure Processing of RTE

Recent Trends

Indira Dey Paul,[1*] *Lakshmi E. Jayachandran,*[2] *and Pradyuman Kumar*[3]
[1]Department of Food Technology, Haldia Institute of Technology, Haldia, West Bengal, India
[2]Faculty of Fisheries Engineering, Department of Food Technology, Kerala University of Fisheries and Ocean Studies, Panangad, Kochi, India
[3]Department of Food Engineering & Technology, Sant Longowal Institute of Engineering & Technology, Longowal, Punjab, India
*Corresponding Author

4.1 INTRODUCTION

According to the 2020 Food and Health survey by the International Food Information Council, the coronavirus disease 2019 (COVID-19) pandemic has brought about significant changes in our food habits (Inanoglu *et al.*, 2022). The fatigue induced by continuous home cooking during lockdown period increased the demand for more and more RTE food products. These products can be in canned, powdered, or frozen form.

Conventional thermal processing is a well-established approach for the processing and preservation of food products. However, the drawbacks of heat processing include a compromised nutrient profile, energy inefficiency, and loss of sensory acceptability. Nonthermal processing technologies employ alternate medium of processing such as pressure, radiation, electricity, sound waves, etc. for processing and preservation purposes pertaining to food products.

4.2 ULTRASOUND TECHNOLOGY

Ultrasound (US) waves in the frequency range of 20 Hz to 20 kHz can be used effectively for elimination of microflora as well as inactivation of spoilage enzymes.

4.2.1 Working Principle

The sound waves vibrating at more than 16 kHz frequency are known as US waves. They propagate by means of a series of expansion and compression (Figure 4.1). Again, the waves propagating at the frequency range of 20–100 kHz are known as power US, which have been widely applied in food processing such as protein structure modification, bioactive compounds extraction, sterilization, etc.

When the local pressure at the rarefaction points drops below the vapor pressure of the fluid (through which the US waves are passing), vapor bubbles form. The bubbles grow in size via the process of rectified diffusion till the local pressure becomes equal to the vapor pressure of the fluid. The moment the local pressure becomes higher than the vapor pressure of the fluid, implosion is

 DOI: 10.1201/9781032622408-4

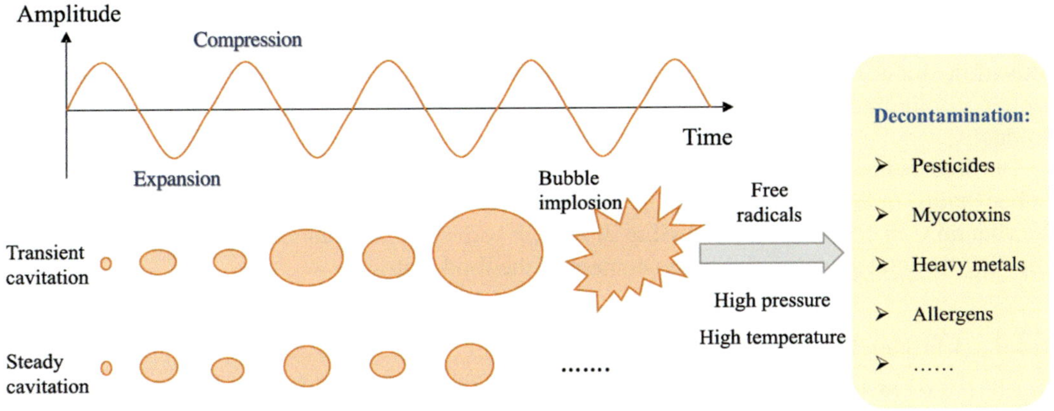

FIGURE 4.1 The phenomenon of cavitation and its application in food processing. (Yuan *et al.*, 2021.)

initiated and the bubbles collapse (Figure 4.1). This phenomenon is termed as cavitation, or precisely transient cavitation (at >1 W/cm^2). Low-intensity US (<1 W/cm^2) produces stable cavitation, where the local pressure is not allowed to exceed the vapor pressure, thereby avoiding implosion.

The transient cavitation results in localized hot spots and high-intensity shear forces and mechanical impacts, which kill the microorganisms or affect the protein configuration. Several physical and chemical changes are induced in the fluid medium due to the cavitation phenomenon. Cavitation is promoted in medium containing impurities or suspended particles. It is associated with the generation of high-temperature zones, high-intensity shock waves, and free radicals which contribute effectively to inactivation of microorganisms. Explosion of the cavitation bubbles may also lead to high-shear microstreaming (Chen *et al.*, 2020). All these mechanisms ensure cell damage and membrane disruption, further resulting in microbial cell death.

4.2.2 US Equipment and Mode of Operation

Ultrasound processing equipment can be classified into bath-type and probe-type based on the configuration of the transducers (Rastogi, 2011). The bath-type sonicator comprises metallic enclosures with in-built transducers and transmits the sound energy to the liquid medium filled inside the enclosure. The product to be processed will be immersed in the liquid medium for uniform distribution of the sound energy. The probe-type sonicators consist of the transducers connected to a probe/horn in series. The design ensures that the probes vibrate with the same frequency as that of the transducer. The probe will be kept in direct contact with the liquid medium and the intensity of treatment will be amplified. It needs to be pointed out that the direct contact of the probe with the food medium may increase the chances of food contamination or leaching of the metal from the probe material. Hence, it should be fabricated with stainless steel or good-quality metal alloys.

4.2.3 Effect on Food Quality Changes

The US can effectively reduce the microbial load in food products by means of cavitation induced cell membrane damage or loss of cell permeability. In addition, prolonged sonication time can lead to temperature buildup, resulting in lethality of spoilage microorganisms. Among the different types of microflora, the bacteria are more susceptible to the lethal effects of US, while the microbial spores are the most resistant to the US. The yeast and mold can be eliminated with an increase in the US power. In contrast to other non-thermal processing techniques, the US treatment leads to irreversible damage to the microbial cells.

The US-induced cavitation generates sufficient energy to break the hydrogen or van der Waals forces of the enzyme structure. The free radicals produced during cavitation alters the activation site of the enzyme structure. The ability of the cavitation phenomenon to disrupt the protein configuration has been exploited in elimination of allergens/toxins in different food products.

The free radicals produced by cavitation hastens the degradation of color pigments, such as anthocyanins, in the fruit and vegetable products. US is also reported to have an adverse effect on the vitamin C in citrus fruits. Processing of food in bath sonicator can lead to leaching out of solids affecting the total soluble solid (TSS) content of the food matrix.

4.2.4 Critical Processing Parameters

Higher power of US is required in viscous food, whereas lower power would suffice in attaining microbial lethality in low-viscosity food. Lower pH is reported to have a significant impact on the enzyme inactivation by US. Presence of sugar in high amounts offers a shielding effect to the microbial population in the food medium.

Research studies reported that the food spoilage bacteria are the most prone to lethality by US mechanism. Among the different types of bacteria, the cocci with large exposure area of the cells are easily eliminated by the cavitation effects. The gram-negative bacteria, due to the thin layer of cell wall, can be easily disrupted than the gram-positive bacteria. However, the microbial spores have been reported to show maximum resistance to the cavitation effects.

Conversely, no definitive trends have been observed in the case of US-induced inactivation of enzymes. Different enzymes show varying responses to the US processing based on the pH of the food, type of food matrix, temperature, etc. Majority of the studies emphasize on the need of coupling US with other mild treatments to enhance the inactivation of enzymes.

Regardless of the type of sonicator used, the frequency, amplitude, and power play a crucial role in altering the treatment intensity. The high levels of amplitude and power often give rise to larger cavitation bubbles, which collapse releasing huge quantity of energy. The cavitation phenomenon is accelerated up to a particular level of frequency (beyond which an increase in frequency further impairs the cavitation).

4.2.5 Application of US in Preparation of Ready-to-Eat Food Products

The US technology exhibits a high versatility in food processing and preservation applications (Figure 4.2). It has been reported to be effective in the sanitization of fresh-cut fruits such as strawberries, bell pepper, lettuce, etc., used for direct consumption (Birmpa *et al.*, 2013; Wang *et al.*, 2022). These products offer a convenient and healthy meal option to busy urban population. US-assisted pasteurization of packaged sausages is also reported to be highly effective in extending the shelf life significantly. Ready-to-drink juices such as apple juice, tomato juice, etc. have been effectively pasteurized by the application of ultrasonication. The US treatment, in combination with the chemical preservatives, has shown interesting results in the shelf-life extension of RTE vegetable salad (Marques *et al.*, 2019). The efficiency of chlorination of winter jujube was improved upon combining with US treatment at 28 kHz. The US pre-treatment was vital in attaining crispy and attractive RTE carrot chips by drying (Konopacka *et al.*, 2017).

Curing and tenderization of meat are enhanced by the application of the US, which is achieved by the changes induced in the myofibrillar components of the meat by the mechanical and physical effects of cavitation (Al-Hilphy et al., 2020). The US processing has been reported to reduce the cooking time of whole brown crab and improve the leaching of the dirt into the cooking water, obviating the need for further cleaning and pasteurization (Condón-Abanto et al., 2018). Promising

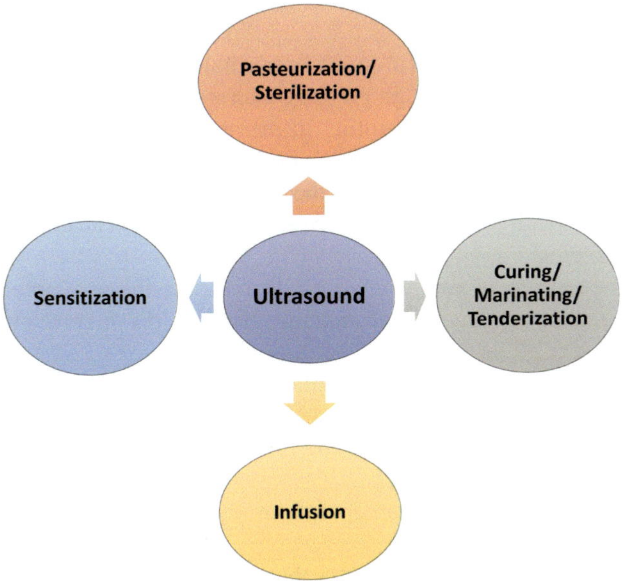

FIGURE 4.2 Major applications of US technology in processing of RTE foods.

results have been reported on the application of US in cooking, brining, and marinating the animal products and seafoods (Gómez-Salazar *et al.*, 2021).

It is worth pointing out the ability of US waves to induce sensitization of allergen/toxins in food. Sonication reduced the levels of trypsin inhibitors in soy bean in addition to enhancing its eating properties (Huang *et al.*, 2008). Sprouting of legumes is also hastened by sonication process (Liu *et al.*, 2021). The sensitization ability of the US technology has also been exploited in removal of allergens in different types of foods such as peanuts and shrimps (Salve *et al.*, 2019; Dong *et al.*, 2020). This technique is also suitable for improving the shelf stability and consistency of ready-to-use non-dairy milk such as peanut milk, coconut milk, almond milk, etc. (Salve *et al.*, 2019).

It can be concluded that the US technology with or without the inclusion of additional hurdles offers a versatile solution to the adverse effects of conventional heat treatment. Elaborate studies need to be performed on individual products to assess the suitability of this approach.

4.3 HIGH-PRESSURE PROCESSING TECHNOLOGY

High-pressure processing (HPP) is a non-thermal food preservation as well as processing method. Traditionally, HPP has been applied in ceramics, steel and super alloy production, crystal, diamond and plastic processing industries. The studies on effect of HPP on foods and microorganisms began in the year of 1889 by Hite (Hite, 1899; Abdel Karim, 2011; Guerrero-Beltran *et al.*, 2004). It is also termed as hyperbaric pressure processing or ultra-HPP or high hydrostatic pressure (HHP) or pascalization. In this method, the food material is exposed to elevated pressures with or without heat application. It gained popularity because it is a green technology, and it causes minimal damage to product's nutritional quality without compromising on its lethal effect against microorganisms. Recently, numerous attempts have been made to apply HPP in RTE food development. Before discussing its utility in RTE food domain, let us shed some light on its working principle, equipment design, and, most importantly, effect on food quality.

4.3.1 Working Principle

HPP involves subjecting food material (in-container or bulk) to elevated pressure ranging from 50 to 1000 MPa. The application of such high pressure exerts shear force on the cell membrane of the microorganisms (present in the food), resulting in membrane disruption, which in turn interrupts their normal cellular function. Under extreme pressurization, the microbial cell compresses, leaching out valuable cell materials. Therefore, although the cell can regain its original size after decompression, it becomes physiologically inactive. Moreover, HPP may also cause localized thermal damage and denaturation of proteins.

Pressure affects the system in two ways, i.e., decrease of the available molecular space and increase of intrachain reactions. Hence, fundamentally, HPP works on three main process principles:

1. *Isostatic principle* states that pressure application is uniform and instantaneous throughout the food sample, regardless of size and shape of the sample (Figure 4.3). It indicates that high pressure will not cause any harm to the food product and upon pressure release, the product will be able to regain its original shape.
2. *Le Chatelier's principle* states that operations leading to volume reduction of food are accelerated by HPP application and vice versa.
3. *Principle of microscopic reordering* states that at constant temperature, increase in pressure increases degree of molecular ordering. As temperature rise increases molecular disorder, temperature and pressure exert antagonistic forces on structure of molecules and in turn on chemical reactions as a whole.

As described by Le Chatelier, HPP favors any phenomenon in food system where volume decrease is involved. However, the reactions involving increase of volume will be inhibited.

4.3.2 HPP Equipment and Mode of Operation

As the method is involved with the application of high pressure, the key components of an HPP system includes four main parts: (1) a pressure vessel, (2) a pressure generation system, (3) a temperature control device, and (4) a material handling system.

The crucial part of the entire HPP equipment is its pressure vessel, where products under treatment are subjected to high-pressure levels. Generally, pressure vessels are made of low-alloy steel. The thickness of the vessel wall is designed based on the maximum pressure to be applied

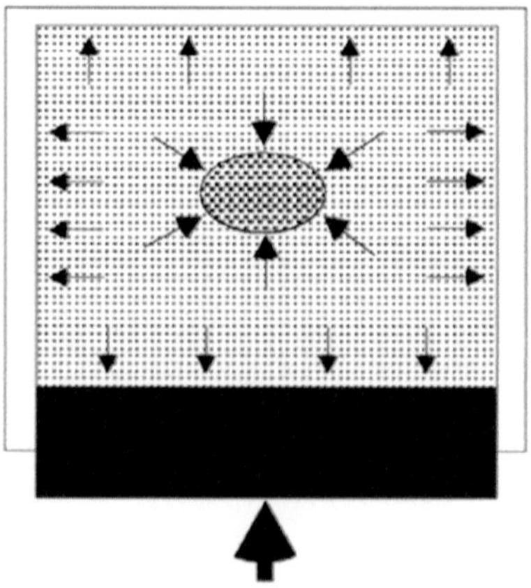

FIGURE 4.3 Isostatic principle of HPP. (Abdel Karim, 2011.)

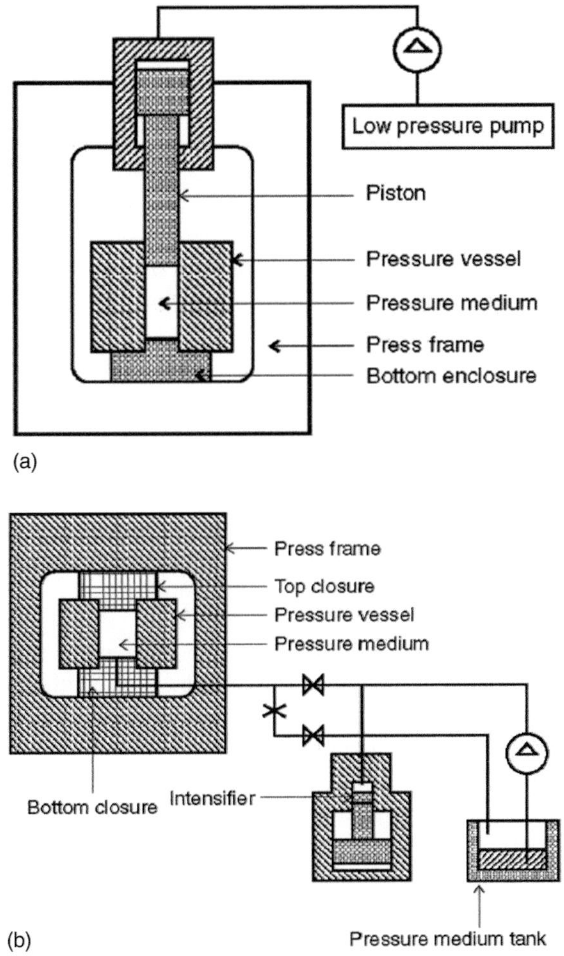

FIGURE 4.4 Schematic diagram of batch HPP employing (a) direct compression and (b) indirect compression. (Yordanov and Angelova, 2010.)

during processing, number of pressurization–depressurization cycles, and volume of the food under treatment. The vessels used for low-pressure treatments (200–400 MPa) are usually made up of forged monolithic wall material, whereas pre-stressed multilayer or wire-wound vessels are suitable for high-pressure (> 400 MPa) operations. Lubricant oils are often used to protect the pressure vessels from corrosion-related damage.

High-pressure processing involves application of enormous amount of pressure, which is generated within the pressure medium in three ways, viz., direct compression, indirect compression, and heating of the pressure medium. Out of these three, direct and indirect compression methods are commonly applied in the food industry (Figure 4.4). Direct compression generates pressure through movement of piston (mostly in batch mode equipment). It allows very fast buildup of pressure; however, its application is limited to laboratory scale and pilot scale due to the requirement of high-pressure dynamic seals between the piston and the pressure vessel surface. In case of indirect compression, the pressure medium from a reservoir is pumped through a high-pressure intensifier into the pressure vessel at desired pressure levels. Intensifier uses oil to purge pressure medium inside the vessel. This method only requires efficient tubing and fittings and is preferred industrially over direct compression method. The third method of compression utilizes heat-induced expansion of the pressure medium, which results in generation of high pressure inside the vessel. Heating of pressure medium is applied when food needs to be treated under combined effect of high pressure and high temperature.

The desired level of pressure is conveyed to the food material through a pressure transmitting medium (PTM) or simply pressure medium. Pure water and inert gases are the most commonly used PTM. Main selection criterion for PTM is its compressibility. The lower the compressibility of a medium, the higher is its pressure transmitting ability. High-pressure application on a highly compressible medium will reduce its volume, because of which the food material within the medium would not experience effective and uniform compression needed for inactivation of microorganisms or enzymes. This may lead to hazardous incidents in the HPP system. As water is 20,000 times less compressible than air, it is preferred as a PTM over gases. When pressure is increased from 0 MPa to nearly 392 MPa at 22°C, water experiences volume reduction by about 5%; this reduction is much smaller compared to that of inert gases. Food material kept within water (acting as pressure medium) is exposed to instantaneous and uniform pressure throughout its surface.

As HPP is primarily a non-thermal treatment, it is necessary to control temperature fluctuations within the equipment. Temperature within the pressure vessel can be maintained by surrounding the vessel with a heating or cooling jacket. The temperature level inside the vessel can be monitored by placing a thermocouple at the desired places of the equipment.

After HPP treatment, the food products need to be disposed carefully to avoid post-contamination hazards. The type of material handling system depends upon the mode of HPP operation.

There are three main categories of HPP operation: batch, semi-continuous, or continuous modes. In the batch mode, the packed food (packed using flexible packaging materials) to be treated is loaded within the pressure vessel. After loading, the vessel is filled up with PTM and pressurization starts. The compression of the PTM is continued till the attainment of the desired pressure level, which is then maintained for a specific time period (holding time or treatment time). After adequate holding, the pressure vessel is depressurized and the treated food product is taken out. Sometimes solid foods are placed within the vessel without packaging. Semi-continuous mode of operation involves more than one pressure vessel. Mostly liquid foods are treated in this setup. Through a low-pressure pump, liquid food material is pumped into one vessel, where a free piston is used to increase the compression. As the vessel fills up, the free piston is pushed up. Then, the inlet port is closed and PTM is introduced behind the free piston via a high-pressure pump to compress the piston from the top. As a result, the liquid product in the first vessel gets compressed. The holding operation occurs in the sequential pressure vessel. After desired holdup, the pressure is released and the free piston gradually moves toward the discharge port, pushing the treated liquid through the outlet into the collection tank. Less work has been conducted on the continuous mode of HPP system, hence its exploration is limited.

4.3.3 Effect on Food Quality Changes

The effect of HPP on food can be clearly understood through the concomitant action of pressure–temperature on the treated products (Yordanov & Angelova, 2010). There are three phases, as shown in Figure 4.5. During the initial compression phase (P_1–P_2; T_1–T_2), the volume of the food decreases along with increase in temperature from T_1 to T_2. Next, the product is held under pressure for a certain period (T_2–T_3; $P_2 \approx P_3$); during this phase, the change in product temperature (T_2–T_3) does not depend on the compression rate, provided there is negligible exchange of heat between the product and the surroundings. The final phase is the decompression phase (P_3–P_4; T_3–T_4) where the pressure is gradually released and the product temperature drops down to a temperature (T_4) slightly lower than its initial temperature (T_1). This decrease in final temperature may be due to the heat losses during the compression phase. For every 100 MPa increase in pressure at room temperature, the

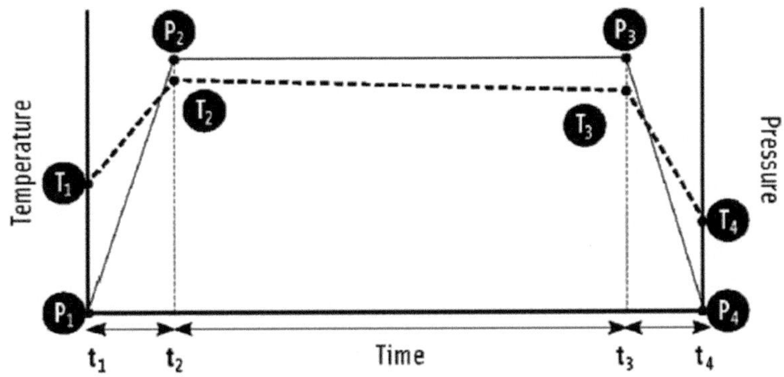

FIGURE 4.5 Pressure–temperature relationship during HPP operation. (Yordanov and Angelova, 2010.)

temperature of the water increases by about 3°C, whereas for fats and oils the temperature rise is about 8 to 9°C.

The quality changes in food due to HPP is primarily brought about by its effect on the cell membrane of the microorganisms and enzymes present in the food. The irreversible changes or reforms in the proteins responsible for the metabolism and replication of the microorganisms is the major cause for their inactivation. The rate of microbial inactivation is significantly influenced by the type of microorganisms (which in turn decides the pressure level), food matrix composition, treatment time, pH, and a_w of the food product. HPP treatment under lower pH and a_w results in greater extent of inactivation. Enzymes are special proteins having catalytic properties. Hence, the mechanism of enzyme inactivation during HPP is similar to the process of protein denaturation. The application of high pressure leads to partial or total unfolding of the native structure of the enzymes, rendering unfavorable modifications of their active sites, which in turn reduces the enzyme activity. Modification of the tertiary and quaternary structures of the enzymes is responsible for the alteration of the hydrogen bonds, and hydrophobic and electrostatic interactions. The primary structure of the enzymes remains unaffected by HPP due to the presence of several covalent bonds.

The nutritional quality of the food encompasses large-molecular-weight compounds such as proteins, nucleic acids, and polysaccharides, as well as the low-molecular-weight compounds such as flavoring agents, pigments, bioactive compounds, and vitamins. The HPP process affects noncovalent bonds and high-molecular-weight compounds, whereas it has no effect on low-molecular-weight compounds. As low-molecular-weight compounds contribute toward the nutritional and sensory characteristics of food, the non-reactivity of HPP toward them helps in retaining the fresh attributes of the processed food.

4.3.4 Critical Processing Parameters

Almost all types of food products are suitable for high-pressure treatment. The technology has been successfully applied to various food items, such as whole fruits, fruit and vegetable juices, milk and milk products, meat, salsa, fish, and rice products. Batch-type systems can be used for both solid and liquid food products, whereas semi-continuous types are mainly suitable for liquid foods (basically pumpable foods). However, choice of packaging materials is limited for batch-type HPP compared to the semi-continuous type HPP. The composition and pH of the food materials can manipulate the efficacy of the HPP treatment toward microorganism inactivation. Again, the level of applied pressure greatly depends on the type of bacterial strains. Spores (> 800 MPa at ambient temperature) are most pressure resistant followed by gram-positive bacteria and then gram-negative bacteria. Germination of the spores increases their susceptibility toward pressure; hence, germinated spores can be killed under 600 MPa at 50 to 70°C. Fungi generally require lesser inactivation pressure.

Treatment under less than 400 MPa for a few minutes is sufficient to inactivate most of the yeasts, whereas pressures between 300 and 600 MPa are sufficient to inactivate most of the molds. The nature of the growth medium also affects the pressure tolerance of the microorganisms (Yordanov & Angelova, 2010). According to Huang *et al.* (2014), fat present in the food can affect the antimicrobial efficiency of the HPP treatment by exerting a protective shield on the microorganisms. In case of enzyme inactivation, the a_w, TSS, amino acids, sugars, and polyols play major roles. Like microorganisms, level of applied pressure is dependent upon type of enzymes and their sources. While studying on the HPP treatment of mango pulp, it was found that polyphenol oxidase (PPO) is most pressure sensitive followed by pectin methylesterase (PME), followed by peroxidase (POD) (Chakraborty & Dhar, 2022).

Although HPP is a non-thermal technique, mild temperature rise can occur because of the high-pressure application (Gay Lussac's Law). The concomitant effects of pressure and temperature enhance the rate of inactivation of microorganisms and spoilage enzymes. Arrhenius equation (eqn. 4.1) is used for expressing the relationship between the inactivation rate of microorganisms/enzymes and temperature variation (at constant pressure, P_{ref}), whereas Eyring's equation (eqn. 4.2) is used for calculating the effect of pressure variation (at constant temperature, T_{ref}) on the inactivation rate of microorganisms/enzymes:

$$\ln(k) = \ln(k_{ref}) + \frac{E_a}{R}\left(\frac{1}{T_{ref}} - \frac{1}{T}\right) \tag{4.1}$$

$$\ln(k) = \ln(k_{ref}) + \frac{V_a}{RT}(P_{ref} - P) \tag{4.2}$$

Here k is the inactivation rate constant, k_{ref} is the reference rate constant, E_a is the activation energy, V_a is the activation volume, R is the universal gas constant (8.314 J/K. mol), T_{ref} is the reference temperature, and P_{ref} is the reference pressure.

4.3.5 Application of HPP in Preparation of Ready-to-Eat Food Products

Over time, the utility of HPP has expanded from microbial inactivation to enzyme inactivation to latest "new product development." In past few years, HPP has been successfully employed in the processing of RTE food products. Stratakos and Koidis (2015) reported that worldwide approximately one-third of HPP machines have been employed for production of RTE vegetable products. This method gained fast popularity because of few reasons, viz., (1) getting approval from regulatory authorities is comparatively easier, (2) as no chemicals are used, no special labeling is required, and (3) proper selection of HPP conditions does not alter the original flavor and texture of the food product (Yordanov & Angelova, 2010).

Generally, shorter HPP pressurization time is required for RTE products than in case of raw meat (Luo *et al.*, 2021). Luo *et al.* (2021) used HPP for enhancing quality attributes of RTE hairtail fish balls. The pre-cooked hairtail (*Trichiurus haumela*) fish balls were subjected to HPP conditions of 250 MPa for 5 minutes or 300–350 MPa for 3 or 5 minutes for producing RTE surimi products. The developed product showed improved textural and sensory properties with reduced total bacterial count (by ≈ 99%) and total volatile basic nitrogen content (by ≈73%). The HPP condition of 300 MPa for 5 minutes was considered to be optimum for manufacturing RTE surimi products with desirable quality attributes.

HPP has been recognized as a post-packaging pasteurization method for low-acid foods by Food and Drug Administration (FDA) and USDA FSIS (United States Department of Agriculture

Food Safety and Inspection Service), especially when spores are not the index microorganisms (Chakraborty *et al.*, 2014). For low-acid RTE foods, mostly HPP treatment is conducted after cooking process. These RTE products do not require specific cooking instructions before consumption. Combined effect of HPP and mild heat treatment was employed by Ates *et al.* (2016) for inactivation of *Listeria monocytogenes* in a model RTE soup. HPP treatment under 600 MPa for 5 minutes at 25°C resulted in more than 6 log reduction of the pathogen, whereas no bacterial regrowth in the soup was observed for 3 weeks (at 4°C and 8°C) under the pressure of 600 MPa for 5 minutes at a temperature above 45°C. Stratakos *et al.* (2015) observed that the combination of HPP (500 MPa for 1 minute) and active packaging (using an active film based on coriander essential oil) on the surface of vacuum-packed RTE chicken breast synergistically reduced the count of *L. monocytogenes* below the detection limit for 60 days during storage at 4°C. Although the said pathogen grew at 8°C, their growth rate was suppressed. The treatments did not have significant effects on the color and lipid oxidation of the finished products. Ferreira *et al.* (2023) studied the effect of pressure (400–600 MPa), time (180–480 seconds), and a_w (0.94–0.96) on inactivation of *Latilactobacillus sakei* (a pressure-resistant lactic acid bacteria) in an RTE meat emulsion model. The response surface methodology results indicated that pressure and time had a significant effect on the inactivation rate, whereas a_w had no influence on the inactivation of the pathogen. Depending on the HPP conditions, the inactivation of the microorganism varied from 0.99 to 4.12 UFC/g.

Toledo *et al.* (2012) recommended treatment of at least 680 MPa for 8 minutes for vacuum-packed Spanish potato omelet to reduce the concentration of intact cells of *Salmonella enterica.* Balamurugan *et al.* (2018) and Rubio *et al.* (2018) successfully enhanced the safety and shelf life of cooked sausages (in-package) by subjecting it to HPP (600 MPa, 3–12.5 minutes) and hot water pasteurization. Vacuum-packed fish fillets of fishes such as hilsa, cod, and salmon have also been processed by HPP under a pressure range of 350–450 MPa for 5–10 minutes at 20–27°C (Chouhan *et al.*, 2015; Arnaud *et al.*, 2018).

The effect of HPP (alone as well as in combination with *Weissella viridescencs* protective culture) against *L. monocytogenes* present in RTE salads was explored by Stratakos *et al.* (2016). The salads were maintained at different pH values (4.32 and 5.59) at storage temperatures 4°C and 12°C. HPP-treated salads under low pH showed lower microbial load than those treated under higher pH. However, the pathogen was able to recover and grow during the later storage period.

Meats are generally marinated with salt and other ingredients for preservation purpose, increased water retention and palatability. The salt in marinated meat is one of the highest sources of dietary sodium. As excessive sodium intake in body leads to high blood pressure problems, kidney failure, and cardiovascular diseases, World Health Organization (WHO) recommended reduction of NaCl intake to 5 g/day. However, lowering salt percentage in meat hampers hydration and binding capacity of water and fat, which are vital for proper texture development of the meat products. Moreover, salt significantly contributes to the taste of the final product. Orel *et al.* (2020) recognized HPP as a potent method for improving physicochemical, sensory, and microbiological quality of sodium-reduced RTE meat products. While working on RTE chicken breasts, they observed that application of HPP in two stages (300 MPa for 5 minutes after tumbling; 600 MPa for 3 minutes to the final product) significantly reduced cooking loss and hardness of the final product, improving the sensory quality of the product. It allowed sodium reduction up to 50%. The shelf life of the sodium-reduced RTE chicken breasts was extended by more than 60 days. Tamm *et al.* (2016) applied HPP at different stages of processing sodium-reduced ham (raw material—100 MPa, after injection or massage—300 MPa, final product—600 MPa) to obtain end products with lower cooking loss, better texture, improved water holding capacity, and desirable color and sensory properties.

Ready-to-drink food beverages have also been extensively treated by HPP. Several researchers conducted comparative study between HPP and thermal treatment of juices of cucumber (Liu *et al.*, 2016), carrot (Picouet *et al.*, 2015), asparagus (Chen *et al.*, 2015), watermelon (Zhang *et al.*, 2011), and a mix of various vegetables (Barba *et al.*, 2010). The HPP-treated juices prepared by Liu *et al.*

(2016) and Barba *et al.* (2010) were observed to retain better color than thermally treated juices. However, Picouet *et al.* (2015) reported that although HPP-treated carrot juice samples had more fresh-like taste, the thermally processed samples retained more color than those treated by HPP; the nutritional quality was same for both types of samples. Regarding functional properties, Chen *et al.* (2015) observed that the ascorbic acid and total phenolic content of HPP-treated asparagus juice samples were significantly higher than thermally treated samples; this result was constant for all the pressure levels (200–600 MPa for 10–20 minutes). However, the antioxidant activity of the juice was more favored by thermal processing (121°C for 3 minutes). While working with watermelon juice, Zhang *et al.* (2011) found HPP (900 MPa for 40 and 60 minutes) to be more effective for enzyme inactivation than thermal treatment (60°C for 5, 20, 40, and 60 minutes).

4.4 CONCLUSION

The popularity of RTE food products among adults and children has witnessed massive growth in last two decades. However, the usage of conventional processing methods for their preparation affects the nutritional quality of the products. The health implications of such practices served as a driving force for the adoption of alternative novel non-thermal technologies such as HPP and US. Through several studies and surveys these technologies have proved their efficacy in terms of higher nutrient retention and lower quality deterioration. From an engineering perspective, these technologies have helped in reducing processing time, which have direct positive effect on energy saving and operational costs. Both US and HPP are considered to be environment friendly as they do not leave any kind of residue after the operation. Despite all these advantages, the usage of US is limited as the process of cavitation exerts certain negative impact on the bioactive components of the food. Moreover, it is inactive against enzymes. Hence, it will be wise to combine US treatment with other non-thermal methods. HPP is effective against both microorganisms and enzymes, but it is mostly suitable for pumpable foods. Although batch-mode HPP can process solid foods, the limited number of HPP suitable packaging materials and efficient seals restrict their wide application. Addressing these issues will help in production of greater number and variety of US and/or HPP-processed RTE products.

REFERENCES

Abdel Karim, P. (2011). *High Pressure Processing as an Alternative Food Preservation Technology and Its Applications for Fruits and Vegetables*. Kansas State University.

Al-Hilphy, A. R., Al-Temimi, A. B., Al Rubaiy, H. H. M., Anand, U., Delgado-Pando, G., & Lakhssassi, N. (2020). Ultrasound applications in poultry meat processing: A systematic review. *Journal of Food Science*, *85*(5), 1386–1396.

Arnaud, C., de Lamballerie, M., & Pottier, L. (2018). Effect of high pressure processing on the preservation of frozen and re-thawed sliced cod (*Gadus morhua*) and salmon (*Salmo salar*) fillets. *High Pressure Research*, *38*(1), 62–79.

Ates, M. B., Rode, T. M., Skipnes, D., & Lekang, O. I. (2016). Modeling of *Listeria monocytogenes* inactivation by combined high-pressure and mild-temperature treatments in model soup. *European Food Research and Technology*, *242*, 279–287.

Balamurugan, S., Inmanee, P., De Souza, J., Strange, P., Pirak, T., & Barbut, S. (2018). Effects of high pressure processing and hot water pasteurization of cooked sausages on inactivation of inoculated Listeria monocytogenes, natural populations of lactic acid bacteria, *Pseudomonas* spp., and coliforms and their recovery during storage at 4 and 10 C. *Journal of Food Protection*, *81*(8), 1245–1251.

Barba, F. J., Esteve, M. J., & Frigola, A. (2010). Ascorbic acid is the only bioactive that is better preserved by high hydrostatic pressure than by thermal treatment of a vegetable beverage. *Journal of Agricultural and Food Chemistry*, *58*(18), 10070–10075.

Birmpa, A., Sfika, V., & Vantarakis, A. (2013). Ultraviolet light and ultrasound as non-thermal treatments for the inactivation of microorganisms in fresh ready-to-eat foods. *International Journal of Food Microbiology*, *167*(1), 96–102.

Chakraborty, S., & Dhar, R. (2022). *Fundamentals of Non-Thermal Processes for Food Preservation*. CRC Press.

Chakraborty, S., Kaushik, N., Rao, P. S., & Mishra, H. N. (2014). High-pressure inactivation of enzymes: A review on its recent applications on fruit purees and juices. *Comprehensive Reviews in Food Science and Food Safety*, *13*(4), 578–596.

Chen, F., Zhang, M., & Yang, C. H. (2020). Application of ultrasound technology in processing of ready-to-eat fresh food: A review. *Ultrasonics Sonochemistry*, *63*, 104953.

Chen, X., Qin, W., Ma, L., Xu, F., Jin, P., & Zheng, Y. (2015). Effect of high pressure processing and thermal treatment on physicochemical parameters, antioxidant activity and volatile compounds of green asparagus juice. *LWT-Food Science and Technology*, *62*(1), 927–933.

Chouhan, A., Kaur, B. P., & Rao, P. S. (2015). Effect of high pressure processing and thermal treatment on quality of hilsa (*Tenualosa ilisha*) fillets during refrigerated storage. *Innovative Food Science & Emerging Technologies*, *29*, 151–160.

Condón-Abanto, S., Arroyo, C., Álvarez, I., Brunton, N., Whyte, P., & Lyng, J. G. (2018). An assessment of the application of ultrasound in the processing of ready-to-eat whole brown crab (*Cancer pagurus*). *Ultrasonics Sonochemistry*, *40*, 497–504.

Dong, X., Wang, J., & Raghavan, V. (2020). Effects of high-intensity ultrasound processing on the physiochemical and allergenic properties of shrimp. *Innovative Food Science & Emerging Technologies*, *65*, 102441.

Ferreira, N. B. M., Rodrigues, M. I., & Cristianini, M. (2023). Effect of high pressure processing and water activity on pressure resistant spoilage lactic acid bacteria (*Latilactobacillus sakei*) in a ready-to-eat meat emulsion model. *International Journal of Food Microbiology*, *401*, 110293.

Gómez-Salazar, J. A., Galván-Navarro, A., Lorenzo, J. M., & Sosa-Morales, M. E. (2021). Ultrasound effect on salt reduction in meat products: A review. *Current Opinion in Food Science*, *38*, 71–78.

Guerrero-Beltran, J. A., Barbosa-Canovas, G. V., Lopez-Malo, A., Palou, E., Welti-Chanes, J., & Bermudez, D. (2004). Fundamentals and applications of high pressure processing to foods. *Novel Food Processing Technologies. Food Science and Technology*, 157–181.

Hite, B. H. (1899). *The Effect of Pressure in the Preservation of Milk: A Preliminary Report* (Vol. 58). The Station.

Huang, H., Kwok, K. C., & Liang, H. H. (2008). Inhibitory activity and conformation changes of soybean trypsin inhibitors induced by ultrasound. *Ultrasonics Sonochemistry*, *15*(5), 724–730.

Huang, H. W., Lung, H. M., Yang, B. B., & Wang, C. Y. (2014). Responses of microorganisms to high hydrostatic pressure processing. *Food Control*, *40*, 250–259.

Inanoglu, S., Barbosa-Cánovas, G. V., Sablani, S. S., Zhu, M. J., Keener, L., & Tang, J. (2022). High-pressure pasteurization of low-acid chilled ready-to-eat food. *Comprehensive Reviews in Food Science and Food Safety*, *21*(6), 4939–4970.

Konopacka, D., Cybulska, J., Zdunek, A., Dyki, B., Machlańska, A., & Celejewska, K. (2017). The combined effect of ultrasound and enzymatic treatment on the nanostructure, carotenoid retention and sensory properties of ready-to-eat carrot chips. *LWT-Food Science and Technology*, *85*, 427–433.

Liu, F., Zhang, X., Zhao, L., Wang, Y., & Liao, X. (2016). Potential of high-pressure processing and high-temperature/short-time thermal processing on microbial, physicochemical and sensory assurance of clear cucumber juice. *Innovative Food Science & Emerging Technologies*, *34*, 51–-58.

Liu, H., Li, Z., Zhang, X., Liu, Y., Hu, J., Yang, C., & Zhao, X. (2021). The effects of ultrasound on the growth, nutritional quality and microbiological quality of sprouts. *Trends in Food Science & Technology*, *111*, 292–300.

Luo, H., Sheng, Z., Guo, C., Jia, R., & Yang, W. (2021). Quality attributes enhancement of ready-to-eat hairtail fish balls by high-pressure processing. *LWT*, *147*, 111658.

Marques, C. S., Grillo, R. P., Bravim, D. G., Pereira, P. V., Villanova, J. C. O., Pinheiro, P. F., ... & Bernardes, P. C. (2019). Preservation of ready-to-eat salad: A study with combination of sanitizers, ultrasound, and essential oil-containing β-cyclodextrin inclusion complex. *LWT*, *115*, 108433.

Orel, R., Tabilo-Munizaga, G., Cepero-Betancourt, Y., Reyes-Parra, J. E., Badillo-Ortiz, A., & Pérez-Won, M. (2020). Effects of high hydrostatic pressure processing and sodium reduction on physicochemical properties, sensory quality, and microbiological shelf life of ready-to-eat chicken breasts. *LWT*, *127*, 109352.

Picouet, P. A., Sarraga, C., Cofan, S., Belletti, N., & Guardia, M. D. (2015). Effects of thermal and high-pressure treatments on the carotene content, microbiological safety and sensory properties of acidified and of non-acidified carrot juice. *LWT-Food Science and Technology*, *62*(1), 920–926.

Rastogi, N. K. (2011). Opportunities and challenges in application of ultrasound in food processing. *Critical Reviews in Food Science and Nutrition*, *51*(8), 705–722.

Rubio, B., Possas, A., Rincón, F., García-Gímeno, R. M., & Martínez, B. (2018). Model for *Listeria monocytogenes* inactivation by high hydrostatic pressure processing in Spanish chorizo sausage. *Food Microbiology*, *69*, 18–24.

Salve, A. R., Pegu, K., & Arya, S. S. (2019). Comparative assessment of high-intensity ultrasound and hydrodynamic cavitation processing on physico-chemical properties and microbial inactivation of peanut milk. *Ultrasonics Sonochemistry*, *59*, 104728.

Stratakos, A. C., Delgado-Pando, G., Linton, M., Patterson, M. F., & Koidis, A. (2015). Synergism between high-pressure processing and active packaging against *Listeria monocytogenes* in ready-to-eat chicken breast. *Innovative Food Science & Emerging Technologies*, *27*, 41–47.

Stratakos, A. C., & Koidis, A. (2015). Suitability, efficiency and microbiological safety of novel physical technologies for the processing of ready-to-eat meals, meats and pumpable products. *International Journal of Food Science & Technology*, *50*(6), 1283–1302.

Stratakos, A. C., Linton, M., Tessema, G. T., Skjerdal, T., Patterson, M. F., & Koidis, A. (2016). Effect of high pressure processing in combination with *Weissella viridescens* as a protective culture against *Listeria monocytogenes* in ready-to-eat salads of different pH. *Food Control*, *61*, 6–12.

Tamm, A., Bolumar, T., Bajovic, B., & Toepfl, S. (2016). Salt (NaCl) reduction in cooked ham by a combined approach of high pressure treatment and the salt replacer KCl. *Innovative Food Science & Emerging Technologies*, *36*, 294–302.

Toledo, J., Pulido, R. P., Abriouel, H., Grande, M. J., & Gálvez, A. (2012). Inactivation of *Salmonella enterica* cells in Spanish potato omelette by high hydrostatic pressure treatments. *Innovative Food Science & Emerging Technologies*, *14*, 25–30.

Wang, J., Huang, K., Wu, Z., & Yu, Y. (2022). Effects of ultrasound-assisted low-concentration chlorine washing on ready-to-eat winter jujube (Zizyphus jujuba Mill. cv. Dongzao): Cross-contamination prevention, decontamination efficacy, and fruit quality. *Ultrasonics Sonochemistry*, *82*, 105905.

Yordanov, D. G., & Angelova, G. V. (2010). High pressure processing for foods preserving. *Biotechnology & Biotechnological Equipment*, *24*(3), 1940–1945.

Yuan, S., Li, C., Zhang, Y., Yu, H., Xie, Y., Guo, Y., & Yao, W. (2021). Ultrasound as an emerging technology for the elimination of chemical contaminants in food: A review. *Trends in Food Science & Technology*, *109*, 374–385.

Zhang, C., Trierweiler, B., Li, W., Butz, P., Xu, Y., Rüfer, C. E., ... & Zhao, X. (2011). Comparison of thermal, ultraviolet-c, and high pressure treatments on quality parameters of watermelon juice. *Food Chemistry*, *126*(1), 254–260.

5 Functional Approaches among RTE Food and Their Stability

Recent Trending Adaptation

Bindu Bazaria,[1] *Ira Tripathi,*[2*] *Ajay Singh,*[3] *and Pradyuman Kumar*[4]
[1]Assistant Professor, Department of Food and Nutrition and Food Technology, Lady Irwin College, New Delhi, India
[2]Assistant Professor, Department of Home Science, K.N. Government P.G. College, Gyanpur, Bhadohi, Uttar Pradesh, India
[3]Assistant Professor, Department of Food Technology, Mata Gujri College, Fatehgarh Sahib, Punjab, India
[4]Professor, Department of Food Engineering and Technology, SLIET Longowal, Punjab, India
*Corresponding Author

5.1 INTRODUCTION

Technological developments have completely changed the convenience food industry, especially in the areas of food processing machinery, procedures, and packaging materials. Constant stress-free lifestyles, dietary requirements, and taste preferences all contribute to the strong demand for convenience food. Instant food is more than just a marketing ploy; it represents a radical shift in how we think about food and life. The idea of convenience or fast meals perfectly captures the spirit of the times we live in.

In a broader sense, convenience food refers to foods that have undergone significant processing by the manufacturer so that, aside from warming, thawing, cooking, frying, diluting, and reconstitution, consumers will need to perform little to no additional processing on them (Paradkar et al., 2007). If a food item satisfies certain requirements, it may be categorized as convenience food.

- It must have had little cooking or processing until the consumer consumes it, and the product must have undergone significant cooking or processing by the producer before it arrives at the store and ultimately the customer.
- There should be little preparation time before it is eaten. Convenience food, also known as tertiary processed food, is intended to save customers time, cut down on spoiling waste, and lower costs by utilizing economies of scale. These dishes only need to be heated, usually requiring very little preparation. According to Balasubramanayam (2005), they are packed to have an extended shelf life with minimal flavor and nutrient loss over time.

In developed nations, the goal of developing convenience foods was to manage the surplus of agricultural products and maintain stability in the food markets. The invention of canning by Nicolas

DOI: 10.1201/9781032622408-5

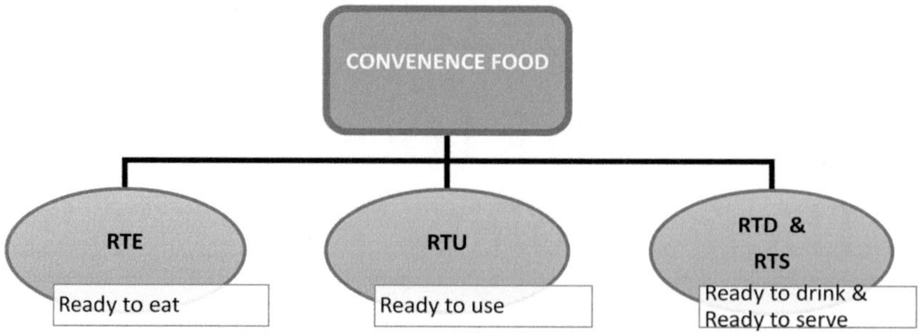

FIGURE 5.1 Classification of convenience foods.

Apppert in the 18th century was the single most significant step in the creation of convenience foods. When it came to feeding the army during the Napoleonic Wars, World Wars I and II, and the Gulf Wars in 1991, canned goods were crucial. Therefore, it is accurate to refer to canned foods as the original mass-produced convenience foods. A lot of these convenient food items are derived from military-developed foods that are meant to be easily prepared, stored, and long-lasting in combat environments. Many commercial food firms had surplus manufacturing facilities after the war. These businesses created new product lines of freeze-dried and canned goods intended for domestic use (Patel and Rathod, 2017).

Convenience foods: With the correct additives, the daily choice can be customized to meet the requirements of the army, airlines, railroads, and even patients. This food category is made up of a diverse range of meals that differ in terms of their nutritional value as well as their size, shape, preparation, and processing techniques. This literally includes anything from straightforward roasted and fried nuts to prepared mixtures, foods in cans and freezers, and elaborate warm-and-serve meals (Manohar et al., 2005). Ready-to-eat foods (RTE-F) are actually a type of convenience food (Manohar et al., 2005) (Figure 5.1).

The US Department of Defense accepted the use of the RTE meals in 1975, and they began to be delivered in 1981. Civilians in Western nations likewise embraced it with great enthusiasm. The Meal, Ready-to-Eat (MRE) is a lightweight, individual field ration that is self-contained and purchased by the US military for its service members. It is intended to be used in battle or other field situations where organized meal facilities are unavailable. The lighter Long Range Patrol (LRP) ration, created by the US Army for special forces and ranger patrol groups in Vietnam, was to be superseded by the MRE in 1981, which took the role of the canned or Meal Combat Individual (MCI) ration meal (Friedl et al., 2020).

RTE-F is, to put it simply, food that has already been packaged, can be quickly and simply prepared, and can be eaten straight out of the package, either with or without heating, thawing, or preparation (Figure 5.2). The term "ready-to-eat" refers to food that is prepared at the point of sale and ready for immediate consumption. It can be eaten hot or cold, raw or cooked, and it does not require additional heating or cooling (Sheshachala et al., 2005). RTE-F is recognized as the "ultimate processed foods" with a notable value addition as they eliminate the laboriousness of cooking meals at home and offer the convenience of "eating off the shelf" (Paradkar et al., 2007). A significant degree of preparation is done by the producer in some RTE-F, such as frozen meals, breakfast cereals, baked goods, canned food, and cooked meat (Singh et al., 2007).

Ready-to-drink (RTD) foods can be directly consumed from the container, many fruit juices and dairy drinks reside in this category. Smoothies is also one among them as RTD. Ready to serve (RTS) varies from that of RTD in terms of where some pre-preparations like dilutions or reconstitution are needed in them before consumption. It includes chicken soup, instant soup powder fruit juice concentrate, and instant juice powder like Rasna, Tang, etc.

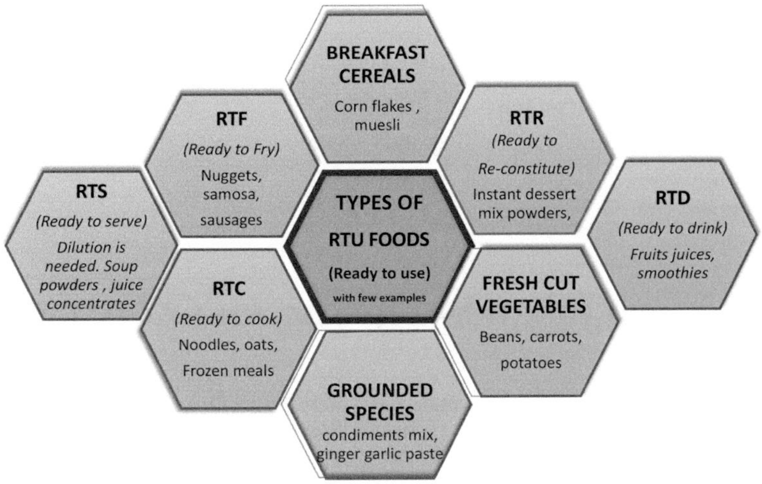

FIGURE 5.2 Classification of RTE foods.

5.2 EVOLUTION IN PROCESSED CONVENIENCE FOODS

The following four key changes highlight the rise of convenience foods:

1. The food processing industry is boosted by the rapid rise of organized retail (Dixit and Ravichandran, 2022).
2. The desire for convenience and "enjoying life" is fueled by changes in age, income, and the proportion of working women (Temgire et al., 2021).
3. The movement worldwide toward outsourcing of goods and services, especially food, to India.
4. The Indian economy has been liberalized and deregulated since 1991 (Manohar et al., 2005).

Reasons to go for RTE foods

The biggest reasons for buying RTE-F are convenience offered (87%) and time scarcity (56%). Following are the other reasons including social and economic demography.

- Changes in socioeconomic conditions:

Marked changes in the lifestyle of people have significantly influenced the growth of ready-to-eat foods. Moreover, in India, there has been an increase in the real income of people, which has promoted attraction toward convenience food.

- Increase in population of working women:

The increased participation of women in the workplace has resulted in less time being available to plan and cook meals. At the same time, double income enables families to purchase convenience food at an affordable price.

a. Scarcity of household labor
b. Lack of time and hectic schedule
c. Changes in meal pattern and food habits: The conventional three meals a-day routine has been replaced by snack-type meals. It compels the consumer to look for:

- Food of convenience
- Easy commercial availability
- Longer shelf life

d. New advances in food technology and packaging.
e. Increasing number of nuclear families and disappearance of the traditional organization of society.
f. Increase in the rate of *urbanization*, working hours, and distance between workplace and home.

5.3 CONTEMPORARY FUNCTIONAL APPROACHES INVOLVED IN RTE-F PROCESSING

The RTE foods require processing for preservation and hence the improved shelf life. The major technologies involved in the processing and manufacturing of RTE products include:

- Preservation by freezing method
- Packaging in retort pouches
- Extrusion cooking and spray drying and hurdle processing
- Microbial safety

5.3.1 Preservation by Freezing Method

With options ranging from frozen desserts to appetizers and dinners, the frozen food and beverage business offers consumers another alternative when they are on the go. The global frozen ready-meal market is expected to reach a value of $89.7 billion by 2032, growing at a compound annual growth rate (CAGR) of 8.2%, according to Future Market Insights (FMI) "Frozen Ready-Meal Market" research.

A wide range of subjects, including the market, preparation and assembly, health and safety, and the significance of product-process-package (PPP) and time-temperature-tolerance (TTT) factors, are covered in relation to frozen ready-meals, a significant segment of the frozen foods market. Because the application of these techniques to ready-meals is relatively new, there is special attention on freeze-chilling and sous vide-freezing of ready-meals and ready-meal components. In wealthy nations, there is a significant demand for frozen-ready meals due to its antimicrobial and fungal properties. Frozen ready meals are becoming more popular in developed countries, where demand is larger than in other nations.

RTE frozen meals typically include components such as vegetables, sauce, protein, and carbohydrates. All of the items can be prepared on-site; alternatively, some ingredients can be prepared on-site and others brought in as frozen goods to be added to the meal when it is assembled. When cooking on-site, the protein ingredients are either cooked through or par-cooked and then added to the tray "as is" or individual quick freezing (IQF) to be added to the trays at a later time. After cooking or par-cooking, carbohydrates are often blast-chilled, whereas vegetables are frequently delivered frozen to the trays. Before the product is placed in frozen storage in the case of frozen assembly, the full trays and product undergo one last freeze. Following a complete freezing process in a blast or spiral freezer while still warm, the sealed trays are transferred to frozen storage.

5.3.2 Packaging in Retort Pouches

The technology chosen for packaging needs to be thoroughly researched and considered to retain both a high nutritional value and a high degree of quality. The combination of packaging technology with cutting-edge materials and functional packaging has made consumers' lives easier. Several novel packaging materials, including microwave, self-heating, active and intelligent, and biodegradable packaging, have been introduced to the ready-meal market. Packaging technology is crucial not only for the protection of food but also for consumer convenience. The drive and inspiration behind the development and refinement of the retorting packaging method emerged from the need that soldiers on the front lines of the armed forces had for light, healthy food that had a guaranteed long shelf life.

"Retort is a process that uses heat and pressure to cook food in a strong sealed package. The packaging is less bulky than traditional cans and jars and food cooks more quickly, providing a better testing product for customer" (Ramakrishna, 2001). The retort packages are available in:

- Pouches
- Roll stock
- Lidding stock

The US Army promoted the concept of flexible retortable pouches for use in combat rations in the 1950s. The idea was to have a lightweight, easy-to-pack, shelf-stable food container to eliminate the heavier traditional can. In 1965 the first commercial retort pouches were produced in Italy. "Retort pouches are flexible laminated packages made from multilayer plastic films with or without aluminum foil as one of the layers". The primary characteristic of these pouches is their construction from heat-resistant plastics, which allows them to withstand temperatures as high as 121°C, which is typically reached during food thermal sterilization. This sets them apart from standard flexible pouches. Its benefit is that it combines the texture and nutritional content of frozen foods with the shelf stability of metal cans.

The packaging material for RTE-F has to satisfy different functional requirements like:

- Stability during sterilization
- Barrier against the permeation of water vapors, oxygen, and flavor compounds.

The organism of major concern is *Clostridium botulinum,* a spore-forming mesophilic bacteria, which, under anaerobic conditions, produces a deadly toxin. Interestingly, this organism cannot grow and produce toxins if the pH of the food is 4.6 or less. All RTE-F with pH >4.6 and water activity >0.85 belong to "Low Acid Canned Foods" (LACF) for which sterilization process (commercial sterilization) is to be designed to eliminate the botulism risk (Rangarao, 2007).

Construction-wise retort pouch is a three-ply laminate consisting of PET + aluminum foil + polypropylene and it is the only combination that is used in India at present.

PET or polyester, with an approximate thickness of 12 µm, serves as the outer layer and provides the necessary strength for the pouch.

- PET may be reverse printed to create eye-catching images by embedding links between the subsequent inner layer and the outer layer due to its great glossy surface and printability.
- A barrier layer of 9–25 µm thick aluminum foil is utilized, and it is this layer that gives the product a shelf life of more than a year. It stops ultraviolet light rays and oxygen from penetrating.

Further materials frequently found in retort pouch structures consist of nylon, ethylene vinyl alcohol (EVOH), and polyvinylidene chloride (PVDC).

- Polypropylene (PP), used as the inner layer, usually about 75 µm thick, provides critical seal integrity, elasticity, durability, and taste and odor integration with a variety of food products.

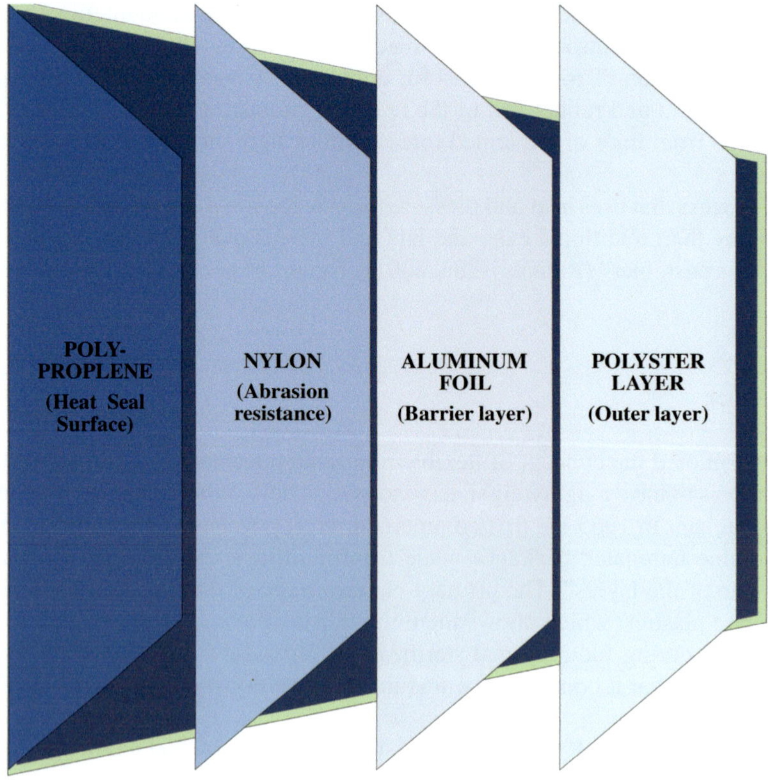

FIGURE 5.3 Retort Pouch Laminate Layers.

- When combined with PP, nylon's poor gas barrier capability makes it suitable for products that do not need to last longer than 3–6 months on the shelf (Ramakrishna, 2001). Figure 5.3 is a sequential description of layer involved in tetrapak packaging.

The pre-formed retort pouch, which is the most commonly used by the food processor, has three sides already sealed by the retort pouch manufacturer. The roll stock laminate is fed through a tensioning device to ensure that the flexible pouch is smooth. A shoulder assembly then folds the laminate along the centerline, exposing the PP surfaces to each other (some roll stock machines bring two separate rolls of laminate together). The seals are formed using a heat-sealing device and the formed pouches are cut off. One style of filler, for a liquid product, uses equipment that forms the pouch, fills, and heat seals (FFS machine) on a production-line basis.

Filling of retort pouches requires accurate measurement and dosing as the quantity of the filled product determines the thickness of the packed pouch and thickness dictates procedure time. Besides, the seal area should not be contaminated with the product as it reduces the seal strength and contributes to pouch failure during retorting. After escaping the sealer, the web of designed pouches is cut by a roller knife to separate each pouch (Rangarao, 2001). Sealing is an important parameter for retort processed foods as they are commercially sterilized. Any failure can happen like the following:

- Inadequate seal strength,
- Formation of micro channel,
- Contaminated seal area.

It may lead to the spoilage of food products in storage. Hence, a suitable continuous heat sealer (hot bar sealing and impulse sealing) that can provide a seal strength of >2 kg/mm after processing is important. Before sealing the pouch, headspace air should be removed to a very low level preferably to <10 cc. The presence of headspace air in the pouch leads to excessive pressure inside the pouch during retorting and can result in the bursting of the pouch. Besides, it interferes with uniform heat transfer which is essential to guarantee sterility. To reduce headspace air in the pouch various techniques are applied:

- Mechanical squeezing of the pouch
- Steam flushing
- Vacuum sealing
- Application of superimposed air-pressure

The sterilization process involves temperature treatment of 110–125°C to eliminate bacterial spores. Retort equipment for retort processed food is divided into steam system and hot water system from the difference of heating medium, stationary system, rotary system, and continuous system from the state of the product to be heated in the retort.

5.3.3 Extrusion Process

The extrusion process has produced a food called "quick small meal" between regular meals, i.e., snack foods, which refers to sweet and spicy/salty ready-to-eat foods and generally includes items such as chips/crisps, extruded snacks, sweets and savory snacks, sugar and chocolate confectionary, etc. (Murughan and Ramesh, 2007). In India, two types of snack foods are very common: packaged snack foods and street-vended traditional snack foods. is The consumption of crisps and chips constitutes the quickest rate of growth in snack food industry. Consumers under the age of 19 are fond of snacks (Murughan and Ramesh, 2007).

Extrusion is a technique used in the production of snack food products that involves molding a plastic or dough-like substance by pressing it through a die or restriction. "The process of extrusion involves forcing a food material through a die that is intended to form and/or puff dry the ingredients under a variety of mixing, heating, and shear conditions" (Choton et al., 2020). The process starts with cleaning the grains, then it involves milling them to remove seed coatings, grinding them into the required flour particle size, and blending them with condiments or spices.

The flour blend is mixed with water and steam in a mixture and extruded through a die in an extruder in the form of noodles or snacks (Debbarma et al., 2017). The process mechanism of extrusion processing is to gelatinize the starch contents by rupturing the particles. As gelatinized starch passes through the extruder, it expands. Figure 5.4 depicts the extrusion process and its impact on to carbohdrates, protein, and fat constituents.

With the use of this technology, food cereals such as wheat, rice, maize, and potatoes, among others, can be expanded into continuous, uniform shapes like pipes, roads, rings, and hexagons, among others (Borah et al., 2019). The pellets that are extruded have a high starch content (65–75%), which causes them to expand greatly when heated in hot air or fried in hot oil. These expanded snack foods are referred to as third-generation snack products, and they are packaged and sold as RTE snacks (Yadav and Sharma, 2005).

There are numerous manufacturers in India and around the globe producing extruded products frequently for the local markets. Although RTE (a meat-based convenience food), the processed meat sector is in its early stages. The major constraints include lack of popularity of meat products, inadequate advertising and campaign to highlight virtues of processed meat products, inadequate processing and cold chain facilities, diversity in regional taste, lack of faith among consumers in the quality of such processed meat products, inadequate distribution and marketing strategy for of

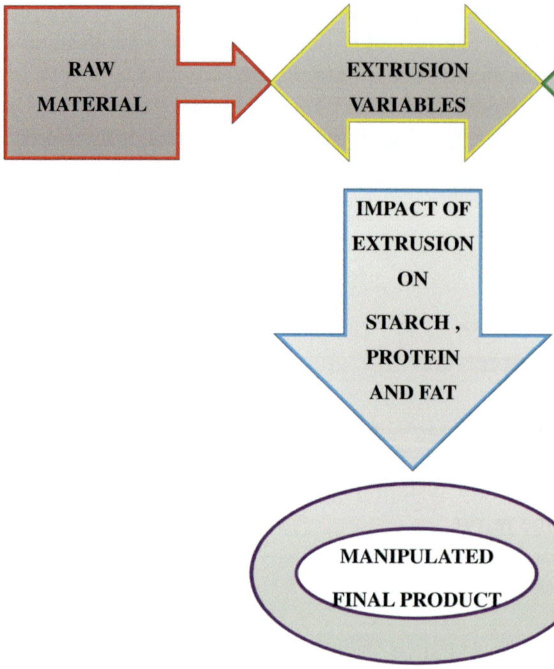

FIGURE 5.4 Extrusion process.

ready-to-eat meat products. The major multi-national companies in the RTE meat market are KFC (Kentucky Fried Chicken), McDonald's, Mary Brown, Pizza Hut, Wimpy, and Domino's.

RTE food market perspective:
The size of the worldwide dinner RTE food market is projected to reach 152.4 billion by 2027, expanding at a CAGR of 6.0% from 2022 to 2027. Dinner RTE foods are traditional frozen RTE meal items, such as frozen pizza, as well as various frozen vegetarian dishes that are kept in extremely cold storage. The increasing efforts of manufacturers to introduce novel flavors and nutritious blends—possibly with the help of plant extracts—are fueling the expansion of the global dinner RTE food market in the midst of the COVID-19 epidemic. Nonetheless, the necessary technical know-how to safely create RTE foods is crucial, and in the lack of such know-how, it is strongly advised to seek appropriate professional help. This is one of the main things impeding the progress of Global Dinner RTE Foods Market. Major global and Indian RTE Foods manufacturing companies are listed in Table 5.1.

In India, like most developing countries, both the industrial and non-industrial sectors are engaged in producing RTE foods.

- The industrial sector faces quality assurance inspections. In India's industrial sector, tonnes of RTE meals are produced annually.
- The non-industrial sector is inherently exempt from legislative regulations and quality assurance. RTE meals are made in the non-industrial sector, such as potato chips and bakery products like bread, biscuits, cakes, and rusks.

According to the Ministry of Food Processing of India (MFPOI), besides premium Indian food retailers such as MTR (Mavalli Tiffin Rooms) Foods, Bikanervala Foods, and Kitchen of India, they have already ventured into the sector are planning to explore it further, following their product preference among consumers in the United States and Europe. The estimated production of RTE foods in retortable pouches covering mostly the popular Indian curries is about 2.5 lakh pouches each day amounting to about 22,500 MT per annum. Instant food makes up between 15% and 20% of both the top retail stores' overall inventory. They claim that throughout the past 10 years, the popularity graph has been continuously rising. The increasing growth in the market of RTE is due to three primary factors: Acceptability, Affordability, and Availability.

Indian snack food market has reached a value of 1530 crores. It is one of the largest snack markets in the world. Euro Monitor International, a market research company, says that the amount of money Indians spend on RTE snacks and food is 5 billion US $ in a year, while overseas Indian or Indian subcontinents spend 30 billion US $ annually.

TABLE 5.1
Leading RTE Players in Global and Indian Market

Brands	Description
Nomad Foods Ltd.	The UK-based frozen food firm Nomad Foods. Across 17 European nations, Nomad Foods manufactures, sells, and distributes RTE meals such as fish fingers, natural fish coated fish, spinach peas, lasagna noodles, pasta, pancakes, and other items.
Bakkavor Foods Ltd	Bakkavor foods is a food production firm established in the United Kingdom that distributes its products to 10 countries in North America, Europe, and Asia. The company's range of goods includes bread, pizzas, dips, desserts, and freshly cooked dishes.
General Mills	US-based General Mills is a food corporation. The product line of the company includes yoghurt, cereals, snacks, and bakery goods.
McCain Foods	McCain Foods is present in over 160 nations worldwide. Vegetables, appetizers such as nuggets, and frozen French fries and patty are among the company's offerings.
PepsiCo,	PepsiCo is a multinational American food, snack, and beverage corporation That makes breakfast Bars, coffee drinks, energy drinks, chips crisps cereals, rice snacks, side dishes, soft drink, biscuits, crackers dips sports nutrition, packaged water, etc.
Premier Foods Group Ltd	Premier Foods is a food company of UK origin. The company produces a range of sauces, soups, sauces, and cakes.
2 Sisters Food Group	It is a UK-based food company. The company merchandise comprises red meat, fish, poultry, and chilled foods.
Greencore Group Plc.	Greencore Group food company was founded in United Kingdon. The manufactured goods range comprises salads, sandwiches, and chilled prepared meals.
Orkla ASA	Orkla ASA is food company from Norway. The company offers ingredients, branded consumer goods, and solutions in its product line.
Nestlé	It is Switzerland-based company. It produces powdered milk, condensed milk, baby food, chocolate goods, chocolates, instant coffee and tea, soups, ice cream, frozen foods, and bottled water, which are the main items offered.
The Kraft Heinz Company	It is an American global food firm. The product line of the company includes meat, cheese, coffee, sauces and toppings, snack nuts beans and pasta dishes, frozen foods, dressings, soups, baby food, and other food and grocery products.
Unilever	The worldwide consumer products firm Unilever is based in Britain. Additionally, their primary products are seasonings, soups, ketchup and dressings, desserts and baking, and sauces and gravies.
ConAgra	The corporation is an American holding company for consumer packaged goods: sauces, side dishes, pizza, frozen meals, snacks, chocolate, items made with tomatoes, spreads, ice cream, toppings, drinks, syrups, protein, and frozen breakfast.

(*continued*)

TABLE 5.1 (Continued)
Leading RTE Players in Global and Indian Market

Brands	Description
Tyson Foods	The company Tyson Foods, Inc. is based in America. It is involved in the processing of food. Its segments include prepared foods, beef, pork, and chicken, and they prepare sales from related items in addition to fresh, frozen, and value-added chicken products.
Amul	The largest food product marketing organization in India is the Gujarat Cooperative Milk Marketing Federation (GCMMF). Amul is well-known for its dairy products, but it has also branched out into the RTE market, selling items like pure ghee, processed cheese, Shrikhand, Nutr-amul, and variety of sweets.
Gits	Variety of popular ready to cook and instant foods that cover a range of ethnic Indian cuisine, and recipes have "Global pallete acceptance".
Haldirams	From a tiny setup, the traditional Indian sweet maker has grown into a full-fledged food manufacturing enterprise that exports its goods to the Western world. It provides a variety of traditional Indian packed snacks, savories, including Chana Masala, Samosa, Alu Tikki, Pao Bhaji, Gol Gappa, and Dhokla. It provides a wide variety of products, including packaged sweets, syrups, namkeens, biscuits, pickles, aloo masala, bhujia, and many more.
Ethnic Kitchens	Offers packaged sweets, syrups, namkeens, cookies, pickles, aloo Masala, Bhujia, Bhelpuri, Chana Dal, Kajui Ladoo, and many more items.
Kitchens of India	The unique flavor of Indian food is now being enjoyed outside of India because of ITC's flagship brand, "Kitchens of India." Experts of Indian cuisine in the United States, United Kingdom, Switzerland, Bhutan, Bangladesh, Hong Kong, Tanzania, Canada, and Australia can now sample these delectable delicacies.
ITC (Indian Tobacco Company)	Products from the ITC stable of India that are sold by Aashirvaad and Kitchens of India include a large variety of ready-to-cook foods and dishes, such as "regular" biryanis, curry pastes, and entrees, as well as Bukhara (an Uzbek recipe) and other exotic cuisines.
MTR (Mavalli Tiffin Rooms)	Ranked among India's top five producers of processed foods, the company says it "markets and exports a wide range of packaged foods to global markets", Currently, MTR's menu offers 22 delectable, authentic Indian curries, gravies, and rice dishes.
Priyafoods	Priya has a range of popular traditional recipes starting from north Indian delicacies along with true southern delicacies. Priya's products are available in United States, Canada, West Indies, UAE, Middle East, Singapore, Malaysia, Australia, UK, New Zealand, etc.
Rajbhog Foods Inc.	Rajbhog is a company that specializes in Indian sweets and snacks in the United States and Canada. It is taking advantage of this market niche by franchising its business model and producing and selling sweets, snacks, savories, frozen meals, ice cream, and beverages.
Rainbow Foods India	They are producers and providers of Indian frozen vegetables, snacks, and meals to the United States and the United Kingdom. They procure meals, snacks, pickles, pastes, fruits, veggies, fruits, and frozen vegetables as well as parathas and Punjabi vegetable curries as well as meals and snacks.
Satnam Overseas Ltd.	A variety of RTE Indian dishes are available at Kohinoor Heat and Eat Indian Curries. Kohinoor boasts that the "Heat and Eat range of curries use the well-established retort technology to offer extended shelf life to the products through steam sterilisation".
Shana	Ready-to-cook ethnic cuisine are the specialties of Shana Frozen cuisine. In the frozen Asian ethnic food sector in the United Kingdom and Canada, Shana is a well-known brand, and recently introduced South Indian line of frozen vadas, idlis, mini uttapams, and masala dosas.
Tasty Bite	Tasty Bite offers a selection of ready meals and entrées. Their remarkable retort pouches were created specifically for the Apollo space programme. Tasty Bite is a favorite among campers, mountain climbers, sailors, and desert safaris because of its retort packing, which has been tested to survive acute temperatures and heights from deep below sea level to as far as the moon.

TABLE 5.1 (Continued)
Leading RTE Players in Global and Indian Market

Brands	Description
Ashoka	ADF Foods Limited, an Indian company that is registered on the BSE, is the owner and manager of the Ashoka brand. The flagship brand, Ashoka, is the top Indian-made ethnic food brand. It is one of the popular ethnic Indian brands. Its selection consists of microwaveable rice, condiment pastes, pickles, chutneys, mango pulp/slices, pappadums, individually quick frozen (IQF) ready-to-cook vegetables, and RTE curries (Heat and Eat).
Vimal Agro Products	With its cutting-edge infrastructure, Vimal Agro Products can produce more than 80 different food products, including RTE meals, fruit slices and pulp, pickles, chutney, canned vegetables, and more. It can also supply these goods to retail establishments in the United States, United Kingdom, and Canada under the brands VIMAL and SWAD.

5.3.4 Microbial Safety of RTE Foods

Listeria monocytogenes poses a serious risk to the safety and security of RTE meat products due to its widespread distribution and capacity to grow at chilled temperatures. *L. monocytogenes* mainly contaminates RTE meat during the slicing and packing processes. Microorganisms have a significant impact on RTE's shelf life. The quantification of *L. monocytogenes'* presence and concentration is crucial in determining how this particular organism is regarded to be controlled in RTE foods (Zhu et al., 2005).

L. monocytogenes is a common environmental pathogen that can be found in sewage, soil, water, and decomposing plants. At refrigeration temperatures, it grows slowly. Every year in the United States, listeriosis causes about 2500 instances of disease and $200 million in financial losses. *L. monocytogenes* is a facultative anaerobic gram-positive rod-shaped bacterium that does not generate spores and is very mobile. According to Ralovich (1992), it can grow in a broad variety of temperature settings (0–42°C) and a pH range of 4.5–9.6. In meats that are close to or above pH 6.0, the organism often develops well; at pH 5.0, it typically grows badly or not at all (Glass and Doyle, 1989). It can withstand nitrite and salt (McClure et al., 1997). It may infect food and result in:

- mild non-invasive gastroenteritis (also known as listerial gastroenteritis).
- Invasive listeriosis: This is a serious illness that has a high risk of death, with a case fatality rate of 20–30%. The following are the primary target populations for listeriosis: Pregnant individuals and fetuses: illnesses throughout pregnancy and infancy, abortion, or stillbirth; immunocompromised individuals and the elderly; cancer patients, especially those with leukemia, patients with diabetes, cirrhosis, asthma, and ulcerative colitis

A Mexican-style soft cheese was the source of a severe listeriosis outbreak in the United States in 1985, which resulted in over 142 instances of illness and 48 fatalities (Linnan et al., 1988).

To varied degrees, *L. monocytogenes* has been found in deli-style salads, sandwiches, cooked crustaceans, smoked seafood, and fresh-cut vegetables and fruits. *L. monocytogenes* in RTE meals has been the subject of three risk evaluations, all of which are great resources for learning the scientifically supported microbiological requirements for this microbe.

The World Health Organization has established guidelines for environmental monitoring and microbiological testing for *L. monocytogenes* in RTE food. A maximum limit has been set for specific food products where the bacteria cannot develop, but in RTE products where proliferation is feasible, zero *L. monocytogenes* are tolerated (Khan and Ruby, 2010).

RTE foods and beverages produced and/or served by vendors and hawkers, particularly in streets and other such public spaces, are referred to as "street foods". Street foods are generally exposed to dust, dirt, soot, fumes, human hands, and other sources of contamination. Thus, the load of different types of bacteria presents is a matter of concern in such foods. Such types of foods are sold on railway counters and are prepared, handled, and served in unhygienic conditions (Senthilnathan and Saxena, 2007). In India, vast varieties of traditional RTE foods are prevalent, e.g. Panipuri, vada pav, cereal-based khakhra, etc. These typical RTE foods are primarily sold on the sidewalks, in crowded public areas, outside of gardens and parks, in marketplaces, at bus stops or railway stations, in movie theatres and other tourist destinations, as well as at upscale restaurants with a good reputation. CART (Center for Analysis, Research and Training), Delhi, showed a study of the microbiological quality of Panipuri sold in Delhi. The study showed that 100% of samples were found bacteriologically unfit for consumption (Senthilnathan and Saxena, 2007). Consumption of such RTE leads to stomach ache, dysentery, and vomiting.

5.4 OTHER ADDITIONAL FUNCTIONAL APPROACHES IN RTE FOODS

Meat and meat products are generally recognized as being components of good nutrition because these are important sources of proteins, vitamins, minerals, and anti-carcinogenic conjugated linolenic fatty acids. However, they also contain inappropriate amounts of fat, fatty acids, cholesterol, salt, and other toxic components like nitrite. So, meat products are designed so as to prevent harmful components or to reduce them to appropriate limits (Sen and Muthukumar, 2007). The designer meat foods have been generally categorized into three groups:

- Low-fat meat products
- Low sodium meat products
- Nitrite-reduced meat products

A product that is needed for defense personnel stationed at high elevations is an appetizer. Altitude sickness, which can result from exposure to high altitude, is known to cause symptoms like headache, nausea, and vomiting. Consequently, the main issues encountered at high elevations are appetite loss and weight loss. There are RTE appetizers in the form of munches, i.e., pepper munch and lemon munch, that have been developed (Wadikar et al., 2010).

Self-heating food, which is one of the essential requirements, is preferred for consumption in hot conditions, except for some cold beverages; ice cream, and similar other products. Self-heating system is one such requirement from consumers, especially service forces during logistic operations as well as the civilian sector during picnics, camping, expeditions, etc. Self-heating food containers do have a real impact as people are able to get hot and tasty meals at any time anywhere without the use of any external energy sources such as electricity and flame. It enhances emergency preparedness and is very useful during emergency situations such as hurricanes, earthquakes, tornados, floods, and snowstorms (Sabapathy and Bawa, 2005).

Self-heating food containers, popularly known as "KITCHEN IN CARTON", are designed to heat and eat on the move. In general, it is a "canincan" type system with one can containing the food product and the other containing the chemical undergoing an exothermic reaction (Bohra et al., 2015). The system developed worldwide for heating RTE retort processed foods in flexible pouches consists of three main parts:

- Chemicals which undergo exothermic reaction
- A reaction chamber where the said chemicals packed in a suitable packaging material are made to undergo exothermic reaction and liberate heat
- The third item is an RTE retort processed food in a flexible pouch (kept in contact with the chemical pad), which has to be heated by the heat liberated during the exothermic reaction.

Most of the self-heating food systems available worldwide utilize any one of the following exothermic chemical reactions:

Hydration of anhydrous calcium oxide: calcium oxide undergoes exothermic reaction with water as per the following equation releasing 0.28 kcal of heat energy per gram of the substance.

$$CaO + H_2O \rightarrow CaO + H_2O$$

This is the reaction used for heating the beverage product in the "sake can" of Japan, "Self-Heating Ready Meals" by Hot Can Auto Alloys Ltd.

Electrochemical reaction of an alloy of magnesium and iron in an electrolytic solution:

Another type of chemical used for undergoing exothermic reaction with water is an alloy of magnesium and iron. The alloy is a patented product of DYMATRON INC.USA known as "Corrodalloy-5". This is a member of a family of super corroding magnesium–iron alloys. It reacts spontaneously at a controlled rate with saline water producing magnesium hydroxide, hydrogen gas, and a large amount of heat. The reaction is a typical galvanic cell reaction in which magnesium acts as an anode and iron as a cathode. The heat evolved during the corrosion reaction of the "Corrodalloy-5" is about 3.76 kcal/g.

$$Mg + 2H_2O \rightarrow Mg(OH)_2 + H_2 + Heat$$

It has the following advantages:

- It has high heat content and is simple to use.
- It produces non-toxic and insoluble end products after the reaction.

Uses: This material is utilized in "Flameless Food Heaters" by ZestoTherm Inc. USA, "Self-Heating Meals" by Hot Pack Inc. USA, "Self-Heating Foods" by Alpine Aire Foods, USA, etc.

Oxidation of polyhydroxyl alcohols in the presence of an oxidizing agent:

Another type of exothermic reaction used for warming RTE retorts processed food in flexible pouches is the controlled oxidation of polyhydroxy alcohols such as glycols, glycerol, etc. These alcohols undergo spontaneous and vigorous oxidation in the presence of oxidizing agents such as potassium dichromate ($K_2Cr_2O_7$), potassium permanganate ($KMnO_4$) sodium peroxide (Na_2O_2), etc. The oxidation is so vigorous and generates so much heat that often it catches fire. Complete oxidation of glycerol to carbon dioxide and water can liberate 4.15 kcal of heat per gram of the substance.

$$C_3H_8O_3 + 3.5\ O_2 \rightarrow 3CO_2 + 4H_2O + Heat$$

Tempara Technology's self-heating system being used in US military MRE is based on the above-described exothermic oxidation reaction in a flexible pouch that wraps around the food container.

DFRL (Defense Food Research Laboratory) at Mysore, India, has developed a self-heating food pouch for the first time in India based on the controlled oxidation of polyhydroxy alcohols in flexible pouches. The system can heat 300 grams of RTE retort processed food products in flexible pouches from ambient temperature, as well as consumption temperature in a reasonable period of time (8–15 minutes) depending on the nature of the product and surrounding temperatures (Bohra et al., 2005).

Some of the commercially available self-heating food containers/packages are Hot Cans manufactured by HOTCAN AUTO Alloys (I and D) Ltd., UK, Zestotherm's {FRH}-Flameless Ration Heater, manufactured by the US army, Alpine-Aire Foods, Hot Pack Self- Heating Meal, manufactured by Canland, Canada and Ontro Inc. Poway, California.

Ohmic Heating: Food materials are heated in this procedure by an electrical current being passed through them, acting as a resistance. Ohmic heating is specifically employed for RTE fruits and for package sterilizing RTE meals, according to research (Sastry et al., 2011). Ohmic heating has certain benefits, according to Ruan et al. (2001). These include the ability to create heat at higher temperatures for foods containing particles, the fact that it minimizes mechanical harm and the loss of nutrition and vitamin value, and the fact that heat transfer is unrestricted during the process of creating heat, unlike traditional methods. A unique bag is used in ohmic heating to accomplish the heating process. Generally, pouches with foil electrodes are made of laminations based on PP that can withstand temperatures used for sterilization (Somavat, 2011).

Irradiation: Irradiation processing, which produces ionizing radiation with a higher energy per atomic passage, is used to preserve food quality and prevent microbiological deterioration (Irawati and Sani, 2012). The benefits of irradiation include improved food microbiological grade while maintaining the food's physical state, prevention of product recontamination, easy destruction of psychotropic microorganisms—a problem for chilled storage—and a reduced risk of disease-causing microorganisms (IAEA, 2003). Based on the target food and particular reasons, the dosage is calculated. For instance, sterilizing ready meals can kill microorganisms, including spore formers, with a maximum dose of 50 kGy. Similarly, prolonging the shelf life of chilled ready meals can be accomplished with a dose range of 0.5–3 kGy, which can harm spoilage bacteria (Wilkinson and Gould, 1996).

Curing: The process of curing, which precedes other drying methods, including salting, fermenting, marinating, and smoking, increases the product's shelf life and develops its texture and aroma. Additionally, curing affects the meat's textural, olfactory, and microbiological qualities (Özbay Doğu and Sarıçoban, 2015). According to Martin (2001), the major ingredients in curing agents are salt, nitrite, nitrate, and sweeteners (sucrose, dextrose, and corn syrup). Not every cured product is fit for consumption; others require additional steps like heating, brining, or dying. Ham, cured salmon, and certain conventional meat products are examples of general cured RTE foods. For instance, fenugreek is used to cure, dry, press, and coat pastırma, a specialty of Turkey (Özbay Doğu and Sarıçoban, 2015). Toldra et al. (2004) state that oxygen is the primary factor influencing the stability of sliced fermented and cured meats. As a result, vacuum packing (VP) and changed atmosphere are popular techniques for packaging cured goods; additionally, active and intelligent packaging is used to serve as a solution for this issue. Furthermore, according to Toldra et al. (2004), the typical gas composition of MAP for cured meats is 20–30% carbon dioxide and 80–70% nitrogen.

5.5 NUTRITION INTERVENTION POTENTIAL

Robin et al. (2013) studied American RTE stuff to evaluate the total sugar, total dietary fiber, and sodium levels in them as these all together emphasize much on to the consumer's health. They found that total sugar and sodium levels decreased 7.6–11.2%, respectively, while dietary fiber rose to 13.4% from the time span of 2005–2011. These positive reforms among RTE breakfast cereals are attributed mainly to the utilization of whole grain cereals.

Dhumal along with his associated researchers (2014) worked on to standardization of process parameters for composite-blended oil-free and microwave-puffed RTE snacks. They analyzed three blended proportions (barnyard millet flour and potato mash), i.e. 50:50, 55:45, and 60:40, subjected to steam for 10, 15, and 20 minutes wherein the most optimal treatment under cold extrusion technique for 55:45 combination steamed for 15 minutes is the best one, taking care of the thickness of the dough sheet to 50 mm thickness.

Jaybhaye and Srivastav (2015) evaluated barnyard millet flour, potato mash, and tapioca powder blend and revealed an optimized combination in a 60:37:3 ratio. A sheeted dough of thin rectangular shape processed under steam-cooked cold extrusion process followed through HTST means as 238°C/39.35 seconds for temperature and time respectively. The obtained puffed product as per its physical attributes in competition with the control sample for moisture content (9%) and expansion ratio (2.05). Immediately after puffing, the obtained mass was oven-toasted at 116.26°C for 20–23 minutes which produced a significant sensory score in comparison with its equivalent control samples.

As per the gathered data from the National Health and Nutrition Examination Survey (NHANES) for RTE cereals snacks among people of different age groups, children are found fond of its consumption (34%), similarly adults (14%), and those 65 years or older (22%). Consumption of RTE-C (cereal) was associated with significantly ($p < 0.05$) higher intake of carbohydrates, dietary fiber, Ca^{2+}, Fe^{2+}, Zn^{2+}, Mg^{2+}, K^{+}, P, along with vitamin A, B_1, B_2, B^3, B^6, B^9, B^{12}, and vitamin D but least in protein and fat consumption. Besides all these, children adapted to RTE-C also had significantly lower intake of sodium and total fat than non-eaters, but on the other side, energy intake was significantly up for RTE-C than non-eaters in adults. In a nutshell, RTE-C eaters have a higher intake of nutrients in competition with non-RTE-C eaters.

A survey was conducted in reference to RTE-C consumers for Hispanic children and Hispanic adults, which reported eaters with 40% and 14%, respectively. Both children and adults who reported consuming RTE-C had significantly higher intakes of whole grains, dairy products, vitamins, and minerals (calcium, zinc, iron, folate, vitamin A, vitamin D, thiamin, riboflavin, niacin, vitamin B12, and vitamin B6,) than non-eaters did. They also had significantly lower intakes of sodium and dietary cholesterol ($P < 0.05$ for all). It has been determined that consuming RTEC improves the quality of our diet, leading to a larger intake of whole grains, dairy products, and various vitamins and minerals in the diets of both Hispanic children and adults.

One of the recently investigated projects on the development of a novel tilapia skin-based RTE snack is by way of infrared drying and air frying using condition parameters of 45°C for 25 minutes and 180°C, respectively. Inclusion of this technique for RTE preparation showed sound records in competition with deep-fat frying, rotary baking, and infrared radiation. Highest puffing time (300 seconds) and medium expansion ratio (292.69%) along with, most interestingly, reduction in fat proportion (14.68%) and elevated sensory outcomes (scored 8.64) were observed.

An investigation for RTE snack standardization through given blends of sorghum, cowpea, and baobab; (45:55:0; 50:45:5; 55:35:5; 55:35:10; 60:25:15; 65:15:20) predicted for their physical, nutritional, and sensory attributes. Popped sorghum, baked cowpea, and pulp powder from baobab were accessed as part of minimal pre-processing operations. In summary, the results showed a larger proportion of iron (4–5g/100g), zinc (1–2g/100g), low crude fat (2–3g/100g), and a high level of crude protein (11–21g/100g). In addition to an increase in vitamin C concentration (8–21g/100g) with the rise in the proportion of baobab, sensory scores of greater than 5 indicate customer approval (Momanyi et al., 2020).

Suri et al. (2020) optimized process parameters to develop iron-rich extruded snacks under the RTE heading, keeping in mind the anemia. Selected ingredients for this work were barnyard millet, defatted soy flour, amla, and rice flour using a blend ratio (barnyard and defatted soy flour), barrel temperature, and amla concentration. Wise selection of these ingredients elevates iron, and vitamin C proportion in them if treated through 6:1 (blend ratio), 114°C (barrel temperature), and (12.25g/100g) amla proportion. These parameters announce 15.71 mg/100 g iron, and 18.91 g/100 g protein upliftment to the final product.

Bender (1978) proposed some general guidelines for creating processed foods that are nutritionally beneficial. These guidelines included making it easy to prepare a full meal, making it simple to meet human needs, placing more emphasis on foods that contain nutrients that are likely to be lacking in diets, and placing less emphasis on foods that contain excess ingredients. Techniques for processing food, like enrichment and fortification, can increase the number of vital nutrients that would otherwise be scarce and change the composition of foods to reduce potentially over-consumed ingredients (Andress et al., 2011; Nickols et al., 2010). Infant cereals fortified with iron, milk supplemented with vitamin D, margarine fortified with vitamin A, processed goods produced with iodized salt (Ohlhorst et al., 2012) cereals fortified with iron and B vitamins, and the most recent fortification of wheat flour with folic acid are a few examples (Weaver et al., 2014). Encouragement of the healthiest food options—that is, those that offer fewer elements to limit and more nutrients to encourage the calories consumed—may be more fruitful in the diet than restricting processed foods. To promote nutritional consumption, more has to be done to select processed foods with reduced levels of added sugars, sodium, and saturated fats (implying a need for a broader definition of nutrient density). Foods that are high in nutrients, whether they are processed or not, increase the likelihood that a diet will satisfy food guidance requirements.

Convenience meals are now developed according to consumer needs, tastes, and nutritional requirements because of technological advancements in food processing equipment, procedures, and packaging materials.

5.6 REGULATIONS FOR RTE FOODS

RTE foods are subject to the same FSS law standards as any other food product. To ensure RTE safety, all inspection protocols, food adulteration laws, sample collection, nutritional labeling, etc., must be followed. A very high quality of hygienic preparation is necessary for the manufacturing of RTE meals. To proceed, certain conditions must be met:

1. Hygiene and Good Manufacturing Practices (GMP), which include:
   - Proper methods for washing and disinfecting equipment;
   - Hygiene standards for the premises;
   - Personal hygiene standards that are high;
   - Reputable vendors for the ingredients.
2. Hazard Analysis and Critical Control Point (HACCP)-based actions that involve separating food that is intended for human consumption (e.g., cooked and raw meat) from non-food items, as well as related personnel and equipment. A system for evaluating the efficacy of hygiene practices and HACCP-based procedures must be in place, and data related to this system must be recorded.

5.7 FUNCTIONAL INNOVATIVE APPROACHES

5.7.1 3D Printing of Food

There are now four main types of 3D printing methods used in the food industry: binder jetting, inkjet printing, selective laser sintering, and extrusion-based printing (Liu et al., 2017). By continuously pressing food through a nozzle, extrusion-based printing produces a replica of food. This technique bears similarities to traditional fused deposition modeling (FDM). While the wire is the starting material for FDM, with low viscosity, the launch medium for extrusion-based printing can be both paste-like and solid. In contrast to FDM, which uses wire as its starting material, extrusion-based printing allows for the use of both solid and paste (soft) with low viscosity. Foods made with this method include cheese and beef paste dough (Pitayachaval et al., 2018). In addition, a process called selective laser sintering uses a power laser to precisely fuse powder particles layer by layer inside a three-dimensional structure. The laser carefully fuses the powder on the exterior of each layer by scanning cross-sections of the material. After scanning each cross-section, the powder bed is lowered, and a fresh layer of powder is applied on top. Until a satisfactory structure is achieved, this procedure is repeated (Liu et al., 2017).

5.7.2 Computer-Aided Food Engineering

The term "computer-aided food engineering" (CAFE) refers to the application of analytical computer software to the design and management of food items, procedures, equipment, or packaging, ranging from the molecular level to the scale of a whole food plant. Rather than unit operations or factory size, which would be a component of the larger computer-aided food manufacturing (CAFM), it concentrates on the food-product scale (Datta et al., 2022). CAFE functions within the overall digital transformation framework that currently defines the food business. Anticipated results span from agriculture to food to health, and they include: identifying hidden patterns, forecasting the future, streamlining processes, developing human resources, enhancing accessibility to information, and automating tasks (Venkataraman et al., 2019).

5.8 CONCLUSION

The nation's economy and public health are significantly impacted by the food trade. The modern consumer is well-informed and diligent. Convenience and nutrition are big priorities for them. Thus, combining nutrition with taste is essential. The emphasis needs to be shifted from processing utility to consumer utility. The current food processing method ought to focus on the convenience and quality of the consumer. The two most important components in drawing on potential customers are packaging and branding. Packaging is becoming a communicator and influencer of purchases. Consumer trust and brand value will increase with trustworthy and honest packaging that includes food traceability. Automation and technology are essential to the food processing industry's thriving expansion. In the present day, product innovation is essential for growth. Applications of technology will lower waste and improve food product quality. The food business is also dedicated to lowering its water and carbon footprints. Technology that is climate-smart and sustainable is warranted today to lessen environmental effects.

REFERENCES

A global AC Nielsen report. (2006). *Consumers and RTE Meals*, pp. 1–8.

Adetola, A., Duodu, K., Taylor, J. (2022). Food-to-food fortification of ready-to-eat cereal-based porridges: A potential sustainable food-based strategy to improve bioavailable iron and zinc in plant-based diets. *Current Developments in Nutrition*, 6(Suppl_1):498.

Alam, S., Kadam, M. (2005). Extrusion technology in food industry. *Processed Food Industry,* 8:28–33.

Allan, G. (2001). Instant food facts and fallacies. *Journal of Food Science,* 122–133.

Andress, E.L., Nickols, S.Y., Peek, G.G., Nickols-Richardson, SM. (2011). Seeking food security: Environmental factors influencing home food preservation and wellness, part II: 1960–2010. *Family and Consumer Sciences Research Journal*, 39(3):233–245. https://doi.org/10.1111/j.1552-3934.2010.02062.x.

Balasubramanyam, B.V., Manohar, S.S. (2005). Convenience foods-growth and prospects. *Indian Food Industry*, 24(4):56–59.

Bawa, A.S., Vibhakara, H.S. (2002). Potential for irradition technique for development of convenience foods in India. *Beverage and Food World*, 21–24.

Bender, A.E. (1978) *Food Processing and Nutrition*. London, United Kingdom: Academic Press.

Bhavyashree, S., Srinivasan, K. (2009). Bioaccesibility of minerals from cereal based composite meals and RTE foods. *Journal of Food Science and Technology*, 46(5):431–435.

Bohra, N. B., Gawit, N. N., Chavan, N. R., Kapse, N. R. a. R. K. D. (2015). An Eco-friendly and reusable heat source for self-heating food packaging. *International Journal of Engineering Research And, V4*(05). https://doi.org/10.17577/ijertv4is050462

Borah, A., Chutia, H., Balasubramanian, S., Mahanta, C.L. (2019). Role of food extrusion in development of healthy food products: A review. *Acta Scientific Nutritional Health Special Issue* 1:39–44. 10.31080/ASNH.2019.S01.0010

Chitov, T., Wangdao, S. (2009). Occurrence of potentially pathogenic *vibrio* species in raw, processed, and RTE sea foods. *Maejo International Journal of Science and Technology*, 3(1):88–98.

Choton, S., Gupta, N., Bandral, J. D., Anjum, N., Choudary, A. (2020). Extrusion technology and its application in food processing: A review. *The Pharma Innovation*, 9(2), 162–168. https://doi.org/10.22271/tpi.2020.v9.i2d.4367

Datta, A., Nicolaï, B., Vitrac, O., Verboven, P., Erdogdu, F., Marra, F., Sarghini, F., Koh, C. (2022). Computer-aided food engineering. *Nature Food*, 3(11):894–904. https://doi.org/10.1038/s43016-022-00617-5.

Debbarma, J., Viji, P., Rao, B. M., Prasad, M. M. (2017). Nutritional and physical characteristics of noodles incorporated with green seaweed (Ulva reticulata) and fish (*Pangasianodon hypophthalmus*) mince. *Indian Journal of Fisheries*, 64(2). https://doi.org/10.21077/ijf.2017.64.2.58918-14

Dhumal, C.V., Pardeshi, I.L., Sutar, P.P., Jayabhaye, R.V. (2014). Development of potato and barnyard millet based ready to eat (RTE) fasting food. *Journal of Ready to Eat Food*, 1(1):11–17.

Dixit, P., Ravichandran, R. (2022). Bridging the skill gap in the Indian food processing sector: A review. *FoodSci: Indian Journal of Research in Food Science and Nutrition*, 9(2), 85. https://doi.org/10.15613/fijrfn/2022/v9i2/217902

Flexible Retort Pouch Defects. (2002). Identification and classification manual. *Canadian Food Inspection Agency.* http://mofpi.nic.in (accesssed March 28, 2022), www.apeda.com (accessed May 14, 2022), www.itc.com (accessed March 28, 2022), www.mtrfoods.com (accessed April 8, 2022), www.foodengineeringmag.com/articles/100260-frozen-ready-meal-market-expected-to-grow (accessed August 15, 2022).

Friedl, K. E., Askew, E. W., Schnakenberg, D. D. (2020). A ration is not food until it is eaten: nutrition lessons learned from feeding soldiers. In *Elsevier eBooks* (pp. 127–146). https://doi.org/10.1016/b978-0-12-820075-9.00007-7

Glass, K. A., Doyle, M. P. (1989). Fate and thermal inactivation of *Listeria monocytogenes* in Beaker Sausage and Pepperoni. *Journal of Food Protection*, 52(4), 226–232. https://doi.org/10.4315/0362-028x-52.4.226

Hwang, A., Haung, L. (2001). Microbiology of RTE foods in RTE foods: Microbial concerns and control measures, 1–40.

International Atomic Energy Agency (IAEA). (2003). Radiation processing for safe, shelf-stable and ready-to-eat food. Proceedings of a Final Research Co-ordination Meeting held in Montreal, Canada, pp.1–260.

Irawati, Z., Sani, Y. (2012). Feeding studies of radiation sterilization ready to eat foods on sprague wawley rats: *In Vivo*. *Natural Science*, 4(2):116–122. https://doi.org/10.4236/ns.2012.42017.

Jaybhaye, R.V., Srivastav, P.P. (2015). Development of barnyard millet ready-to-eat snack food: part II. *Food Science*, 6(2):285–291.

Kadidlova, H., Paveb, H. (2010). The effect of long term storage on amino acid content of RTE entrees. *International Journal of Food Science and Technology*, 45:966–970.

Khadel, N., Rao, M. (2005). Development and acceptability of RTE snack for the institutional elderly. *Journal of Food Science and Technology*, 42(2):176–179.

Khan, A.K., Ruby. (2010). Synchronization of global food regulations for RTE foods. *Beverage and Food World*, 51–55.

Linnan, M.J., Mascola, L., Lou, X. D., Goulet, V., May, S., Salminen, C., Hird, D. W., Yonekura, M.L., Hayes, P., Weaver, R. (1988). Epidemic listeriosis associated with Mexican-style cheese. *The New England Journal of Medicine*, 319(13), 823–828. https://doi.org/10.1056/NEJM198809293191303

Liu, Z., Zhang, M., Bhandari, B., Wang, Y. (2017). 3D printing: Printing precision and application in food sector. *Trends in Food Science and Technology*, 69:83–94. https://doi.org/10.1016/j.tifs.2017.08.018.

Manohar, S. S., Balasubramanyam, B. V., Sheshachala, C. (2005). Indian food industry. *Association of Food Scientists and Technologists,* 24(4):56–59.

Martin, M. (2001). Chapter 20: Meat curing technology. In: Young, O.A., Rogers, R.W., Hui, Y.H., Nip, W.K. (Eds), *Meat Science and Applications*, CRC Press. https://doi.org/10.1201/9780203908082.ch20.

McClure, P. J., Beaumont, A. L., Sutherland, J. P., Roberts, T. A. (1997). Predictive modelling of growth of Listeria monocytogenes. The effects on growth of NaCl, pH, storage temperature and NaNO2. *International Journal of Food Microbiology*, 34(3), 221–232. https://doi.org/10.1016/s0168-1605(96)01193-2

Mckinsey Report. (2000). Retailing in India: The emerging revolution CII retail report.

Mishra, H.N., Sinija, V.R. (2008). Food technology to meet the changing needs of urban consumers. *Comprehensive Reviews in Food Science and Food Safety*, 7:358–368.

Momanyi, D., Owino, W., Makokha, A. (2020). Formulation, nutritional and sensory evaluation of baobab based ready-to-eat sorghum and cowpea blend snack bars. *Scientific African*, 7:e00215.

Motey, R., Lele, S. (2003). Plastic films for processed foods: Special requirement. *Packaging India*, 35(5):151–158.

Murughan, M., Ramesh, P. (2007). Snack food in India—A healthy eating. *Beverage and Food World*, (18):36–38.

Nadanasabapathy, S., Ramakrishna, A. (2001). Current status and potential for retort processed foods in India. *Indian Food Industry*, 20(3):67–68.

Nickols, S.Y., Andress, E.L., Peek, G.G., Nickols-Richardson, S.M. (2010). Seeking food security: Environmental factors influencing home food preservation and wellness, part I: 1910–1959. *Family and Consumer Sciences Research Journal*, 39(2):122–136. https://doi.org/10.1111/j.1552-3934.2010.02051.x.

Ohlhorst, S.D., Slavin, M., Bhide, J.M., Bugusu, B. (2012). Use of iodized salt in processed foods in select countries around the world and the role of food processors. *Comprehensive Reviews in Food Science and Food Safety*, 11(2):233–284. https://doi.org/10.1111/j.1541-4337.2011.00182.x.

Özbay Doğu, S., Sariçoban, C. (2015). Meat drying technology and some dried meat products consumed in the World. *Journal of Food and Health Science*, 109–123. https://doi.org/10.3153/jfhs15011.

Paradkar, M., Sharma, K., Mallya, R. (2007). RTE: Technological process and Indian perspective. *Indian Food Industry*, 39–44.

Patel, D., Rathod, R.M. (2017). "Ready to eat Food Perception, Food Preferences and Food Choice: A Theoretical Discussion."

Pattan, N., Yenagi, N. (2001). Product quality of traditional RTE Madeliand its changes during storage. *Journal of Food Science and Technology*, 38(4):371–373.

Pitayachaval, P, Sanklong, N., Thongrak, A. (2018). A review of 3D food printing technology. *MATEC Web of Conferences*, 213:1–5.

Potty, V.H. (2003). Indian food industry: Will the sun ever rise? *Indian Food Industry*, 21(1):12–20.

Radhakrishnan, K., Das, H. (2001). Preservation of mutton as RTE curry by hurdle technology. *Journal of Food Science and Technology*, 38(3):287–289.

Ralovich, B. (1992). Data for the properties of Listeria strains (a review). *Acta Microbiologica Hungarica*, 39(2), 105–113.

Rane, J. (2011). Street vended foods in developing world: Hazard analysis. *Indian Journal of Microbiology*, 4:18–22.

Rangarao, G.C. (2001). RTE Indian foods in retort pouches: The second wave. *Indian Food Industry*, 21(1):12–18.

Rangarao, G.C. (2007). Room for innovation. *Times Food Processing Journal*, 5(10):23–28.

Rangarao, G.C., Ashwini, B. (2004). Shelf stable RTE as HMR: A pilot survey on consumer perception and preferences. *Indian Food Industry*, 23(1):21–28.

Rao, M. (2005). Extruded foods in India: Emerging technology and exciting products. *Indian Food Industry*, 24(2):56–58.

Rao, P., Jonnalagedda, R.P. (2004). Type, extent and use of colors in RTE food prepared in non-industrial sector: A case study from Hyderabad. *International Journal of Food Science and Technology*, (39):125–131.

Robin, G.T., Pamela, R.P., Jaspreet, K.C., Ahuja, E.S., Kevin, B.M. (2013). Recent trends in ready-to-eat breakfast cereals in the U.S. *Procedia Food Science*, 2:20–26.

Ruan, R., Ye, X., Chen, P., Doona, C.J., Taub, I. (2001). Ohmic heating. In: Richardson, P. (Ed), *Thermal Technologies in Food Processing*, CRC Press.

Sabapathy, S.N., Bawa, A.S. (2005). Self heating system for RTE foods. *Indian Food Industry*, 24(1):54–57.

Sanghvi, A., Saraswati, G. (2010). Development of dehydrated and RTE Brahmi Chutney. *Beverage and Food World*, 19–24.

Sastry, S.K., Shynkaryk, M., Somavat, R. (2011). Ohmic and moderate electric field processing: developments and new applications.11th International Conference on Engineering and Food, May 22–26, 2011, Athens Greece, NFP121.

Sen, A.R., Muthukumar, M. (2007). Convenience meat product: Indian perceptive. *Processed Food Industry*, 10:48–51.

Senthilnathan, P., Saxena, S.K. (2007). Safety evaluation of Indian RTE foods: Panipuri and samosa. *Processed Food Industry*, 17:29–33.

Sharma, G.K., Bawa, A.S. (2003). DFRL in the services of specific consumers convenience foods and operational rations. *A Souvenir of IFCON*, 101–102.

Shiby, V.K., Sinija, V.R. (2007). RTE health foods: A promising concept. *Indian Food Industry*, 47–54.

Singh, J., Kaur, L., McCarthy, O. (2007). Factors influencing the physico-chemical, morphological, thermal and rheological properties of some chemically modified starches for food applications—A review. *Food Hydrocolloids*, 21(1):1–22. https://doi.org/10.1016/j.foodhyd.2006.02.006

Somavat, R. (2011). Applications and effects of ohmic heating: Sterilization, influence on bacterial spores, enzymes, bioactive components and quality factors in food. Dissertation, The Ohio State University, pp. 4–5.

Suri, S., Dutta, A., Shahi, N.C., Raghuvanshi, R.S., Singh, A., Chopra, C.S. (2020). Numerical optimization of process parameters of ready-to-eat iron rich extruded snacks for anemic population. *LWT-Food Science and Technology*, 134:110164.

Temgire, S., Borah, A., Kumthekar, S., Idate, A. (2021). Recent trends in ready to eat/cook food products: A review. *The Pharma Innovation*, 10(5), 211–217. https://doi.org/10.22271/tpi.2021.v10.i5c.6207

Toldra, F., Gavara, R., Lagaron, J.M. (2004). Fermented and dry-cured meat: Packaging and quality control. In: Hui, Y.H., Meunier-Goddik, L., Hansen, A.S., Josephsen, J. (Eds), *Handbook of Food and Beverage Fermentation Technology*, Marcel Dekker, pp. 445–458..

Upidi, S.A., Kakkar, D.A. (2002). Microbiological quality of RTE meat and meat products sold in Mumbai city. *Journal of Food Science and Technology*, 39(3):299–303.

Venkataraman, H., Westerhof, K., Olson, S. (2019). *The Digital Transformation of the Food Industry*. Lux Research; https://web.luxresearchinc.com/hubfs/Press%20Release%20Assets/Lux%20Research%20-%20The%20Digital%20Transformation%20of%20the%20Food%20Industry.pdf

Wadikar, D., Nanjappa, C., Bawa, A.S. (2010). Development of RTE appetizers based on pepper and their quality evaluation. *Journal of Food Science and Technology*, 47(6):638–643.

Wadikar, D. D., Nanjappa, C., Premavalli, K. S., & Bawa, A. S. (2010). Development of ginger based ready-to-eat appetizers by response surface methodology. *Appetite*, 55(1), 76–83. https://doi.org/10.1016/j.appet.2010.04.004

Wang, L., Chen, W., Zhou, R., Ren, Y., Wang, K., Jiang, N., Gao, R. (2022). Physicochemical and sensory properties of a tilapia skin-based ready-to-eat snack prepared by infrared drying and air frying. *Applied Food Research*, 2(2):100–155.

Weaver, C. M., Dwyer, J., Fulgoni, V. L., 3rd, King, J. C., Leveille, G. A., MacDonald, R. S., Ordovas, J., Schnakenberg, D. (2014). Processed foods: Contributions to nutrition. *The American Journal of Clinical Nutrition*, 99(6), 1525–1542. https://doi.org/10.3945/ajcn.114.089284

Wilkinson, V.M., Gould, G.W. (1996). *Food Irradiation: A Reference Guide,* Butterworth-Heinemann, p. 10.

Yadav, B.S., Sharma, A. (2005). Extruders for snack foods. *Beverage and Food World*, pp. 50–53.

Zhu, M., Du, M., Cordray, J. (2005). Control of *Listeria Monocytogenes* contamination in RTE meat products. *Comprehensive Reviews in Food Science and Food Safety*, (4):34–42.

Zhu, N., Jain, N., Normington, J., Michno, J., Holschuh, N., Smith, J. (2021). Consumption of ready-to-eat cereal and its associations with nutrient intake and nutrition adequacy in the United States, NHANES 2017–2018. *Current Developments in Nutrition*, 5(2):1115.

Zhu, Y., Smith, J., Vanage, V., Jain, N., Sharma, M., Holschuh, N., HermetetAgler, A. (2019). Associations between consumption of ready to eat cereal and dietary intake and diet quality in Hispanics. *Current Developments in Nutrition*, 3(1):18–52.

6 RTE Production and Consumption

Statistics at a Glance

Mohammad Raziuddin,[1] *Shaikh Adil,*[2*] *Ajay Singh,*[3]
Ashok Bhosale,[4] *and Vijay Kele*[2]
[1]Department of Livestock Products Technology, COVAS, MAFSU, Udgir, Maharashtra, India
[2]Department of Dairy Technology, PIT, PARUL University, Vadodara, Gujarat, India
[3]Department of Food Technology, Mata Gujri College, Fatehgarh, Punjab
[4]Department of Veterinary Microbiology, COVAS, MAFSU, Udgir, Maharashtra, India
*Corresponding Author
Shaikh Adil
Assistant Professor, PIT, Parul University, Vadodara, Gujarat, India

6.1 INTRODUCTION

The Indian ready-to-eat (RTE) market was projected to increase at a compound annual growth rate (CAGR) of 20.64% between 2022 and 2027. According to projections, the market will expand by USD 1,064.32 million. As of 2017, the market was assessed to be worth USD 263.4 million (PR Newswire, New York, April 14, 2023). The growing consumer preference for RTE foods is primarily attributable to their ease of preparation and storage, as well as consumer allure factors including value, texture, convenience, and attractive appearance (Harper, 1981). RTE products are defined by the Food Standards Agency (FSA, UK) as "any food intended for consumption without additional heating or processing." This definition applies to both unwrapped and pre-wrapped RTE products, regardless of whether the food can be ingested in a chilled condition or not. The phrase "additional heating or processing" does not encompass food preparation tasks performed by the consumer in lieu of consuming an RTE food item, including light washing, slicing, dicing, portioning, marinating, or preserving (Food Standards Agency, 2011; Mankani & Tata, 2022).

Variations in lifestyle are directly correlated to alterations in dietary patterns. The consumption of convenience foods and the culture of "eating out" has developed in conjunction with the transformation of eating habits in modern Indian society. These are commercially available and are formulated for simple ingestion. Although they meet this requirement, restaurant meals are seldom referred to using this term. Convenience foods consist of prepared items such as RTE foods, frozen foods, and prepared mixes, including chutney, dosa, and upma mixtures, among others. Considerable evolution has occurred in the composition, function, and visual presentation of a diverse range of RTE foods throughout history. This transformation can be attributed to technological progress and shifts made in usage and customs that have characterized numerous epochs (Scholliers, 2015).

Although India is a global leader in food production, with the organized sector contributing USD 34,827 (Commodity Online, 2007) to the total food output, only a small percentage of its agricultural

DOI: 10.1201/9781032622408-6

products undergo processing to create goods with added value. For example, despite ranking first in milk production, second in fruits and vegetable production, and third in livestock production, processing of less than 2–5% of the country's output is observed. This highlights the tremendous investment potential in the processed food industry for machinery, packaging, and infrastructure. In reality, India requires an investment of USD 28 billion to grow its food processing plants by 8–10%. In 15 years, both Indian cuisine and style of life have undergone substantial transformations. These transformations can be attributed to a variety of factors, including, among others, the function of the media, marketing pressure, globalization, dual income, and remote-living spouses.

As a result of the demands imposed by modern society, consumers today prefer short clear food preparation methods to prolonged, lengthy procedures. Due to the scarcity of basic materials and the difficulty encountered in preparing masala, individuals opt for RTE items of this nature. These products' market potential is not restricted to a specific category. Numerous multinational corporations that capitalize on, foster, and develop swift transformations in human lifestyles have established their operational hubs in India. A substantial number of individuals are relocating to urban areas in search of employment and higher education. They perceive RTE items as a more comfortable substitute for dining in establishments. The majority of dual-income families, due to time constraints, would rather forego the inconveniences associated with cooking. They prefer dining out and spending time with their families on the weekends, whereas their demanding work schedules compel them to purchase these items during the week. An added rationale for promoting heightened utilization is the extensive variety and convenient accessibility of these items. The widespread availability of RTEs in supermarkets has caused a significant and profound change. Furthermore, the accessibility of these packages enables one to prepare most international cuisines for personal consumption. These foods have a high added value despite being the most processed, as they enable consumers to "eat off" the shelf without having to engage in the time-consuming activity of preparing a supper from the beginning.

6.2 DIFFERENT TYPES OF FOODS

6.2.1 High-Risk Foods

The term "high-risk foods" refers to foods that possess the capacity to promote the proliferation of pathogenic bacteria, which are also notably associated with higher occurrences of food poisoning (Beshiru et al., 2022). They are typically protein-rich, necessitate stringent temperature regulation and contamination prevention, and comprise the following:

1. Meat and poultry that have been cooked, including beef, pork, ham, lamb, chicken, turkey, and duck.
2. Prepared dishes that require cooking, including meat pies and pasties, pate, meat stock and gravy, and cook-chill foods.
3. Dairy products, including but not limited to matured soft and molded cheeses, cream, artificial cream, custards, and products containing unpasteurized milk.
4. Egg-based products, including quiche, homemade ice cream, mayonnaise, and mousse; cooked eggs; and quiche and products comprising uncooked or lightly cooked eggs.
5. Oysters in their uncooked state, mussels, cockles, cooked prawns, and other seafood
6. Farinaceous entrees, such as boiled couscous, pasta, and rice.

6.2.2 Low-Risk Foods

Low-risk foods include bread, cookies, cereals, crisps, and cakes (but not dairy cakes), which remain stable at room temperature. The following foods are extremely unlikely to contribute to food poisoning:

1. Dried products, those that contain minimal amounts of moisture, such as bread, flour, and biscuits.
2. Foods that have been preserved, for example, smoked or salted fish.
3. Acidic foods, for example, pickled foods, vinegar, and fruit.
4. Fermented products such as salami and pepperoni.
5. Foods with high sugar/fat content, for example, jam or chocolate.
6. Canned food. (Source: Farber et al., 2021)

6.2.3 Acidic Foods

According to US Food and Drug Administration (FDA) Code of Federal Regulations (CFR 2), acidic foods are those that have a pH of 4.6 or below. "Low-acid foods" are foods, other than alcoholic beverages, with a final equilibrium pH greater than 4.6 and a water activity (a_w) greater than 0.85. Acidified foods are low-acid foods to which acid(s) or acid food(s) are added. These products might be called "pickles" or "pickled" and include beans, cucumbers, cabbage, artichokes, cauliflower, puddings, peppers, tropical fruits, fish, vinegar, fruits, cheeses, and other fermented products such as tempeh, miso, sauerkraut, kefir, kombucha, and kimchi (Barron et al., 2012).

6.3 CONSUMER BEHAVIOR TOWARD THE PURCHASE OF READY-TO-EAT FOOD

The subject of consumer behavior concerning the acquisition of RTE food is complex, involving a range of psychological, social, and cultural elements that impact the decisions of individuals in the swiftly expanding convenience food industry. Recent years have witnessed an increase in the demand for RTE foods as a result of shifting dietary preferences, rapid urbanization, and a growing desire for time-saving meal options. A comprehensive comprehension of the complex interplay between consumer preferences, motivations, and perceptions is imperative for organizations operating in this sector to successfully customize their marketing approaches (Bae et al., 2010).

Taste, packaging, nutritional value, and brand support are all significant determinants in influencing consumer choices. Upon reviewing the most recent scholarly works, one discovers an expanding collection of investigations that scrutinize the changing trends in consumer behavior within this field. These studies examine various facets, including the influence of digital marketing on perception formation, health consciousness, and sustainability (Siddiqui et al., 2022). Regarding the various RTE foods they commonly procure from grocery stores, the findings indicate (Figure 6.1) that among the non-price factors considered significant by the participants in their pre-cooked international food purchases, taste was ranked as the foremost consideration. Following "taste" with an average rating of 4.66 out of 5, the subsequent most significant factors were food safety (4.65/5), quality (4.61/5), flavor (4.57/5), and authenticity (4.10/5). Brand familiarity received the lowest rating of 3.68 out of 5. Therefore, regardless of the culinary variety, the subsequent non-price variables held varying degrees of significance for every consumer: taste (ranking highest), quality, flavor, food safety, authenticity, and brand recognition (ranking lowest).

6.4 DEMAND OF RTE FOOD

The rapid growth of the RTE food market can be attributed directly to the achievements of food companies and their competitors in providing convenience. Distinctive technologies have collaborated to make it possible for consumers to receive RTE and ready-to-prepare foods at their preferred time and location. The proliferation of RTE food products brought about by an increase in domestic deliveries has driven the expansion of the RTE food market. Utilizing research, packaged food manufacturers in the RTE market have introduced more nutritious ingredients and improved packaging, including

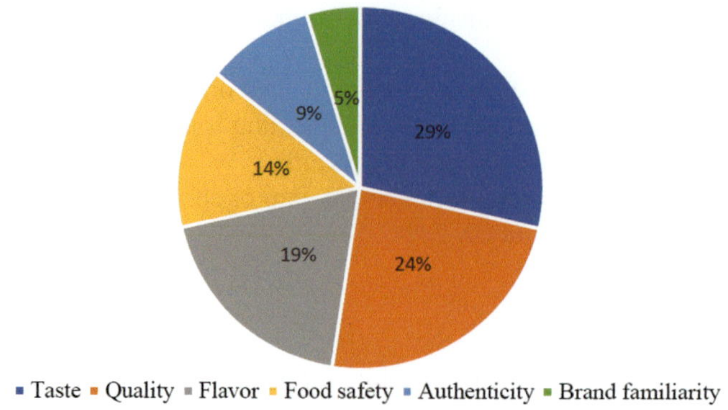

FIGURE 6.1 Key Non-Price Factors Important to Consumers. (Bhuyan & Govindasamy, 2020.)

those derived from exotic plants. The food and beverage industry benefits significantly from the increasing prevalence of online delivery platforms. The advent of advanced indoor farming technologies in urban areas has opened up novel opportunities. Key product categories include cereals and microwave breakfast foods. Meat and poultry products, instant broth and snacks, ready-made meals, and backed goods. Prominent distribution channels include e-commerce platforms, food specialty stores, hypermarkets, and convenience stores (RTE Food Market, 2023).

6.5 READY-TO-EAT FOOD MARKET: KEY TRENDS

A major development in the RTE food market is the increasing demand for convenience foods among time-crunched consumers. Lifestyle changes have strengthened the trend toward consumption on the go, thereby increasing RTE's acceptability. Health and wellness becoming more prevalent in food consumption patterns have stimulated innovation in the RTE food market. Rapid urbanization in a number of regions has emerged as an important factor in millennials' adoption of novel food consumption trends. Developed countries have already witnessed substantial demand in the global RTE food market due to the increasing prevalence of packaged foods. Researchers are currently investigating novel alternatives, particularly for the wealthier class who favor protein-based constituents in RTE products (RTE Food Market, 2019).

In recent years, consumers in developing regions have gained more knowledge not only of the flavor but also of the nutritional composition of the packaged foods they purchase, thereby creating new revenue opportunities for the RTE food market. The emphasis on contemporary food trends has further stimulated the emergence of novel prospects for RTE manufacturers (Smigic et al., 2023).

6.6 CHALLENGES FOR READY TO EAT FOOD

Increasing obesity and diabetes-related health concerns denote a significant obstacle to the expansion of the RTE food market in India. Nowadays, the prevalence of obesity has been rising in India. Specific ingredients incorporated into widely ingested foods directly contribute to the rise in obesity rates. For example, obesity and diabetes can result from an excessive intake of sugar (Hurt et al., 2010). To prevent the consumption of trans fats, most consumers stop consuming RTE products, including packaged and frozen foods. Experts advise avoiding cake-containing products as they may trigger allergic reactions. Manufacturers face a significant challenge in substituting them with non-allergenic ingredients (Gulati & Misra, 2014).

Prohibiting the presence of trans fats in RTE products presents a significant obstacle for RTE food manufacturers. Substitutes for trans fats influence the qualities of products such as flavor, consistency, longevity, and steadiness. Detrimental lipids significantly elevate cookie dough. Due to the predominantly high-fat content of RTE products, excessive consumption may result in obesity, overweight conditions, and additional health complications. They are therefore not favored by health-conscious consumers. These variables could impede the market's expansion throughout the projected timeframe (Mozaffarian & Stampfer, 2010).

6.7 VENDOR OVERVIEW

Vendors employ a variety of strategies, such as geographical expansion, product and service launches, strategic alliances and partnerships, mergers, and acquisitions, to bolster their market presence.

Furthermore, the report on the RTE food industry contains comprehensive analysis of the market's competitive landscape and details about 15 vendors, such as:

- Bikaner Express
- Conagra Brands Inc.
- Curry Culture LLP
- Dabur India Ltd.
- General Mills Inc.
- Gits Food Products Pvt. Ltd.
- Haldiram Foods International Pvt. Ltd.
- Heritage Foods Ltd.
- ITC Ltd.
- Kohinoor Foods Ltd.
- Maples
- McCain Foods Ltd.
- Nestle SA
- Nomad Foods Ltd.
- Orkla ASA
- Pristine Organics Pvt. Ltd.
- Savola Group
- The Kraft Heinz Co.
- Tyson Foods Inc.
- Vadilal Industries Ltd.

To aid clients in conceptualizing the larger business climate and the strengths and weaknesses of key market participants, qualitative and quantitative vendor analyses have been performed. Vendors are classified as pure play, category-focused, industry-focused, and diversified based on qualitative analysis of the data; dominant, leading, strong, tentative, and weak vendors are identified through quantitative analysis (Indo Asian News, 2023).

6.8 READY-TO-EAT FOOD MARKET DISTRIBUTION CHANNEL IN INDIA

In India, supermarkets and grocery stores became the most popular channels of distribution for RTE food items. The growth of grocery stores and the flourishing retail sector have both played a role in the segment's increased sales. These retail establishments provide an extensive assortment of RTE products, enabling customers to conveniently locate all the necessary items in a single location. Additionally, the inclusion of price comparisons on store displays influences consumer preference toward RTE products at hypermarkets and supermarkets. It is anticipated that the opening of new locations by major retailers, such as Reliance Retail and D-Mart, will stimulate consumer interest in RTE food products in the coming years (FNB News, 2023).

Furthermore, convenience and grocery stores are well-liked for their adaptability, vast selection of products, and extended business hours. Consumers are able to compare various RTE brands in these stores prior to making a purchase. Moreover, they offer convenient amenities such as virtual payment alternatives and physical ATMs, which appeal to a wide range of customers. In addition, offline distributors are implementing promotional offers and same-day delivery as service delivery innovations. Consequently, it is expected that the RTE food products sector will undergo a substantial annual expansion rate throughout the projected timeframe (Market Research Reports, 2023).

6.9 SEGMENT OF READY-TO-EAT FOOD MARKET IN INDIA

The rising demand for frozen food items in India is driving the RTE food market. In 2017, the frozen food segment was worth USD 146.23 million, and its growth was anticipated to continue until 2021. Frozen food items comprise a wide variety of products, such as RTE dishes, fruits, vegetables, meat, and poultry, which are subjected to rapid freezing and can remain frozen until they are consumed. Typically, these items have a shelf life of not less than six months. Consumers' busy lifestyles and long hours of work have generated a need for readily prepared food alternatives that are convenient in nature. The growing female workforce and the availability of frozen food products both help to meet this demand. In response to the increasing market demand, vendors are extending their distribution networks within India. Major companies in India that provide RTE frozen food products are Conagra Brands Inc., McCain Foods Ltd., and Nestle SA. The rising demand for frozen foods will drive the expansion of the RTE food market in India throughout the forecast period (Market Research Reports, 2023).

6.10 CONCLUSION

The projected CAGR for the RTE food market in India is 20.64% between 2022 and 2027, which corresponds to an estimated increase in market size of USD 1,064.32 million. The increase in demand for RTE foods can be due to their increasing convenience, ease of preparation, and appealing features, including flavor and texture.

Urbanization and the demand for rapid food preparation techniques are factors that are influencing the transformation of dietary habits in India. India, a leading global producer of food, processes a negligible proportion of its agricultural output, suggesting considerable investment potential in the processed food sector. In place of price, consumers are increasingly considering non-price variables such as flavor, quality, and safety when purchasing RTE foods. Trends suggest a rising need for convenience, advancements in packaging technology, and the influence of health and wellness on the consumption of RTE foods. However, challenges arise in the form of health concerns associated with obesity and diabetes. The vendor overview provides an inventory of significant market participants, while the distribution channels, encompassing hypermarkets, supermarkets, and online platforms augment the overall comprehension of the sector.

Finally, shifting consumer preferences and lifestyle dynamics indicate that the Indian RTE food market has ample opportunities for substantial expansion. Successfully managing obstacles associated with health considerations and capitalizing on emergent trends will be indispensable for long-term prosperity in this ever-changing sector.

REFERENCES

Bae, H. J., Chae, M. J., & Ryu, K. Consumer behaviours towards ready-to-eat foods based on food-related lifestyles in Korea. *Nutrition Research and Practice*, 4(4), (2010): 332–338.

Barron, F. H., Fraser, A. M., & Innocenzo, M. Acidified foods: food safety considerations for food processors. *Food Industry*, (2012): 231–239.

Beshiru, A., Okoh, A. I., & Igbinosa, E. O. Processed ready-to-eat (RTE) foods sold in Yenagoa Nigeria were colonized by diarrheagenic *Escherichia coli* which constitute a probable hazard to human health. *Plos One*, 17(4), (2022): e0266059.

Bhuyan, S., & Govindasamy, R. Factors affecting consumption of RTE international food in the United States. *International Journal of Food and Agricultural Economics*, (1), (2020): 257–269.

Commodity Online, 2007. www.rediff.com/money/report/food/20071115.htm

Farber, J. M., Zwietering, M., Wiedmann, M., Schaffner, D., Hedberg, C. W., Harrison, M. A., ... & Gummalla, S. Alternative approaches to the risk management of *Listeria monocytogenes* in low-risk foods. *Food Control*, 123, (2021): 107601S.

FNB News. Innovations and opportunities for RTE food market. 2023. www.fnbnews.com/FB-Specials/innovations-and-opportunities-for-rte-food-market-44979

Food Standards Agency. E. coli O157 control of cross-contamination guidance for food business operators and enforcement authorities. (2011), https://webarchive.nationalarchives.gov.uk/ukgwa/20131206104225/http://www.food.gov.uk/multimedia/pdfs/publication/ecoliguide0211.pdf

Gulati, S., & Misra, A. Sugar intake, obesity, and diabetes in India. *Nutrients*, 6(12), (2014): 5955–5974.

Harper, J. M. Extrusion of starches and starches materials. In: Harper, J.M. (ed) *Extrusion of food*, 2nd edn. CRC Press, Boca Raton, pp. 41–60, (1981).

Hurt, R. T., Kulisek, C., Buchanan, L. A., & McClave, S. A. The obesity epidemic: challenges, health initiatives, and implications for gastroenterologists. *Gastroenterology & Hepatology,* 6(12), (2010): 780.

Indo-Asian News Service. (2023). https://ians.in/pr-wire-detail/clay-market-size-to-grow-at-a-cagr-of-52-from-2022-to-2027-active-minerals-international-llc-austin-white-lime-coltd-burgess-pigment-co-and-more-among-the-key-companies-in-the-market-technavio-01-12-2023

Mankani, R., & Tata, K. A Study on Consumer Behaviour of Ready-to-Eat Food with Special Reference to Women in Mumbai. *Journal of Positive School Psychology,* (2022): 4549–4562.

Market Research Reports – Industry Analysis Size & Trends – Technavio. (n.d.). Technavio, www.technavio.com/, All Right Reserved 2023. www.technavio.com/report/ready-to-eat-food-market-in-india-industry-analysis

Mozaffarian, D., & Stampfer, M. J. Removing industrial trans-fat from foods. *British Medical Journal*, (2010): 340.

PR Newswire. New York (April 14, 2023). www.prnewswire.com/news-releases/ready-to-eat-food-market-size-in-india-to-grow-by-usd-1-064-32-million-between-2022-and-2027-growth-driven-by-growing-preference-for-convenience-food-products---technavio-301797125.html

Ready-to-Eat (RTE) Food Market (2023). Ready to Eat Food market Report. IMARC Group. www.imarcgroup.com/ready-to-eat-food-market. Accessed on 24/03/2024.

Scholliers, P. Convenience foods. What, why, and when. *Appetite*, 94, (2015): 2–6.

Siddiqui, S. A., Zannou, O., Bahmid, N. A., Fidan, H., Alamou, A. F., Nagdalian, A. A., ... & Arsyad, M. Consumer behavior towards nano packaging new trend in the food industry. *Future Foods*, 6, (2022): 100191S.

Smigic, N., Ozilgen, S., Gómez-López, V. M., Osés, S. M., Miloradovic, Z., Aleksic, B., ... & Djekic, I. Consumer attitudes and perceptions towards chilled ready-to-eat foods: a multi-national study. *Journal of Consumer Protection and Food Safety*, (2023): 1–14.

7 Consumer Behavior and RTE

Market Adaptability and Likeness Preference Terms

Sourav Sharma
Assistant Professor, Department of Management Studies, Mata Gujri College, Fatehgarh Sahib (An Autonomous College), Fatehgarh Sahib, Punjab, India

7.1 INTRODUCTION

Consumer behavior refers to the study of how individuals, groups, or organizations make decisions related to the selection, purchase, use, and disposal of goods, services, experiences, or ideas to satisfy their needs and wants. When examining consumer behavior in the context of ready-to-eat (RTE) products, it involves understanding the factors that influence consumers' choices, perceptions, attitudes, and actions toward RTE foods.

7.2 HOW THE CONCEPT OF CONSUMER BEHAVIOR IS RELEVANT IN REFERENCE TO RTE PRODUCTS

- **Understanding Consumer Needs:** By studying consumer behavior, RTE product manufacturers and marketers can identify the specific needs and desires of their target audience. This knowledge helps in developing products that align with consumers' preferences, leading to higher satisfaction and increased demand.
- **Segmenting the Market:** Consumer behavior analysis allows RTE product providers to segment the market based on factors such as demographics, lifestyle, and psychographics (Graca, et al., 2015). This segmentation enables targeted marketing strategies to reach different consumer groups effectively.
- **Product Development and Innovation:** By understanding consumer behavior, companies can uncover insights into what consumers are looking for in RTE products. This information can drive product development and innovation to meet evolving consumer demands and preferences.
- **Influencing Purchase Decisions:** Knowledge of consumer behavior helps in crafting effective marketing messages and promotions that resonate with consumers. Understanding factors like convenience, taste, health, and packaging allows companies to highlight the aspects that matter most to their target audience, influencing purchase decisions positively.
- **Brand Loyalty and Advocacy:** Building brand loyalty among consumers is crucial for RTE product companies. Understanding consumer behavior helps identify the elements that create brand loyalty and advocacy, leading to repeat purchases and positive word-of-mouth recommendations.
- **Pricing Strategies:** Consumer behavior insights aid in determining appropriate pricing strategies for RTE products. Companies can assess consumers' price sensitivity and willingness to pay for convenience and quality, optimizing pricing for profitability and customer value.

 DOI: 10.1201/9781032622408-7

- **Packaging and Presentation:** Packaging plays a vital role in attracting consumers to RTE products. Understanding consumer preferences and perceptions about packaging can lead to effective designs that catch consumers' attention and communicate the product's attributes clearly (Hartmann and Siegrist, 2017).
- **Addressing Health and Nutrition Concerns:** As consumers become more health-conscious, understanding their preferences for healthier RTE options can help companies develop and promote suitable products that align with these preferences.
- **E-Commerce and Online Shopping:** In the digital age, consumer behavior plays a significant role in e-commerce. Understanding how consumers browse, compare, and purchase RTE products online helps optimize website design, user experience, and online marketing strategies (Food Industry Association, 2020).
- **Consumer Feedback and Improvement:** Analyzing consumer behavior allows companies to collect feedback and reviews on RTE products. This feedback is valuable for continuous improvement, addressing concerns, and enhancing product offerings.

Studying consumer behavior in the context of RTE products is highly relevant for companies aiming to thrive in this competitive market. By understanding consumers' motivations, preferences, and decision-making processes, RTE product providers can tailor their offerings, marketing strategies, and overall brand experience to meet consumer needs effectively. This, in turn, leads to increased customer satisfaction, brand loyalty, and business success.

7.3 CONSUMER BEHAVIOR AND CONSUMPTION BEHAVIOR

Consumer behavior and consumption behavior are related concepts, but they have distinct meanings and implications in the context of RTE products.

Consumer Behavior	Consumption Behavior
Consumer behavior refers to the study of how individuals or groups of people make decisions about purchasing, using, and disposing of products and services. It encompasses a wide range of factors that influence these decisions, including psychological, social, cultural, and economic factors. In the context of RTE products, consumer behavior analysis might involve studying why individuals or households choose to buy and consume these products.	Consumption behavior, on the other hand, specifically refers to the actions and patterns of consumption once a product or service has been acquired. It focuses on how individuals or groups use and interact with the product after purchase.
Factors Influencing Consumer Behavior	**Factors Influencing Consumption Behavior**
• Convenience • Taste and Quality • Price • Health and Quality • Branding and Packaging • Social and Cultural Factors	• Frequency of Consumption • Portion Sizes • Heating and Preparation • Satisfaction and Feedback

From the above table it is very much clear that while consumer behavior focuses on the broader process of decision-making, including why consumers choose RTE products, consumption behavior delves into how these products are used and consumed once they are in the hands of the consumer. Both aspects are important for businesses in the RTE industry to understand and cater to the needs and preferences of their target market.

7.4 STEPS INVOLVED IN THE BUYING PROCESS OF RTE PRODUCTS

Buying behavior, also known as consumer behavior, refers to the process that individuals and organizations go through when making purchase decisions. It can be a complex process influenced by various factors. The following are the typical steps in buying behavior in the context of RTE Products:

- **Recognition of Need**:
  The buying process often begins with the recognition of a need or a problem (Graca, et al., 2015). This need could be triggered by internal factors (e.g., hunger, desire for a new gadget) or external factors (e.g., advertising, friends' recommendations).
- **Information Search**:
  Once the need is identified, consumers seek information to help them make a decision. This information can be gathered through various sources, such as online research, product reviews, asking friends and family, or visiting physical stores.
- **Evaluation of Alternatives**:
  Consumers typically consider multiple options to satisfy their needs. They evaluate these alternatives based on factors like price, quality, brand reputation, features, and personal preferences. This process can involve creating a pros and cons list for each option.
- **Purchase Decision**:
  After evaluating the alternatives, consumers make a decision to purchase a particular product or service (Harris, et al., 2009). This decision can be influenced by a variety of factors, including budget constraints, emotional factors, and external influences.
- **Purchase**:
  This is the point at which the consumer actually buys the chosen product or service (Rozin, et al., 1999). The purchase can occur in a physical store or online, and it often involves payment.
- **Post-Purchase Evaluation**:
  After making the purchase, consumers evaluate their decisions and their satisfaction with the product or service. If the product meets or exceeds their expectations, it leads to a positive post-purchase evaluation. If not, it may result in dissatisfaction and potentially lead to returns or complaints.
- **Post-Purchase Behavior**:
  Depending on the level of satisfaction experienced, consumers may engage in various post-purchase behaviors (Seiders, et al., 2015). If they are satisfied, they may become loyal customers and recommend the product or service to others. If dissatisfied, they may seek refunds, exchange the product, or leave negative reviews.
- **Repeat Purchase and Loyalty**:
  If consumers have a positive experience with a product or brand, they are more likely to make repeat purchases and become loyal customers. Building customer loyalty is an important goal for businesses.
- **External Influences**:
  Throughout the buying process, external factors such as social influences, cultural norms, marketing efforts, and economic conditions can play a significant role in shaping buying behavior.
- **Decision-Making Heuristics**:
  Consumers often use mental shortcuts or decision-making heuristics to simplify the buying process. Common heuristics include brand loyalty, price as an indicator of quality, and recommendations from trusted sources.

7.5 BUYING BEHAVIOR MODELS

By understanding and analyzing buying behavior for RTE products, several consumer behavior models can be applied to gain insights into how consumers make purchasing decisions. The prominent models often used in this context are:

1. The Consumer Decision-Making Process Model:
   This model breaks down buying behavior into a series of stages that consumers typically go through when making purchasing decisions. It can be applied to RTE products as follows:
   - **Problem Recognition:** Consumers recognize a need for convenient and quick meal solutions, such as RTE products, often due to time constraints or a desire for convenience.
   - **Information Search:** Consumers may seek information about different RTE products available in the market, their features, prices, and nutritional values.
   - **Evaluation of Alternatives:** Consumers assess the various RTE product options, considering factors like taste, brand reputation, price, and nutritional content.
   - **Purchase Decision:** After evaluating the alternatives, consumers make a decision to purchase a specific RTE product.
   - **Post-Purchase Evaluation:** After consumption, consumers evaluate their satisfaction with the product. Positive experiences may lead to brand loyalty and repeat purchases.
2. The Theory of Planned Behavior (TPB):
   TPB suggests that consumer behavior is influenced by three key factors: attitude toward the behavior, subjective norm, and perceived behavioral control. In the context of RTE products:
   - **Attitude toward Behavior:** Consumers' attitudes toward RTE products, including their taste, convenience, and healthiness, can influence their likelihood of purchasing them.
   - **Subjective Norm:** The opinions of friends, family, or peers regarding RTE products can influence consumer choices. For example, if a person's friends frequently consume RTE products, they may be more inclined to do so as well.
   - **Perceived Behavioral Control:** Factors such as accessibility to RTE products, budget constraints, and cooking skills can affect consumers' perceived control over their decision to buy and consume RTE products.
3. The Hierarchy of Effects Model:
   - This model suggests that consumers go through a series of stages when making buying decisions, starting from awareness and ending with action (purchase). In the context of RTE products:
   - **Awareness:** Consumers become aware of RTE products through advertising, word of mouth, or in-store displays.
   - **Interest:** Consumers develop an interest in RTE products, perhaps due to their convenience or unique features.
   - **Desire:** A desire for RTE products may be created through effective marketing and positive reviews.
   - **Action:** The final stage is when consumers decide to purchase RTE products based on their previous experiences and the information they've gathered.
4. The Fishbein Model:
   This model quantifies the consumer's attitude toward a product based on beliefs about the product attributes and the importance or evaluation of those attributes. For RTE products:

   Consumers may have beliefs about attributes like taste, convenience, healthiness, and price. They assign importance or evaluation scores to each attribute based on personal preferences and priorities. These scores can be used to predict consumer attitudes and, subsequently, their likelihood of buying RTE products.

5. The Psychoanalytic Model:
   The psychoanalytic model of consumer behavior is rooted in Sigmund Freud's psychoanalytic theory, which explores the role of unconscious thoughts, desires, and motivations in shaping human behavior. While psychoanalysis is primarily associated with psychology and therapy, it has been applied to marketing and consumer behavior to understand the deeper, often subconscious, influences that drive consumers' choices and preferences. This model suggests that consumer behavior is not solely rational and conscious but is also affected by hidden psychological processes. Here are the key components of the psychoanalytic model of consumer behavior:

- **Unconscious Motivations**:
  - According to psychoanalysis, individuals have deep-seated, unconscious motivations that influence their decisions. These motivations can include desires, fears, unresolved conflicts, and repressed memories.
    - **Id, Ego, and Superego**:
    - Freud's theory includes three parts of the psyche: the id, ego, and superego.
    - **Id**: This represents the primal, instinctual part of the mind, driven by desires and pleasure-seeking tendencies.
    - **Ego**: The ego is the conscious, rational part of the mind that seeks to balance the id's desires with the constraints of reality.
    - **Superego**: The superego represents the internalized moral and societal standards that guide behavior.
    - **Defense Mechanisms**:
    - Freud introduced the concept of defense mechanisms, which are psychological strategies people use to protect themselves from uncomfortable thoughts and feelings. In the context of consumer behavior, these defense mechanisms can influence how individuals perceive and react to marketing messages and products.
    - **Symbolism and Dream Analysis**:
    - Psychoanalysis places a strong emphasis on symbolism and the interpretation of symbols and imagery. Marketers often use symbolism in advertising to tap into consumers' unconscious desires and fears.
    - **Emotions and Desires**:
    - This model suggests that emotions and desires play a significant role in consumer behavior. These emotions and desires may not always be apparent on the surface but can be uncovered through psychoanalytic techniques.
    - **Subliminal Messaging**:
    - The psychoanalytic model has led to the exploration of subliminal messaging in advertising. It posits that advertisers can appeal to consumers' unconscious minds by using subtle cues and imagery that are not consciously perceived but can still influence behavior.
    - **Brand Loyalty and Attachment**:
  - Consumer loyalty to brands and products can be analyzed through the lens of attachment theory, which is related to psychoanalysis. Brands can become associated with deep emotional needs, providing consumers with a sense of security or fulfillment.

It is important to note that the psychoanalytic model of consumer behavior has received criticism and skepticism within the field of marketing and psychology. Critics argue that it lacks empirical evidence and scientific rigor, making it difficult to apply in a consistent and reliable manner. Many contemporary consumer behavior models, such as cognitive psychology-based models and social influence theories, have gained more widespread acceptance due to their empirical support and practical utility in marketing research and practice. Nevertheless, the psychoanalytic perspective remains an intriguing and thought-provoking lens through which to explore consumer motivations and behavior.

7.6 IMPORTANT FACTORS IN SHAPING CONSUMER CHOICES

RTE products are a category of convenience foods that are pre-cooked, pre-packaged, and require minimal or no preparation before consumption[3]. Consumer decision-making in the context of RTE products can be impacted by a variety of factors (Chen and Lee, 2020). Here are some key factors that play a role in shaping consumers' choices:

- **Convenience:** Convenience is one of the primary drivers for choosing RTE products. Busy lifestyles and time constraints lead consumers to opt for RTE options, as they offer quick and hassle-free meal solutions.
- **Taste and Quality:** Consumers are drawn to RTE products that are not only convenient but also taste good and maintain their quality. Positive experiences with a particular brand or product can lead to repeat purchases.
- **Brand Reputation:** Established and trusted brands tend to have an advantage in the RTE market. Consumers often rely on well-known brands to ensure product safety, consistency, and reliability.
- **Packaging and Presentation:** Attractive and informative packaging can influence consumers' perceptions and purchasing decisions. Eye-catching designs, clear product information, and convenient portion sizes can make RTE products more appealing.
- **Health and Nutrition:** Increasing health consciousness has led consumers to consider the nutritional content of RTE products. Products that are perceived as healthier, such as those with low salt, sugar, or fat content, are likely to be preferred.
- **Ingredients and Transparency:** Consumers are becoming more discerning about the ingredients used in RTE products. They may look for natural, organic, or ethically sourced ingredients and appreciate brands that are transparent about their sourcing practices.
- **Price:** Cost is a significant factor in consumer decision-making. RTE products need to offer perceived value for money, as consumers compare the convenience and quality against the price.

4. **Advertising and Marketing:** Effective advertising and marketing campaigns can influence consumer perceptions and create a desire to try or repurchase RTE products.
5. **Word-of-mouth and Online Reviews:** Recommendations from friends, family, or online reviews can influence consumers' choices, as they rely on the experiences of others to inform their decisions.
6. **Availability and Accessibility:** RTE products need to be readily available and easily accessible in stores or online platforms to encourage consumer purchases.
7. **Cultural and Regional Influences:** Consumer preferences for RTE products can vary based on cultural norms, regional tastes, and dietary habits. Manufacturers may need to customize products to suit local preferences.
8. **Sustainability and Environmental Impact:** With growing environmental concerns, consumers may seek RTE products from brands that prioritize sustainability and eco-friendly packaging.

Consumer decision-making for RTE products is impacted by a combination of convenience, taste, quality, brand reputation, health considerations, price, packaging, and various social and cultural factors (Seiders, et al., 2015). Understanding and addressing these factors are crucial for manufacturers and marketers to effectively cater to consumer needs and preferences in this competitive market segment.

7.7 MARKETING AND ADVERTISING STRATEGIES FOR RTE PRODUCTS

Marketing and advertising strategies for RTE food products are essential to create brand awareness, attract customers, and increase sales. Here are some effective strategies specifically tailored for RTE products:

- **Highlight Convenience:** Emphasize the convenience aspect of RTE products in all marketing materials. Use catchy slogans and messages that convey the time-saving benefits of these products, especially for busy individuals and families.
- **Visual Appeal:** Invest in high-quality food photography and videography to showcase the appetizing appearance of RTE products. Visual content can be used in social media posts, advertisements, and packaging to entice potential customers.
- **Influencer Marketing:** Collaborate with food bloggers, influencers, and social media personalities who can create engaging content featuring RTE products. Influencers can provide honest reviews and recipe ideas to reach their followers effectively.
- **Product Demonstrations:** Organize in-store product demonstrations or sampling events to let consumers try the RTE products. This hands-on experience can create a positive impression and encourage purchase.
- **Health and Nutrition Messaging:** Highlight the nutritional benefits of RTE products, such as low calories, reduced sodium, or the use of natural ingredients. Clear labeling of nutritional information can appeal to health-conscious consumers.
- **Promote Unique Flavors and Varieties:** If your RTE product line offers a range of flavors or unique combinations, use this as a selling point in your marketing efforts. Highlighting variety can attract adventurous eaters and those seeking new tastes.
- **Packaging Innovation:** Create eye-catching and functional packaging that stands out on store shelves. Clear labels, eco-friendly materials, and packaging that maintains product freshness can make a significant impact on consumers.
- **Online Presence:** Establish a strong online presence through a well-designed website and active social media channels. Engage with your audience by sharing recipes, customer testimonials, and behind-the-scenes content.
- **Seasonal Promotions:** Use seasonal themes and occasions to create limited-time promotions and discounts. Tie RTE products to holidays or special events to increase consumer interest.
- **Loyalty Programs:** Implement an attractive loyalty program to reward frequent customers and encourage repeat purchases. Offer incentives, discounts, or exclusive access to new products for loyal customers.
- **Collaborations and Cross-Promotions:** Partner with other food brands or retailers for cross-promotional opportunities. Joint campaigns can expand the reach of RTE products to new audiences.
- **Video Content:** Utilize video marketing to demonstrate RTE products being prepared and served, along with customer testimonials. Videos can be shared on social media platforms, websites, and even in-store displays.
- **Location-Based Marketing:** If you have physical retail locations, use location-based advertising to target nearby consumers. Offer special deals to nearby customers through mobile marketing.
- **Packaging Inserts and Coupons**: Include coupons or promotional offers inside the RTE product packaging to incentivize repeat purchases and increase customer retention.
- **Sustainability Messaging:** Communicate your brand's commitment to sustainability and eco-friendly practices, which can resonate with environmentally conscious consumers.

By combining these marketing and advertising strategies, RTE product brands can effectively connect with their target audience, drive demand, and build a loyal customer base. The key is to focus on the unique selling points of RTE products, understand consumer preferences, and tailor the marketing approach accordingly.

7.8 MARKET ADAPTABILITY OF RTE PRODUCTS

The market adaptability of RTE products in the context of the entire world can vary based on several factors. Although RTE products have gained popularity in many regions due to their convenience and time-saving benefits, their acceptance and success can be influenced by the following factors:

- **Cultural Differences:** Different cultures have unique food preferences and dietary habits (Hetherington and Blundell-Birtill, 2018). The success of RTE products in a particular region depends on how well these products align with local tastes and traditions. For example, certain RTE products may be more readily accepted in Western countries where fast-paced lifestyles are common, while in other regions, traditional home-cooked meals may hold greater significance.
- **Food Safety and Regulations:** Food safety regulations can vary significantly from one country to another. RTE products must comply with local food safety standards and labeling requirements to be accepted in international markets.
- **Distribution and Infrastructure:** The availability of RTE products can be affected by the distribution infrastructure in different regions. Access to cold storage, reliable transportation networks, and retail distribution channels all play a crucial role in reaching consumers in various parts of the world (Harris, et al., 2009).
- **Local Competition:** The RTE market in each country or region may already be saturated with local and international competitors. The success of RTE products will depend on their ability to stand out from the competition and offer unique value to consumers.
- **Health and Nutrition Awareness:** As consumers become increasingly health-conscious worldwide, the market adaptability of RTE products may depend on their nutritional content and the perception of being healthy and balanced meal options.
- **Price Sensitivity:** Price is a significant factor affecting consumer behavior in any market. RTE products need to be priced competitively to gain wider acceptance, especially in regions with price-sensitive consumers.
- **Brand Recognition:** Established RTE brands may find it easier to enter new markets due to their existing reputation and trust among consumers. Building brand recognition in new regions can be a challenge for lesser-known brands.
- **Packaging and Product Localization:** Adapting product packaging, labeling, and messaging to suit local languages and cultural nuances can enhance the appeal and acceptance of RTE products in international markets.
- **Dietary Restrictions and Preferences:** RTE products must cater to various dietary preferences, including vegetarian, vegan, gluten-free, and halal options, to resonate with diverse consumer groups.
- **Trade and Import/Export Regulations:** RTE products may face import/export restrictions and tariffs, affecting their market adaptability in different countries.

Despite the challenges, RTE products have shown promising growth in many regions as consumers increasingly seek convenience and time-saving solutions for their busy lifestyles (Smith and Johnson, 2021). As globalization continues to shape consumer preferences and habits, RTE product manufacturers can explore opportunities to adapt their offerings to meet the needs of various markets worldwide. In doing so, understanding and adapting to local preferences, regulations, and cultural differences will be essential to achieve market success on a global scale.

7.9 DISTRIBUTION STRATEGIES FOR RTE PRODUCTS

Distribution channel strategies for RTE products are critical to ensure efficient and widespread availability to consumers. The aim is to get the products from manufacturers to end consumers in a timely and cost-effective manner. Here are some distribution channel strategies commonly used in the context of RTE products:

- **Retail Stores:** One of the most traditional distribution channels for RTE products is through retail stores, including supermarkets, convenience stores, and grocery chains. This channel offers a wide reach to a diverse consumer base.
- **Online Retailers:** With the rise of e-commerce, selling RTE products through online retailers and marketplaces has become increasingly popular. This channel allows consumers to order RTE products online and have them delivered to their doorstep (Kumar and Gupta, 2017).
- **Specialty Food Stores:** Targeting specialty food stores that focus on organic, health-conscious, or gourmet products can be advantageous for premium RTE products catering to specific consumer segments.
- **Departmental Stores:** RTE products can be strategically placed in departmental stores that have dedicated food sections, attracting consumers with impulse purchases.
- **Quick Service Restaurants (QSRs):** Collaborating with QSRs or fast-food chains to offer RTE products as part of their menu can expose the products to a wider audience and capitalize on existing foot traffic.
- **Food Trucks and Mobile Vendors:** For on-the-go RTE products, utilizing food trucks or mobile vendors at high-traffic areas or events can be a targeted distribution approach.
- **Institutional Sales:** Selling RTE products to institutions such as schools, hospitals, corporate cafeterias, and airlines can provide bulk sales opportunities and a steady revenue stream.
- **Meal Kit Services:** Partnering with meal kit delivery services can allow RTE products to be included in meal kit packages, expanding their reach to consumers who prefer home-cooked meals with minimal effort.
- **Online Direct-to-Consumer (D2C):** Some RTE product manufacturers may opt for direct-to-consumer sales through their websites, offering exclusive products or subscription-based services.
- **Export and Global Distribution:** For international markets, RTE products can be exported through distributors or retailers, considering the unique regulations and preferences of each country.
- **Distributors and Wholesalers:** Working with distributors and wholesalers can streamline the supply chain and help RTE products reach a broader network of retail outlets.
- **Vending Machines:** Placing RTE products in vending machines at strategic locations can cater to consumers seeking quick and convenient snacking options.
- **Subscription Services:** Offering RTE products through subscription-based services can create a loyal customer base and provide recurring revenue.
- **Food Delivery Apps:** Partnering with food delivery apps can make RTE products accessible to consumers looking for delivery options.

Each distribution channel strategy comes with its advantages and challenges, and the choice of channels will depend on factors such as the target market, product positioning, shelf life, packaging, and the overall business model. A well-balanced and diversified distribution approach can help RTE product manufacturers maximize their market reach and profitability.

7.10 CONSUMER PREFERENCES AND LIKINGS FOR RTE PRODUCTS

Consumer preferences and liking for RTE products can be influenced by various factors. Here are some terms related to consumer preferences and likings commonly associated with RTE products:

- **Convenience:** Consumers appreciate RTE products for their convenience and time-saving benefits, allowing them to enjoy a quick meal without extensive preparation.
- **Taste:** The flavor and taste of RTE products plays a crucial role in consumer liking. Products with delicious and appealing tastes are more likely to be preferred.
- **Quality:** Consumers seek RTE products that maintain their quality, texture, and taste even after being packaged and stored for a certain period.
- **Variety:** A wide range of RTE product varieties, including different flavors and options, appeals to consumers who enjoy exploring different tastes and cuisines.
- **Healthiness/Nutrition:** Consumer preference for RTE products is impacted by their perception of being healthy and nutritious. Products with low salt, sugar, and fat content, and natural ingredients are often favored.
- **Packaging:** Attractive and functional packaging that preserves freshness and enhances the product's visual appeal can positively impact consumer liking.
- **Brand Reputation:** Consumers often choose RTE products from trusted brands with a positive reputation for quality and safety.
- **Pricing:** The price of RTE products is a significant factor in consumer preferences. Affordable pricing can attract a wider range of consumers.
- **Eco-friendly/Sustainability:** Consumers increasingly prefer RTE products from brands that prioritize eco-friendly packaging and sustainable sourcing practices.
- **Portion Size:** Consumers may prefer RTE products with portion sizes that suit their individual or family needs, reducing food waste.
- **Dietary Preferences/Restrictions:** Products that cater to specific dietary preferences, such as vegetarian, vegan, gluten-free, or halal, are favored by consumers with these restrictions.
- **Brand Loyalty:** Repeat purchases and brand loyalty can be influenced by positive experiences with RTE products, leading to continued consumer likeness for a particular brand.
- **Word-of-mouth:** Recommendations and positive reviews from friends, family, or online influencers can influence consumer preferences for RTE products.
- **Online Reviews and Ratings:** Consumer reliance on online reviews and ratings to assess product quality and likability before making a purchase.
- **On-the-Go Snacking:** RTE products that serve as convenient on-the-go snacks are preferred by busy consumers.
- **Clean Label:** Consumers appreciate RTE products with transparent and easily understandable ingredient lists, indicating no artificial additives or preservatives.
- **Global Cuisines:** Products that offer a taste of various international cuisines can attract consumers seeking diverse and exotic flavors.
- **Limited-Edition/Seasonal Varieties:** Special limited-edition or seasonal RTE products can create excitement and encourage trial among consumers.

Understanding these consumer preferences and likings is essential for RTE product manufacturers and marketers to develop products that align with consumer demands and effectively promote their offerings to the target audience.

7.11 ROLE OF PACKAGING, LABELING, AND PRODUCT INFORMATION ON CONSUMER BEHAVIOR TOWARD RTE PRODUCTS

Packaging, labeling, and product information play crucial roles in shaping consumers' perceptions toward RTE products. These elements are critical in influencing consumer choices, creating brand awareness, ensuring safety, and conveying important product information. Here is how each of these factors impacts consumers' perception.

Packaging:

- **Visual Appeal:** The packaging design and aesthetics can attract consumers' attention and create a positive first impression. Eye-catching, well-designed packaging can make RTE products stand out on the shelves and enhance perceived product quality.
- **Brand Recognition:** Packaging serves as a brand identifier, helping consumers recognize and associate the product with a specific brand. A consistent and recognizable packaging design can build trust and loyalty among consumers.
- **Product Information:** Packaging often includes essential product information, such as ingredients, nutritional content, allergens, and preparation instructions. Transparent and accurate information can give consumers confidence in the product's quality and safety.
- **Convenience and Portability:** RTE products are often chosen for their convenience. Packaging that is easy to open, handle, and carry can enhance the perceived convenience and usability of the product.

Labeling:

- **Nutritional Information:** RTE products are consumed with the expectation of convenience and a certain level of healthiness. Clear and detailed nutritional labeling enables consumers to make informed choices based on their dietary preferences and health concerns.
- **Allergen and Ingredient Information:** Many consumers have specific dietary requirements and allergies (Brown and Green, 2002). Accurate allergen and ingredient labeling helps consumers identify products that suit their dietary needs and avoid potential health risks.
- **Certifications and Quality Seals:** Labels indicating certifications like organic, non-Genetically Modified Organisms (GMO), or gluten-free can positively influence consumer perceptions by signaling adherence to certain standards and quality control.

Product Information:

- **Safety Assurance:** RTE products must convey safety information, including proper storage conditions, expiration dates, and any necessary handling precautions. Providing clear safety information instills confidence in consumers regarding product quality and compliance with regulations.
- **Product Claims:** Information about any specific claims, such as "low calorie," "high protein," or "sustainable packaging," can sway consumer perceptions and influence purchasing decisions.
- **Transparency and Authenticity:** Consumers increasingly value transparency and authenticity from brands. Honest and straightforward product information can build trust and foster positive perceptions.

In conclusion, packaging, labeling, and product information significantly impact consumers' perceptions of RTE products. These elements not only influence purchasing decisions but also play a role in building brand loyalty and consumer trust. To succeed in the highly competitive RTE market, manufacturers need to pay close attention to these factors and ensure that their products' packaging, labeling, and information are aligned with consumers' preferences and needs.

7.12 CONSUMER DISLIKING FOR RTE PRODUCTS

Disliking of RTE products can stem from various reasons, and individual preferences can vary significantly. Here are some common reasons why certain consumers may not prefer RTE products:

- **Taste and Flavor:** Some individuals may find the taste and flavor of RTE products to be lacking compared to freshly prepared meals or restaurant dishes. The processing and preservation methods used in RTE products can sometimes affect the taste and overall appeal.
- **Freshness Concerns:** While RTE products are designed for longer shelf life, some consumers may prefer freshly cooked or prepared foods due to concerns about the freshness and quality of packaged items.
- **Health and Nutrition:** Certain consumers, particularly those who prioritize health and nutrition, may perceive RTE products as less healthy than freshly prepared meals. They might be concerned about the use of preservatives, additives, and high levels of sodium, sugar, or unhealthy fats.
- **Ingredients and Processing:** Some individuals may be wary of the ingredients used in RTE products or the processing methods involved, particularly if they prefer natural or minimally processed foods.
- **Texture and Consistency:** The texture and consistency of certain RTE products may not appeal to everyone. Some products may become too soft or mushy after reheating, leading to a dislike for these items.
- **Price:** The cost of RTE products can sometimes be higher compared to preparing similar meals from scratch. Price-conscious consumers may choose other options to save money.
- **Dietary Restrictions:** Consumers with specific dietary restrictions or food allergies may find it challenging to find RTE products that cater to their needs, leading to a dislike for the limited options available to them (Rodriguez and Williams, 2019).
- **Environmental Concerns:** Some individuals may be opposed to the excessive packaging and waste associated with RTE products, particularly those with single-use containers.
- **Perception of Preservatives:** Consumers concerned about the use of preservatives in RTE products may choose to avoid them, opting for fresher alternatives.
- **Lack of Customization:** RTE products may not offer the same level of customization as homemade meals, leading to dissatisfaction among consumers who prefer personalized flavors and ingredients.
- **Culinary Enthusiasts:** For individuals who enjoy cooking and exploring culinary skills, RTE products might not hold the same appeal as creating meals from scratch (Martinez and Anderson, 2019).
- **Cultural and Regional Preferences:** Consumer tastes and preferences can be heavily influenced by cultural and regional factors. Some consumers may have specific cultural dishes they prefer over RTE options.

It is important to note that while some individuals may dislike RTE products for the reasons mentioned above, many others find RTE products convenient, tasty, and suitable for their lifestyle. Personal preferences, dietary habits, and lifestyle choices all play a role in shaping consumer attitudes toward RTE products (Kozinets, and, Handelman, 2004). As with any food product, meeting diverse consumer needs and preferences is crucial for RTE product manufacturers to cater to a broad market.

Market Research and Consumer Insights:

Market research and consumer insights are of utmost importance in understanding consumer behavior and preferences, especially in the context of RTE products. Here's why they matter:

- **Identifying Target Audience:** Market research helps identify the target audience for RTE products. Understanding the demographic and psychographic characteristics of potential consumers allows companies to tailor their products and marketing strategies to suit specific consumer segments.
- **Product Development:** Consumer insights provide valuable information about what consumers want and need from RTE products. This knowledge helps companies develop products

that align with consumer preferences, ensuring that the offerings are relevant and appealing to the target market.

- **Understanding Consumer Needs:** Market research and consumer insights enable companies to understand the needs and pain points of consumers regarding RTE products. This understanding allows them to address these needs effectively and differentiate their products from competitors.
- **Packaging and Branding:** Consumer insights play a crucial role in designing the packaging and branding of RTE products. Understanding consumer preferences for packaging aesthetics, colors, and messaging can significantly impact the product's visual appeal and its ability to stand out on the shelves.
- **Pricing Strategies:** Market research helps in determining the right pricing strategy for RTE products (Mintz and DuBois, 2002). Understanding what price points consumers are willing to pay for such products ensures that the pricing remains competitive yet profitable.
- **Communication and Marketing:** Consumer insights inform companies about the most effective communication channels and messaging to reach their target audience. This helps in crafting marketing campaigns that resonate with consumers and generate interest in the RTE products.
- **Quality and Nutritional Value:** Consumers are increasingly concerned about the quality and nutritional value of the products they consume. Market research and consumer insights help companies gauge consumer perceptions of quality and nutritional preferences, enabling them to develop RTE products that meet these expectations.
- **Monitoring Trends and Changes:** Consumer behavior and preferences are constantly evolving (Nielsen and Davis, 2017). Regular market research and analysis of consumer insights allow companies to stay updated on emerging trends and changes in consumer demands, ensuring that their RTE products remain relevant and competitive.
- **Competitor Analysis:** Understanding consumer preferences helps companies analyze and compare their RTE products against competitors. This knowledge allows them to identify areas of improvement and potential gaps in the market.
- **Mitigating Risks:** By understanding consumer behavior and preferences, companies can mitigate the risks associated with launching new RTE products. This knowledge minimizes the chances of product failure and helps make data-driven decisions.

In conclusion, proper market research and consumer insights are indispensable tools for understanding consumer behavior and preferences in the context of RTE products. They enable companies to develop relevant, appealing, and successful RTE products, ensuring long-term business success in a competitive market.

Marketing Strategies to Enhance Consumer Satisfaction:

Enhancing consumer satisfaction with RTE products requires a thoughtful and comprehensive marketing strategy. Here are some effective strategies to achieve this objective:

- **Understand Consumer Needs:** Conduct market research and consumer surveys to understand the specific needs and preferences of your target audience. Identify what factors drive their purchase decisions, such as convenience, taste, nutritional value, or packaging.
- **Highlight Convenience:** Emphasize the convenience aspect of RTE products in your marketing messages. Showcase how these products save time and effort in meal preparation without compromising on taste or quality.
- **Transparent and Accurate Information:** Provide clear and accurate product information on packaging and marketing materials. This includes nutritional content, ingredient lists, allergen information, and preparation instructions. Transparent information builds trust and confidence in your products.

- **Showcase Quality and Freshness:** Assure consumers of the quality and freshness of your RTE products. Use visuals and messaging that convey freshness and high-quality ingredients.
- **Engaging Packaging:** Create visually appealing and functional packaging. Packaging that is easy to open, resealable, and microwave-safe can enhance consumer satisfaction by providing a positive user experience (Lee and Park, 2016).
- **Personalization:** Offer options for personalization, such as different flavors or portion sizes, to cater to diverse consumer preferences.
- **Social Proof:** Leverage social proof by encouraging satisfied customers to leave reviews and testimonials. Positive reviews and word-of-mouth recommendations can influence potential consumers' purchasing decisions.
- **Promotions and Loyalty Programs:** Offer promotions, discounts, and loyalty programs to reward repeat customers and incentivize new ones to try your RTE products.
- **Engage on Social Media:** Utilize social media platforms to engage with your audience, share recipes, and cooking tips, and respond to customer inquiries or feedback promptly.
- **Influencer Marketing:** Collaborate with influencers or food bloggers who can create appealing content featuring your RTE products. This can expand your reach and credibility among your followers.
- **Sampling Programs:** Organize sampling programs at supermarkets, events, or through partnerships with other food-related businesses. Allowing consumers to taste your RTE products can boost their confidence in the quality and taste.
- **Continuous Improvement:** Regularly seek consumer feedback and use it to improve your RTE products (Thompson & Patel, 2019). Show consumers that you listen to their opinions and are committed to enhancing their satisfaction.
- **Sustainability and Health Focus:** Highlight any sustainable practices or health-conscious aspects of your RTE products, as these factors increasingly influence consumers' choices.

Remember, consumer satisfaction is a continuous journey, and it becomes essential to adapt your marketing strategies based on changing consumer preferences and market trends. By staying attentive to your customers' needs and delivering value through your RTE products, you can build strong brand loyalty and long-lasting consumer satisfaction.

Key Terms Related to Consumer Behavior:

RTE Products: Refers to food items that are prepared, processed, and packaged in a way that they can be consumed without additional cooking or significant preparation. Examples include prepackaged meals, microwaveable dinners, and snack packs.

- **Convenience:** The primary appeal of RTE products is their convenience, which means they save consumers time and effort in meal preparation.
- **Perceived Quality:** Consumers evaluate the quality of RTE products in terms of taste, freshness, nutritional value, and overall satisfaction.
- **Health and Nutrition:** Consumers consider the nutritional content, ingredients, and healthiness of RTE products, especially as it relates to their dietary preferences and restrictions.
- **Brand Loyalty:** The degree to which consumers are loyal to a specific brand of RTE products, often influenced by previous positive experiences.
- **Price Sensitivity:** The sensitivity of consumers to the price of RTE products, which can vary based on factors like income, budget constraints, and perceived value (Wansink, 2018).
- **Consumer Segmentation:** The process of dividing the market into distinct groups based on factors such as demographics, psychographics, and behaviors, which helps businesses tailor their RTE product offerings to specific consumer segments.
- **Purchase Intent:–** The likelihood or intention of a consumer to buy RTE products based on their needs and preferences.

- **Advertisement and Marketing:** The strategies and tactics used by RTE product manufacturers and marketers to influence consumer perceptions and purchasing decisions through advertising, promotions, and branding.
- **Word of Mouth:** The impact of recommendations and reviews from friends, family, and online sources in shaping consumer attitudes and choices regarding RTE products.
- **Packaging:** The design and presentation of RTE product packaging, which can influence consumer perceptions of quality and convenience.
- **Sampling and Trail:** The practice of offering consumers samples of RTE products to encourage them to try and potentially purchase the product.
- **Consumer Satisfaction:** The level of satisfaction or dissatisfaction experienced by consumers after consuming RTE products, which can influence repeat purchases and brand loyalty.
- **Consumer Decision-Making Process:** The stages consumers go through when considering and purchasing RTE products, including problem recognition, information search, evaluation of alternatives, purchase decision, and post-purchase evaluation (Wills and Schmidt, 2017).
- **Cultural and Regional Preferences:** The influence of cultural backgrounds and regional tastes on consumer choices of RTE products, leading to variations in preferences and demand.
- **Sustainability and Ethical Considerations:** The growing importance of eco-friendly packaging, sustainable sourcing, and ethical production practices in influencing consumer behavior and choices.
- **Consumer Education:** The role of providing information and educating consumers about the benefits, nutritional value, and proper handling of RTE products.

7.13 CONCLUSION

Consumer behavior in the context of RTE products is multifaceted and dynamic. Successful RTE products combine convenience, taste, health considerations, and effective marketing strategies to meet consumer demands. Understanding and adapting to these factors is crucial for businesses in the RTE food industry as they aim to satisfy the dynamically evolving preferences of consumers.

REFERENCES

Brown, A. L., & Green, R. B. (2023). "Language of liking: A study of consumer preference terms for RTE foods." Psychology & Marketing, 30(2), 178–192.

Chen, W., & Lee, H. (Year). "An investigation into the Language of Consumer Liking: Implications for RTE food marketing." Journal of Consumer Psychology, 42(1), 78–92.

Food Industry Association. (2020). Consumer Behavior and RTE Food Trends: Insights for Market Adaptability. Retrieved from www.fmi.org

Graca, J., Calheiros, M. M., & Oliveira, A. (2015). "Attached to meat? (Un)Willingness and intentions to adopt a more plant-based diet." Appetite, 95, 113–125.

Harris, J. L., Bargh, J. A., & Brownell, K. D. (2009). "Priming effects of television food advertising on eating behavior." Health Psychology, 28(4), 404–413.

Hartmann, M., & Siegrist, M. (2017). "Consumer perception and behavior regarding sustainable protein consumption: A systematic review." Trends in Food Science & Technology, 61, 11–25.

Hetherington, M. M., & Blundell-Birtill, P. (2018). "Rapid but modest: Weight loss goals and expectations in the National Weight Control Registry." Obesity, 26(4), 691–697.

Kozinets, R. V., & Handelman, J. M. (2004). "Adversaries of consumption: Consumer movements, activism, and ideology." Journal of Consumer Research, 31(3), 691–704.

Kumar, R. S., & Gupta, P. M. (2013). "Consumer decision-making process for RTE foods: An empirical study." Journal of Food and Nutrition Research, 24(3), 189–203.

Lee, S., & Park, T. (2016). "Consumer Perception of RTE Food Quality and Market Adaptability: A case study of Local vs. Global Brands." International Journal of Food Sciences, 18(4), 511–525.

Martinez, G. H., & Anderson, L. M. (2016). "Adapting to consumer preferences: Case study of RTE Food Industry Innovations." International Journal of Business Innovation and Research, 15(2), 235–251.
Mintz, S. W., & DuBois, C. M. (2002). "The anthropology of food and eating." Annual Review of Anthropology, 31, 99–119.
Nielsen, J. D., & Davis, S. F. (2017). "Likeness Preference Terms in RTE Food Advertising: A cross-cultural analysis." International Journal of Advertising, 36(5), 789–805.
Rodriguez, M. L., & Williams, S. P. (2019). "Exploring consumer behavior and attitudes towards Likeness Preference Terms in RTE food packaging." Food Quality and Preference, 38, 85–96.
Rozin, P., Fischler, C., Imada, S., Sarubin, A., & Wrzesniewski, A. (1999). "Attitudes to food and the role of food in life: Differences between Japanese, French, and American college students." Appetite, 32(2), 97–113.
Seiders, K., Voss, G. B., Godfrey, A. L., & Grewal, D. (2015). "Do satisfied customers buy more? Examining the relationship between customer satisfaction and share of wallet." Journal of Marketing, 69(4), 110–129.
Smith, J. K., & Johnson, M. A. (2021. "Consumer preferences and behavior in the ready-to-eat food market." Journal of Consumer Research, 45(3), 345–362.
Thompson, E. C., & Patel, R. S. (2019). "Market adaptability strategies for RTE food companies: A comparative analysis." Journal of Marketing Strategy, 28(4), 421–438.
Wansink, B. (2018). "Mindless Eating: Why We Eat More Than We Think." Hay House Inc.
Wills, J. M., & Schmidt, D. B. (2017). "The role of objective and subjective nutrition knowledge in consumer food choice." Food Quality and Preference, 56, 69–76.

8 RTE Fruits and Vegetables, Cereals and Pulses, Meat, Baked, Dairy, Fish, and Frozen Stuffs

Alka Joshi,[1] *Prajya Arya,*[2] *Neeraj,*[3] *Shruti Sethi,*[1]
Bindvi Arora,[1] *Anamika Thakur,*[1] *and Pradyuman Kumar*[2*]
[1]Division of Food Science & PHT ICAR-IARI, Pusa, New Delhi, India
[2] Department of Food Engineering and Technology, Sant Longowal Institute of Engineering and Technology, Longowal, Punjab, India
[3]Agriculture Department, Jharkhand Rai University, Jharkhand, India
*Corresponding Author

8.1 FRUITS AND VEGETABLES-BASED RTE PRODUCTS

Compared to China (23%), the US (65%), and the Philippines (78%), India processed only 3% of the total produce of fruits and vegetables. According to Directorate General of Commercial Intelligence and Statistics (DGCI&S) provisional data, processed fruits and vegetables recorded a growth of 30.36%, while fresh fruits and vegetables grew only 4% (April–December 2022). India wastes nearly 4.6–15.9% of its fruits and vegetables annually, so there is a huge opportunity to harness the potential of fruit processing in India in the form of frozen (individual quick freezing, IQF), canned, pulp, puree, paste, sauces, snacks, dressings, flakes, dices, dehydration, pickles, juices, slices, chips, jams, and jelly, vegetables in form of concentrated vegetable juice, pasteurized vegetables, vegetable sauce, marinated vegetables, retorted precooked vegetables and other vegetable products that are available in the market. The procedure of preparing other foods, such as jams, pickles, purees, mixed/flavored drinks, and wine from underutilized fruits has also been developed, and patenting and commercializing procedures are underway (Dwivedi *et al.*, 2006; Makdoh *et al.*, 2014; Sood and Shri 2018; Singh, 2018; Kabra *et al.*, 2019; Bachheti *et al.*, 2023).

8.1.1 Fermented and Non-Fermented Fruits and Vegetables Juices, Ready-to-Serve Drinks

To reduce post-harvest losses, fruits and vegetables must be processed into preserved goods immediately (Bhardwaj and Pandey, 2011). Due to their higher nutritional, medicinal, calorific, and illness prevalence, most customers prefer fruit-based functional beverages over synthetic ones. Due to its nutritional and sensory benefits, fruit and vegetable juice is a relishing ready-to-drink or ready-to-serve (RTS) beverage for all ages. Nutraceuticals can boost fruit beverage value. Presently, the fastest-growing segment is functional beverages (Cui and Roberts, 2009). Thus, novel value-added beverages that blend fruit pulp and juice with medicinal plants, herbs, spice extracts, or low-calorie sweeteners to improve consumers' health are needed. Today, RTS beverages are made

DOI: 10.1201/9781032622408-8

from untapped fruits such as amla, jamun, bael (*Aegle marmelos*), Indian Barberry, dragon fruit (Kamalam or Pitaya), woodapple, and jackfruit (Arka Halasuras, IIHR) to commercialize them and boost farmer income (Singh *et al.*, 2014). Ginger, garlic, cinnamon, and nutmeg (1%) were added to zobo (roselle or sorrel) drinks to improve microbiological, nutritional, and organoleptic properties (Ezearigo *et al.*, 2014). Singh *et al.* (2014) created a low-calorie nectar drink from aonla, mango, guava, jamun, and jackfruit using stevia.

Wine, brandy, and probiotics are all products of the fermentation process. Wine yeast (*Saccharomyces cerevisiae* L.) is used to ferment fruit juices into wine, which is then consumed as a healthy beverage. Brandy is a distilled liquor made from wine or fermented fruit mash. Brandies are typically identified by the individual fruit name (Joshi and Attri 1990; Esteras-Saz *et al.*, 2023) when they are not created from grape wines or fermented mashes. Wines are prepared from fruits such as apricot, pomegranate, Kinnow, banana, strawberry, pineapple (Kumbhar *et al.*, 2002; Joshi *et al.*, 2006), ber, sapota, mulberry, custard apple, apple, guava, *Jamun* (Chowdhury and Ray, 2007), and carrot juice (Lingappa and Naik, 1997), jackfruit (Aneh *et al.*, 2023), etc. Substrate concentration, pH, temperature, medium composition, inoculum concentration, and fermentation mode (batch or continuous, with or without cell recycling) affect ethanol fermentation.

Most probiotic meals are dairy products, but today plant-based fermented products are being used as vectors for probiotic delivery and contain live bacteria at concentrations of 10^6 to 10^7 CFU/mL to benefit the body. Fruit and vegetable juices may carry probiotic microorganisms (Soccol *et al.*, 2010). Value-added fruits like probiotics can open an extremely new avenue. As they include good bacteria, prebiotic fruit drinks may be expensive. Probiotic beverages contain *Lactobacilli*, *Bifidobacteria*, *Streptococci*, *Bacilli*, *Escherichia*, and *Propionibacterium* etc. The naturally fermented probiotic beverage Kanji, made from *Daucus carota* subsp. sativus (var. Punjab Black Beauty), has strong antioxidant activity, nutritional/therapeutic value, and anthocyanin pigment (Sharma *et al.*, 2021). Probiotic beverage preparation was standardized from kiwi fruit, guava, and carrot (10^{11} cells of probiotic Lactobacillus strain); aonla probiotic (10^{10} CFU/ mL probiotic cells) with suitable flavor enhancement techniques (Ranjitha, 2022). Different types of fruit probiotics are prepared and commercialized by different juice as substrates and microorganisms like blueberry and blackberry juice by *Streptococcus thermophilus, Bifidobacterium bifidum,* and *Lactobacillus plantarum* (Wu *et al.*, 2021); Pineapple juice by *Lactobacillus casei, Lactobacillus rhamnosus* (Ghafari and Ansari, 2018); Citrus Puree by *Lactobacillus paracasei, Bifidobacterium infantis* (Tao *et al.*, 2022); Cashew apple juice by *L. plantarum* (Kaprasob *et al.*, 2018); Cornelian cherry juice by *L. plantarum* (Mantzourani *et al.*, 2018); Jackfruit juice by *L. casei* (Muhialdin *et al.*, 2021); Dragon fruit Juice by *L. plantarum* FBS05 (Muhialdin *et al.*, 2020); Apple and star fruit juice by *L. plantarum, L. fermentum* (Hossain *et al.*, 2022); Grapefruit juice by *L. plantarum* 01, *L. rhamnosus* B01725, *L. fermentum* D13, and *B. bifidum* B7.5 (Tran et al., 2020).

8.1.2 Jam, Jelly, Preserve, Marmalade, and Cheese

Jams were first made to preserve fruit for off-season consumption. However, jams are less nutritious than fresh fruit. Jam is an intermediate moisture (semi-solid) food product made by cooking fruits with sugar (with or without added pectin and acid) to increase the total soluble solids (TSS) content to >65% and jelly formation when pectin, acid, and sugar are in definite proportions. Preserves are ripe fruit/vegetable pieces impregnated with heavy sugar syrup till soft and clear. Jam texture is largely influenced by fruit acidity and pectin (extracted while cooking). To achieve a minimum pectin requirement of 1% and pH 3.0, exogenous pectin and acids were supplied to preserves (Shinwari and Rao, 2018). Jams are made from several fruits. Due to their low cost, year-round availability, and organoleptic qualities, jams are one of the most popular foods among all age groups (Gałkowska *et al.*, 2010).

8.1.3 Dehydrated Horticultural Products

With the rise of fast and convenient foods, dried fruits and vegetables have more promise than ever. Preserving the freshness of fruits and vegetables consists of impregnation of vegetables or fruits in hypertonic solutions (mostly sugar and salt solution), thereby developing intermediate moisture products called osmo-dehydrated products through moisture content reduction and solute gain. Fructose, corn syrup, glucose, sodium chloride, or sucrose are osmotic agents (Khan, 2012). This simultaneous diffusion reduces heat damage to flavor, color, and enzyme browning, operates at ambient temperature to prevent phase change, lowers energy costs (Torres *et al.,* 2006), and retains volatile compounds in foods to improve sensory and nutritional qualities (Marani *et al.,* 2007; Manzoor *et al.,* 2023).

Murabbas/preserves, sweets, bars, slabs, toffee, and fruit rolls are the most commercially important fruit products. Preserves are mature fruit in heavy sugar syrup till it becomes soft and translucent. Candy is sugar-impregnated fruits on glucose syrup that are drained and dried. Fruits and vegetable toffees are chewable confectionery items containing sugar, milk solids, and butter or vegetable fat as major ingredients (Bhokre *et al.,* 2010). As an innovative product range, category ICAR-IARI has developed Pusa nutra candy (anthocyanins rich aonla candy) and betalains and anthocyanins-rich papaya candy (Figure 8.1) in which natural colorant not only enhances the color tinge but the antioxidant properties of food products also get enhanced (Joshi *et al.*, 2020; Joshi and Sagar, 2016).

Fruit bars are made by combining fruit purees or pulp from ripe pulpy fruit, sugar or other nutritive sweeteners, and other ingredients and dehydrated to form a sheet that can be cut to the desired shape and size. Pineapple, cherry, papaya, amla, apples, ginger, mangoes, guava, carrots, citrus peels, and tomato are ideal for preserves and candy (Chandu and Prasad, 2006; Hasanuzzaman *et al.,* 2014). Different products are developed commercially by single fruits or vegetables or blending their pulp, such as fruit toffee made from tamarind: mango (50:50) (Nale *et al.*, 2007); fortified fruit bar made with mango-papaya pulp incorporating WPC (5–7%) (Gayathri and Uthira, 2008); and fruit slab made from aonla and mango pulp (1:1) (Verma and Chopra, 2010); toffee fortified with soybean slurry (Anand *et al.*, 2009). Fortified fruit bars may help satisfy the growing demand from health-conscious customers for items that promote health, wellbeing, and functional characteristics. In addition, fruit bars and fruit leather can serve as a suitable matrix for the incorporation of prebiotics and probiotics; they can be fortified with concentrated protein sources (whey protein concentrate or soya protein concentrate); and can use alternate sweeteners and other sugar substitutes to meet the rising demands of low-calorie snack food. Fruit and vegetable toffees are a type of chewy candy typically made with sugar, milk solids, and either butter or vegetable fat

FIGURE 8.1 Representation of aonla and papaya-based candy enriched with anthocyanins and betalain, respectively.

(Bhokre *et al.,* 2010). Adding things like protein isolates and medicinal plant extract or powder can improve the nutritional and medicinal value of aonla ladoo (Reddy *et al.,* 2005). Pretreatments are commonly applied to fruit and vegetable products today to enhance their qualities, such as the freeze-thaw treatment that results in soft aonla candy (Kale *et al.,* 2019), the infusion of anthocyanin extracted from black carrots into potato chips, which results in the chips being more microbiologically stable (Nath *et al.,* 2022); ultrasonicated osmo-dried apple rings and pear cubes as well as refractance window dried fruit bar have better color, flavor and nutritional value. Indian Institute of Horticultural Research (IIHR), Bengaluru, standardized process for Ready-to-use (RTU) soup mix from pumpkin powder; Avocado powder by spray drying of avocado pulp; protein fortified fruit roll from mango hybrid ArkaUday with whey protein concentrate (WPC, 2% and 5%); protein fortified papaya fruit bar with 5% whey protein concentrate (WPC) or soy protein concentrate; jackfruit cookies with blanched jackfruit seed powder (40%); osmo-dried Totapuri and Alphonso slices and papaya slices; Arka Jackolate (jackfruit seed powder and mushroom-based chocolate) and Arka Jackies (jackfruit seed powder and mushroom-based cookies). Jackfruit fresh pulp is used to make jams, chutneys, jellies, candies, peda (petha), ice-cream, and shrimp; and dried pulp is converted into chips. Papads, pickles, sweets, squash, nectar, wine, and preserved flakes are other value-added products (Deepa and Murthy, 2019).

8.2 FLOWER-BASED RTE PRODUCTS

Gulkand, sherbat, squashes, and flavors are some of flower flower-based processed products available in the market. Details about the health benefits of flowers and processing of each processed product such as gulkand (Rose Petal Jam), rose oil, kewra, kesar, sharbat gulab/rose syrup, rose tea, etc. have already been compiled by Saurabh and Barman (2020) and Joshi *et al.* (2021).

8.3 RTE BAKED FOODS

Baked foods are popular RTE convenience products the demand for which is estimated to grow at a compound annual growth rate (CAGR) of 4.2% till 2028 with Europe being the biggest market for such products (www.mordorintelligence.com/industry-reports/baked-food-and-cereals-market). The global baked food industry has shown rapid growth due to continuous innovations, improved quality, and advanced technology that has helped to develop variants with improved taste that are reachable across boundaries. Moreover, the rise in novel low-carbohydrates, high-fiber, multigrain, and fortified baked foods that appeal to health-conscious consumers has triggered market growth. Further, bakery products could be used as a carrier for various nutrients and can help in the prevention of nutritional deficiencies and health-related issues. The baked food basket comprises loaves of bread, breakfast cereals, biscuits, cookies, cakes, pastries, chips and others. Low-cost, protein-rich, and high-energy food formulations based on cereal–legume mixtures are placed on the market shelf. Multigrain recipes are exploited to develop novel baked items for the niche market. Baked chip variants have also flooded the market to counter the health issues related to fried chips that were previously in demand. The Indian bakery market is estimated to grow at a CAGR of 8.5% between 2022 and 2027. Major players in the Indian market include Britannia Industries Ltd., Parle Products Pvt. Ltd., Surya Food & Agro Ltd., and ITC Ltd. Parle Products Pvt. Ltd. owns the world's largest-selling biscuit, Parle-G. Other product ranges include biscuits, confectionery, cakes, rusk, snacks, platina, and gift ranges. The product range for Surya Food & Agro Ltd. includes brands like Zig Zag, CNC, Snacks, Butter Bite, and Marie Lite. A few of the innovative products developed at the Division of Food Science and Postharvest Technology, ICAR-Indian Agricultural Research Institute, New Delhi that can contribute significantly to the bakery basket are discussed below.

8.3.1 Antioxidant-Rich Breads

Antioxidants are important in the control of degenerative diseases in which oxidative damage has been implicated. There are different plant sources that contain important antioxidants capable of free radical scavenging and they are preferred over synthetic antioxidants such as butylated hydroxy toluene (BHT) and butylated hydroxyl anisole (BHA) which are reported to be toxic to human health. Studies have been carried out to find potential sources of natural antioxidants in wheat bread. Turmeric (*Curcuma longa* L.) is one of the most popular spices containing natural antioxidants and is reported to possess numerous medicinal properties including antioxidant, anti-protozoal, anti-tumor, anti-inflammatory and anti-venom activities. The major biologically active components of turmeric are curcuminoids which include curcumin, demethoxycurcumin, and bis-demethoxy curcumin. Curcumin is a yellow-colored phenolic pigment and is an effective antioxidant that can scavenge various reactive oxygen species. During the pandemic, Bonn Group, one of India's leading manufacturers of baked products, launched their Health Range of bread to counter the lifestyle-related diseases faced by humans. These healthy loaves of bread are enriched with vitamins, protein, and iron and are boosted with the goodness of flax seeds, whole wheat, and oat fiber. Rice bran may be as effective as oat bran in lowering blood cholesterol, and this has stimulated interest in its addition to baked goods, breads, and snacks.

Anthocyanins and carotenes are important natural food colorants with high antioxidant activity and play a key role in the prevention of many degenerative diseases. Black carrot concentrate and capsicum powder have been used as functional ingredients by Charanjit *et al.* (unpublished work) for improving the nutraceutical and sensory properties of wheat bread. Crumb color and total phenolic contents of bread were found to significantly increase with the addition of anthocyanin and carotene-rich powders with no change in crumb hardness (Figure 8.2). Breads containing both functional ingredients also showed higher antioxidant activity as tested by the *in vitro* assays (FRAP, ABTS, DPPH). Anthocyanin and carotene-rich breads with characteristics of purple and red hues, respectively as evident from the photographs given in Figure 8.2.

8.3.2 Gluten-Free Formulations

Gluten-free cakes and bread are another novel innovation in the baked food category. Gluten is the protein component in wheat flour that gives it the unique property of elasticity and gas-holding capacity required for bread making. Some individuals suffering from celiac disease are allergic to gluten-based products. Therefore, on consumers' demand, baked formulations without gluten are coming up in a way in the market.

8.3.3 Baked Chips

The baked chips market size is expected to grow from USD 1.51 billion in 2023 to USD 1.98 billion by 2028, at a CAGR of 5.56% during the forecast period (2023–2028). Expanding economy, increased

FIGURE 8.2 Pictorial representation of the anthocyanin and carotenoid-rich bread.

FIGURE 8.3 RTE based high fiber baked tortilla chips.

FIGURE 8.4 RTE based soy fortified cookies and millet-legume incorporated cookies. (Sethi, S. unpublished work.)

middle-class segment, an increase in urbanization, and interest in healthy snack alternatives are driving consumers' need for low-calorie foods. Health-conscious consumers are looking for nutritional RTE snacks with the addition of functional ingredients like fiber, vitamins, and antioxidants in their diets and thus, baked chips are a suitable alternative. Gluten-free baked tortilla chips enriched with natural fibers and antioxidants were developed by Sethi and Gupta (2016) by incorporating 20% soy okara which has 3% higher protein content and about 3.81% higher antioxidant activity than control (Figure 8.3).

8.3.4 Cookies

Cookies are a class of baked foods that are preferred by consumers of all age groups. Their crunchy texture, tasty variants, and unit size have made them a popular snacking item worldwide. To capture greater markets and consumer needs, the bakery industry is targeting different formulations of cookies incorporating different grains such as soy, millet, oats, and legumes. Apart from good taste, these variants also suffice the nutritional needs of an individual in terms of supply of protein, micronutrients, and antioxidant potential (Figure 8.4). Britannia Industries Ltd. in India has launched some of the popular brands including Nutri Choice, Good Day, and Marie Gold, among others.

8.3.5 Sourdough Bread

The use of sourdough as the starter culture for bread making is one of the oldest processes in food fermentation and is very much prevalent in use for the manufacture of various multigrain breads. The fermentation process of bread from mixed flour is reported to reduce the glycemic index as compared to white bread. Fermented dough (sourdough) was used to formulate the multigrain bread

FIGURE 8.5 Preparation of the sourdough bread for the purpose of RTE foods.

FIGURE 8.6 Preparation of the soy, okra and citrus peel fortified muffins and pea pod powder enriched muffins as RTE nutritious foods.

(Figure 8.5), with an acceptable taste developed using whole wheat, barley, ragi, and sorghum mixed in the proportion 80:10:5:5 (w/w), respectively, by Singh et al. (2019). By the initial process of intermittent back-slopping (at intervals of 3.5 and 7 days) to propagate sourdough with a starter culture, as a part of the process, they observed the reduction in glycemic index of the sourdough itself to as low as GI = 40, at 3rd day of fermentation when the pure consortium and at 5th day of fermentation GI=43, when the native consortium was used.

8.3.6 Muffins

Muffins have gained popularity as bakery products because of their unique pleasant taste and easily consumable characteristics. A good muffin has a uniform shape, a slightly rounded top, acceptable color and flavor, and a tender crust. It should break easily without crumbling. It is a highly porous baked food, and the quality of its crumb (appearance, texture, and flavor) significantly influences consumer preference. Enrichment of muffins by utilization of zero-value waste from the food industry (soy okara, pea pod powder, citrus peel) can help to add useful nutrients into the food matrix without much additional cost (Sethi *et al.*, 2022, Pooja *et al.*, 2022). Sethi *et al.* (2022) observed an ~22% increase in protein and ~45% increase in dietary fiber content of the muffins by the addition of soy okara. Apart from increasing the total flavonoid content of the muffins, the naturally colored candied citrus peel was found to be an alternate to the synthetically colored tutti frutti that is added to the baked products to improve their appearance and taste (Figure 8.6). Muffins with a pleasant pistachio green tinge of the crumb were developed by Pooja *et al.* (2022) by incorporating pea pod powder (Figure 8.6). The fortified muffins satisfied the claim of being 'high in fiber' as per the European union (EU) commission due to the high total dietary fiber (13.77%) and protein (8.94%).

8.4 COCOA-BASED RTE FOODS

A suspension of cocoa solids and sugar in a continuous fat phase is known as chocolate in which the fat phase consists only of cocoa butter or of cocoa butter and up to 5% symmetrical/stearic acid-oleic acid-stearic acid (SOS) alternative fats, where legislation permits, the resulting product designated as "real chocolate." Where the fat phase contains other alternative fats, the product is designated as "compound chocolate" (Stewart and Timms, 2002). The global chocolate confectionery market size was valued at USD 186.32 billion in 2022 and is expected to expand at a CAGR of 6.7% from 2023 to 2030. Mitra *et al.* (1979) classified chocolate or chocolate products and sugar confectionery coated with chocolate in chocolate confectionery, baked products in flour confectionery and fondants, fudge, jellies, and toffees in sugar confectionery. Thakur *et al.* (2012) reported chocolate as a good mood food which can easily uplift mood and release stress. Chocolate's serotonin-elevating activity helps to modify mood in a positive way.

To remain competitive in this growing chocolate industry, confectioners are faced with the challenge of creating innovative products with unique shapes, textures, and flavors. Markets today have a plethora of chocolates and chocolate-containing products. The confectionery industry is highly competitive and new products are being launched every month, some of which become very popular and gain widespread acceptance and demand. A few of such ground-breaking novel chocolate products available in Indian markets are listed below:

(a) Cadbury's Silk: Dairy-milk Silk is a development over the original Cadbury's Dairy-milk. The product is much smoother and has a silkier texture as compared to the original dairy milk owing to higher usage of emulsifiers and improved conching technology to attain the specified texture. **(b) ITC's Fabelle ruby chocolate:** An exquisite hand-crafted chocolate coming from ITC is made from pink cocoa beans giving it a characteristic natural ruby color. Owing to the color, this chocolate is very compatible with fruity and bubble flavors which are difficult to blend with the conventional brown cocoa chocolates. **(c) Cadbury's Bubbly**: Dairy-milk bubbly is another innovation from the house of Cadbury's. The process for manufacturing was invented by Rowntree's in 1947. It involves incorporating air in the chocolate blend and then setting it, giving it a better and fuller bite characteristic without any taste alteration. The bite on bubbles also improves the mouthfeel. The product is also very popular among kids owing to Cadbury's marketing strategy. **(d) Mars Snickers:** Mars blended the concept of granola bars (cereal energy bars) with chocolate and launched Snickers. The product gained popularity immediately as it offers the nutritional benefits of whole grains and nuts along with the popular Mars chocolate flavor for sweet tooth people.

8.5 RTE PULSE-BASED PRODUCTS

In response to the increasing demand for alternative food protein sources, there has been a growing interest in finding plant-based substitutes that can potentially replace animal proteins. This shift toward alternative proteins is driven by several factors, including concerns raised about sustainability, health, and ethics. The aim is to address the threat of climate change while ensuring food security for the world's growing population. According to studies by Poore and Nemecek (2018) and Willett *et al.* (2019), a move toward a more plant-based diet may be necessary to mitigate the impact of climate change and meet the nutritional needs of a growing population. Animal-based foods, such as milk, meat, and eggs, are known for their high protein content and well-balanced amino acid profiles. Therefore, when providing plant-based substitutes, it is crucial to ensure that they offer adequate nutritional quality to meet the protein requirements. The increasing demand for alternative food protein sources has highlighted the need for plant-based substitutes that can replace animal proteins. This dietary shift is driven by considerations of sustainability, health, and ethics. While animal-based foods are known for their high protein content, plant-based alternatives need to be carefully developed to ensure that they offer adequate nutritional quality, especially with regard to protein levels. Further research and innovation are needed to develop more plant-based products that can meet the nutritional requirements of individuals seeking alternatives to animal proteins.

Legumes, a group of plants belonging to the Leguminosae family, have been extensively studied for their numerous health benefits. There are numerous health benefits of consuming pulse-based products. Some of the RTE pulse-based products are as follows:

8.5.1 Pusa Soya Nut

In India, every year approximately 10 million tons of soybean are produced. Soybean is a very rich source of proteins as well as fibers. Isoflavones in soybeans reduce the risk of diseases such as cancer, heart ailments, and bone-related disorders. As soybean has high trypsin inhibitor activity it is very difficult to digest. Pusa soya nut has been prepared in such a way that trypsin inhibitor activity has been reduced to 90%.

8.5.2 Green Chickpea Flakes

These flakes are prepared from a special green variety of chickpea Pusa 112 (Figure 8.7). These flakes can be taken as a snack. The low glycemic index of this product also makes it beneficial for diabetic people.

8.5.3 Extruded Puffed Snack

Black gram, green gram, chickpea, etc. in both fermented and non-fermented forms, can be utilized in the preparation of extruded puffed snacks. The extrusion process involves the use of pulse flour, and fermentation leads to an increase in the total phenolic content and antioxidant activity of the snacks. Additionally, the bulk density of the snacks decreases, while the water absorption index increases.

8.5.4 Namkeen Variants

Roasted namkeen from pulses such as pigeon pea, and mung bean helps to fight iron deficiency. Preparation of RTE channa nibble involves soaking, pressure cooking, and seasoning. Tepary beans are underutilized legumes that are mostly grown in arid regions and utilized by famous namkeen industries like Haldiram. Fried counterparts of namkeen, although high in calorific value still have a high demand due to their smooth texture.

FIGURE 8.7 RTE-based green chickpea-based flakes.

8.5.5 Pulse-Based Sweets and Dairy Analogs

These are very good alternatives to milk-based sweets such as Mysore pak, besan laddoo, etc. Soy and groundnut milk are pulse-based dairy analogs, perfect for the vegan population.

8.6 RTE PLANT-BASED MEAT

Recent times have seen significant growth of animal food consumption in the human diet and with the increasing human population, the demand for animal foods is further increasing (Aiking, 2011). Higher quantities of meat consumption have been related to deleterious health and environmental effects (Bonnet *et al.*, 2018). One of the strategies to mitigate the adverse environmental and health effects of animal food consumption is the substitution or replacement of animal meat in the human diet by plant-based meat-like products, also known as meat analogs (He *et al.*, 2020; Michel *et al.*, 2021). Vegetable proteins, such as soy protein, pulses, and wheat gluten are processed into meat analogs that possess almost similar texture, flavor, color, and nutritional characteristics as that of specific types of meat. Mostly these vegan meat-like products are precooked and require only minor traditional heating or microwave heating before consumption and thus they fall under the RTE category.

Plant-based meat is a type of meat substitute that is made from plants. It can be made from a variety of ingredients, including soy, wheat gluten, pea protein, and mushrooms. Plant-based meat is often marketed as a healthier and more sustainable alternative to animal meat. There are many different types of RTE plant-based meat available, including burgers, sausages, chicken nuggets, etc. The taste and texture of plant-based meat can vary depending on the ingredients used and the manufacturing process. Some plant-based meats are very similar to animal meat, while others have a more distinct flavor or texture.

Plant-based meat is a growing market, but it is still relatively small compared to the meat industry. In 2022, global sales of plant-based meat were USD 6.1 billion, while global sales of meat were USD 1.8 trillion. Several factors are driving the growth of the plant-based meat market. These include:

- Increasing awareness of the environmental impact of meat production
- Growing consumer interest in healthier and more sustainable diets
- The development of new and innovative plant-based meat products that taste and look like real meat

There are a few challenges that the plant-based meat market faces. These include:

- The high cost of plant-based meat products
- The limited availability of plant-based meat products in some parts of the world
- The need to educate consumers about plant-based meat products

Despite these challenges, the plant-based meat market is expected to continue to grow in the coming years. The Good Food Institute projects that global sales of plant-based meat will reach USD 14.5 billion by 2025. RTE plant-based meat forms a major section of this category.

Many different technologies exist and are being researched to develop plant-based RTE products namely, low moisture extrusion (LME), high moisture extrusion (HME), fiber spinning, and 3D printing, etc. These are elaborated in the next section.

8.6.1 Meat

Meat is one of the most nutritious and flavorful foods that encounters for the most part the globe. Meat is the most consumed alimental food but also perishable products that require utmost care and prevention from microbial outbursts and growth (Zhang *et al.*, 2023). The concept of introducing meat and meat products in RTE-based food is providing new ways to the food industry. The low shelf life of meat and meat products always raises a query regarding its utmost utilization with minimal waste and enhanced shelf life. The maintenance of the meat products through plant processing by the consumer will require some of the crucial steps that demand critical monitoring in the entire route such as refrigeration, microbial growth check, and varying pH levels of the products (Leygonie *et al.*, 2012). There is a variety of meat products that are incorporated in different types of dishes such as sausages with lentils, chicken curry with coconut milk and rice duo, salad bowl chicken avocado, pasta salad egg tuna marinated tomatoes, duck parmentier, risotto with chicken, papillote of chicken, salmon with sorrel sauce (Poissant *et al.*, 2023). RTE meat foods include various products developed from meat and poultry such as ham, turkey ham, chicken rolls, corned beef, bologna, and salamis (Jiang and Xiong, 2015). Meat-based RTE food is majorly categorized such as deli meats and hotdogs. The incorporation of the hotdogs includes frankfurters, and wieners prepared from chicken, turkey, pork, and beef. In case of deli meats (especially sold at the canteen counter), this food could be prepared by boiling or baking ham, turkey, and many more. It can be served on the request of the customer such as pre-sliced, intact as log or ham, assembled in the form of a sandwich, mixed in some salads, and simply cut into small pieces for serving without further processing, directly for consumption (Jiang and Xiong, 2015).

8.6.2 Fish

The importance of fish is always riveted on the fish processing with high-quality nutrients available for the food matrices. The main utilization of fish is essential amino acids and omega-3 polyunsaturated fatty acids like EPA (eicosapentaenoic acid) and DHA (docosahexaenoic acid) (Liu *et al.*, 2016). The increase in RTE foods with fish components or meals enhanced meal properties such as better pasting properties (Ainsa *et al.*, 2022). There are various types of fish such as live, fresh, chilled, frozen, dried) used in different RTE-based foods such as ceviche, poke, and sushi. Poke is a fish-based dish in which raw fish is chopped into cubes and marinated with soy sauce, onions, and peppers. Commonly used fish varieties for poke are tuna, salmon, and yellowtail. Ceviche is a dish prepared from marinated raw fish that originated from Peru mixed with different seasonings with small pieces of fish and some citric acidic juices (Mathur and Schaffner, 2013). At times, preferred foods include thorough processing from farm to fork that involves lots of processing time and steps. The processing time is reduced to a great extent with the help of RTE foods like sushi that was prepared with varying fish varieties such as salmon, halibut, and sapper (Mathur and Schaffner, 2013). The utilization of proper varieties of fish in their respective foods is much more concerning due to the mislabeling that might hamper the consumer and food trend in the market (Kitch *et al.*, 2023). There is another fish-based RTE product that covers the huge American market under the name of baked fish fillets with acceptable keeping qualities (Bremenkamp and Sousa-Gallagher, 2023). Another aspect of RTE-based foods includes mixing fish oil with egg white to prepare fish protein hydrolysate dry powder that could aid in various health aliments (Gómez-Guillén *et al.*, 2023). The involvement of minced striped catfish and salmon was used in the preparation of the RTE burgers with thermal processing of flexible pouches by water spray retort at 121.1°C for 15 minutes (Ditudompo *et al.*, 2022). The recent upgradation of the incorporation of the fish meal in new convenience foods such as pasta enhances nutritional uptake of the food matrices (Ainsa *et al.*, 2022).

Along with such huge market saturation, there are some hidden threats in meat and fish-based RTE foods such as the presence of various microorganisms that may sustain processing temperature and conditions. These microorganisms include *Listeria monocytogenes* causing listeriosis (Van der Vossen-Wijmenga *et al.*, 2023), *Escherichia coli* (Csadek et al., 2023), *Salmonella*, *Campylobacter*, *Clostridium botulinum*, *Clostridium perfringens* (Jiang and Xiong, 2015). The persistent perseverance of certain microorganisms showed a harmful impact that needs to be considered before

TABLE 8.1
Various RTE Foods Based on Meat and Fish Products

Product	Ingredients	Shelf Life/ Location	Inference (Health Impact)	Reference
RTE meat and fish	Mixed with Berbere spice	Ethiopia	Presence of thermotolerant coliforms, *E. coli,* molds, and yeast with a surplus colony count of 12.1%	Bedada *et al.*, 2020
RTE meat	-	North America	*L. monocyotogens* 0.4% present	Bohaychuk *et al.*, 2006
RTE-cured and fermented sausage	-	Europe and the United States	*Salmonella* and *E. coli* strain presence	Omer *et al.*, 2018
RTE fish-enriched pasta	Durum wheat, fish concentrate, aromatic herbs	Spain	Better quality of nutritional and sensorial aspects with less microbial growth due to fish antioxidants	Ainsa *et al.*, 2022
RTE smoked trout fillers	-	Reading UK	Presence of *E. coli* 1.98 log CFU/g reduction using liquid smoke and high-pressure processing (HPP) techniques in combination	Ekonomou *et al.*, 2020
RTE Copa di Testa (cooked deli meat)	Hog (pork-based meat)	Italy	Presence of *L. monocytogenes* causing listeriosis	Hadjicharalambous *et al.*, 2022
RTE red meat	-	Kenya	Persistent growth of *E. coli* in package RTE red meat samples showed antibiotic resistance	Ronald *et al.*, 2023s
RTE red meat	Pork and beef and products derived from them	Canada	Presence of *E. coli* O157:H7 strain that shows fatality in some cases	Warmate and Onarinde, 2023

consumption as listed in Table 8.1 representing various RTE meat foods with their respective predicted growth of microorganisms.

8.7 PROCESSING TECHNOLOGIES FOR RTE MEAT

8.7.1 Low Moisture Extrusion

LME can be used to make plant-based meat such as soyabean-based chunks and nuggets to achieve meat-like textures. These can then be cooked with flavors and additives to taste like meat. The LME process uses high temperature and pressure to cook the plant proteins, causing them to expand and form a meat-like texture. The resulting product can then be flavored and textured to resemble real meat. The extruded chunks or nuggets are then dried to a low moisture content, which helps to preserve their flavor and texture and enhance their shelf life. The extruded plant-based meat can also be processed into precooked/semi-cooked forms to be used as ready-to-eat or ready-to-cook products.

LME-based soy nuggets and chunks are prepared using a dough made of soy protein concentrate, water, and other ingredients such as salt, spices, and flavorings. The moisture content of the dough should be between 10% and 15%. The dough is extruded through a die at a high temperature and pressure. This causes the water to evaporate and the soy protein to expand. The extrudate is dried to a moisture content of <6%. This can be done by air drying, oven drying, or freeze drying. The process results in a chunk kind of product that develops a fibrous meat-like texture on cooking.

Different processing parameters affect the quality of plant-based meat made using LME.

- The moisture content of the dough: The lower the moisture content of the dough, the more the nuggets will expand during extrusion. However, the dough should not be too dry, else it will be difficult to extrude.
- The temperature and pressure of extrusion: The higher the temperature and pressure, the more water will evaporate and the more the soy protein will expand. This will result in nuggets that are lighter and more porous.
- The drying method: The drying method will affect the texture and flavor of the nuggets. Air drying is the most common method, but oven-drying and freeze-drying can also be used.

LME is a fairly common technology to produce meat-like chunks from plant protein but there are certain limitations. LME results in a dry product that can be cooked through boiling/steaming/microwaving etc. to be used as meat extenders. Yet, LME-TVPs have not been successful as meat replacers and have been reported to have limitations including sponge-like texture, and an undesirable 'toughness' that does not meet consumer expectations (Sun *et al.*, 2022). These LME-TVP products as in the case with soy-based meat analogs are mostly utilized as meat extenders or ground meat substitutes, but fail to imitate the appearance, fibrous texture, and chewiness of muscle meat (Vatansever *et al.*, 2020).

8.7.2 High Moisture Extrusion

HME is a food processing technique that uses a screw extruder to cook and shape a mixture of food ingredients under high pressure and temperature. The moisture content of the mixture is typically between 40% and 80%, which is significantly higher than the moisture content of the mixture used in LME. HME works by subjecting the food mixture to a combination of mechanical shear, heat, and pressure. The mechanical shear breaks down the food particles and proteins, while the heat denatures the proteins and causes them to unfold. The pressure helps to cook the food and create a cohesive mass. The resulting mass can then be shaped into different meat forms and processed into RTE plant-based meats (Schmid *et al.*, 2022). The formation of anisotropic structures of plant protein concentrates and isolates during the HME occurs by two mechanisms, namely (i) unfolding and aligning of protein molecule in the direction of flow (ii) phase separation of biopolymer mixtures (protein–protein or protein–polysaccharide) that is further stabilized through covalent bonds. These processes texturize the plant proteins such that they mimic the textural behavior of meat. The process parameters, such as the temperature, pressure, and shear rate, can also be adjusted to achieve different textures and flavors. For example, a higher temperature and pressure will produce a firmer product with a more meat-like texture.

Texturization of plant proteins using HME could alleviate the problem of dry, rubbery, and sponge-like textures of extruded products. As protein functionality is directly associated with its secondary and tertiary structures, induced denaturation and protein-protein interactions may be used to generate plant protein extrudates with meat-like textures having the desired fiber orientation and porosity. HME is a versatile process that can be used to produce a variety of plant-based meat analogs, such as burgers, sausages, and nuggets. It is a relatively efficient process that requires less energy than other food processing methods. Additionally, the products produced by HME are typically shelf-stable.

8.7.3 Fiber Spinning

Fiber spinning can be used to create products from plant-based proteins that have a similar texture and appearance to meat. This technology can create long thin strands of material from plant proteins and fibers. The end products are RTE as they are already processed to be directly heated and consumed.

Fiber spinning may employ an extruder through which a mixture of plant proteins, water, and other ingredients is forced through a small opening (a thin die), which creates long, thin strands of material. These strands can then be molded, intermeshed, or woven together into various shapes, such as nuggets, sausages, or burgers.

Another way to create vegan meat using fiber spinning is to use a process called electrospinning. In electrospinning, a liquid solution of plant proteins is sprayed through a small opening into a strong electric field. The electric field causes the liquid to form into long, thin fibers that can then be collected, and processed to create vegan meat products.

Fiber spinning is a promising technology for creating vegan meat that is both nutritious and has a similar texture and appearance to meat (Angonese *et al.*, 2023). However, the technology is still in its early stages of development, and there are some challenges that need to be addressed before it can be widely adopted. One challenge is to create plant proteins that have the same functional properties as animal proteins. For example, animal proteins are typically very elastic, which is why they can be used to create products like sausages and burgers. Plant proteins, on the other hand, are often not as elastic. This means that these proteins need to be modified to have the same functional properties as animal proteins. Thus, proteins may need pre-processing prior to fiber spinning. Despite associated challenges, fiber spinning is a promising technology for creating vegan meat that is both nutritious and has a similar texture and appearance to meat. As the technology continues to develop, it is likely that vegan meat products made using fiber spinning will become more widely available and affordable.

8.7.4 3D Printing

3D printing is a rapidly developing technology that has the potential to revolutionize the way we produce food. In the context of vegan meat, 3D printing can be used to create plant-based products that have a similar texture and appearance to meat. Using 3D printing plant-based materials such as proteins, fats, and carbohydrates can be layered onto each other to create a 3D structure. The structure can then be processed to create RTE vegan meat products with the desired texture and flavor (Wen *et al.*, 2023).

Another way to create vegan meat using 3D printing is to use a process called scaffold printing. In scaffold printing, a printer is used to create a 3D structure that acts as a scaffold for plant-based cells. The cells are then seeded onto the scaffold and allowed to grow, thereby creating a vegan meat product that is similar to muscle tissue.

3D printing is a promising technology for creating vegan meat that is both nutritious and has a similar texture and appearance to meat. However, the technology is still in its early stages of development, and there are some challenges that need to be addressed before it can be widely adopted. Despite challenges, 3D printing is a promising technology for creating vegan meat that is both nutritious and has a similar texture and appearance to meat.

Some of the potential benefits of using 3D printing for vegan meat:

- It can create products with a more complex and realistic texture than traditional vegan meat products made by extrusion.
- It can be used to create products with a variety of shapes and sizes, which can appeal to a wider range of consumers.
- It can be used to create products that are more nutritious than traditional vegan meat products.
- It can be used to create products that are more sustainable than traditional vegan meat products.

8.8 REGULATORY FRAMEWORK FOR PLANT-BASED MEAT

The regulatory conditions for vegan meat vary from country to country. In the United States, vegan meat is regulated by the Food and Drug Administration (FDA) guidelines. The FDA defines vegan meat as "a food that is made from plant-based ingredients and does not contain any animal products." Vegan meat products must meet the same safety and labeling requirements as other food products. There are no specific regulations issued by the FDA for vegan meat. However, the FDA has issued guidance documents that provide information on how to comply with the existing regulations. In the European Union, vegan meat is regulated by the European Commission. The European Commission defines vegan meat as "a food that is made from plant-based ingredients and is intended to be used as a substitute for meat." Vegan meat products must meet the same safety and labeling requirements as other food products.

Vegan meat products must be safe for human consumption. This means that they must be free from harmful bacteria and other contaminants. As mostly these products are RTE and require heating/microwaving before consumption, food safety is of paramount importance. Vegan meat products must be labeled accurately. The label must clearly state that the product is vegan and must list all of the ingredients and the product's nutritional information.

In India, the Food Safety and Standards Authority of India (FSSAI) has issued regulations for vegan food, which came into effect on March 1, 2023. These regulations define vegan food as any food or food ingredient that does not contain any animal products or ingredients that have been used in their production or processing. Animal testing is also prohibited for vegan products unless it is required by a regulatory authority. The regulations also set out labeling requirements for vegan food. All vegan food must be labeled with the following statement: "This food is vegan and does not contain any animal products." In addition, the food must be packaged in a way that distinguishes it from non-vegan food. The FSSAI has also established a logo for vegan food, which must be displayed on all vegan food products. The logo is a green circle with the word "Vegan" written in white. These regulations are a major step forward for the vegan food industry in India. They will help to ensure that vegan food is safe and that consumers are aware of its contents.

In a nutshell, it can be summarized that India's RTE food market has a lot of potential for development and growth. Companies can enter this growing market by catering to consumer demands for convenience, real flavors, and excellent quality. The sector will continue to grow and be an essential part of our fast-paced lifestyles if there is constant innovation, strategic distribution partnerships, and a continuous focus on consumer expectations (Reshi *et al.,* 2023; Ungarov, 2023). To make shelf stable, extend the shelf life, and provide extraordinary sensory advantages RTE processed products are found to be loaded with synthetic preservatives, where interventions of food technologists and microbiologists should be dragged upon. Natural alternatives of these additives such as natural colorants, natural preservatives, and active packaging are the possible ways by which full or partial reduction of additives can be possible. As a tremendous and future market is looking for versatile as well as health-booming RTE products in the marketplace and therefore, healthier counterparts can compete and be sustained in a better way.

8.9 CONCLUSION

The upgradation of the market that is loaded with certain nutritious amounts of RTE foods in different categories that are widely entered into the majority of the households. The acceptance of RTE foods depends on the power of quenching hunger, providing a satiety feeling, and at last, requiring minimum time for cooking or preparing a meal. The fast-moving consumer goods (FMCG) market has prevailed as a good platform for RTE foods including all age groups. The concept of using various bioactive compounds, antioxidants, and nutritional supplements is becoming a more lucrative factor in consumers' minds for attracting more concrete nutrition from a small bowl of serving. This chapter rivets over various aspects of RTE-based food items that include vegetable, plant, and meat sources including red meat, fish, poultry, green chickpeas, soya, bread, muffins, baked chips, cookies, chocolate items, dried fruits, and vegetables. There are certain usages of techniques that fulfill market requirements and technological developments. Along with these good aspects, some microbial developments occurred especially in the meat-based RTE foods that must be looked after in packaging and before consuming. RTE-based foods, the most upcoming food categories are constantly, responsively removing former foods items from the shelves of the food markets by claiming more nutritious foods in the packaged items.

REFERENCES

Aiking, Harry. "Future protein supply." *Trends in Food Science and Technology* 22, no. 2–3 (2011): 112–120. doi: 10.1016/j.tifs.2010.04.005

Ainsa, Andrea, Sergio Roldan, Pedro L. Marquina, Pedro Roncalés, José Antonio Beltrán, and Juan B. Calanche Morales. "Quality parameters and technological properties of pasta enriched with a fish by-product: A healthy novel food." *Journal of Food Processing and Preservation* 46, no. 2 (2022): e16261. 1–10. doi: 10.1111/jfpp.16261

Anand, V., Bhat, A., Kaul, R. K. and Gupta, A. "Preparation and evaluation of Peach-Soy toffees." *Beverage and Food World* 36, no. 9 (2009): 47.

Aneh, Aneh Phillins, Pride Ndasi Ngwasiri, Wilson Agwanande Ambindei, Makebe Calister Wingang, Ngwa Martin Ngwabie, and Martin Benoit Ngassoum. "Enzyme assisted juice extraction from Dacryodes macrophylla as a potential bio-resource for wine production." *Heliyon* 9, no. 6 (2023): 1–12. doi: 10.1016/j.heliyon. 2023.e16443

Angonese, Mariana, Amanda G. A. Sá, Gabriel Emiliano Motta, Débora de Oliveira, Marco Di Luccio, and Bruno AM Carciofi. "Methods for physically structuring meat analogs: Challenges and perspectives." *Food Reviews International* 39, no. 9 (2023): 6805–6833.

Bachheti, Archana, Deepti, Rakesh Kumar Bachheti, Anjita Singh, Meseret Zebeaman, Yilma Hunde, and Azamal Husen. "Bioactive constituents and health-promoting compounds of underutilized fruits of the northern Himalayas of India: a review." *Food Production, Processing and Nutrition* 5, no. 24 (2023): 1–21. doi: 10.1186/s43014-023-00140-5

Bedada, Tesfaye L., Tatek K. Feto, Kaleab S. Awoke, Firehiwot A. Derra, Samson G. Gebre, Waktole G. Sima, Tigist Y. Negassi, and Yosef Beyene. "Microbiological and public health status of cooked meat and fish in Ethiopia." *The Open Microbiology Journal* 14, no. 1 (2020): 123–129. doi:10.2174/1874285802014010123.

Bhardwaj, Raju Lal, and Shruti Pandey. "Juice blends—a way of utilization of under-utilized fruits, vegetables, and spices: a review." *Critical Reviews in Food Science and Nutrition* 51, no. 6 (2011): 563–570. doi: 10.1080/10408391003710654

Bhokre, C. K., A. B. Rodge, P. U. Ghatge, and P. M. Chilkawar. "Preparation and evaluation of banana toffee." *Beverage and Food World* 37, no. 9 (2010): 50–51.

Bohaychuk, V. M., G. E. Gensler, R. K. King, K. I. Manninen, O. Sorensen, J. T. Wu, M. E. Stiles, and L. M. McMullen. "Occurrence of pathogens in raw and ready-to-eat meat and poultry products collected from the retail marketplace in Edmonton, Alberta, Canada." *Journal of Food Protection* 69, no. 9 (2006): 2176–2182. doi: 10.4315/0362-028x-69.9.2176

Bonnet, Céline, Zohra Bouamra-Mechemache, and Tifenn Corre. "An environmental tax towards more sustainable food: Empirical evidence of the consumption of animal products in France." *Ecological Economics* 147 (2018): 48–61. doi: 10.1016/j.ecolecon 12.032

Bremenkamp, Ina, and Maria J. Sousa-Gallagher. "Design and development of an edible coating for a ready-to-eat seafood product." *Preprints* 12 (2023), 1–14. doi: 10.20944/preprints202312. 1033.v1

Chandu, N., and M. Prasad. "Development of guava candies." *Journal of Food Science and Technology-Mysore* 43, no. 2 (2006): 210–212.

Chowdhury, P. R. C. R., and R. C. Ray. "Fermentation of Jamun (*Syzgium cumini* L.) fruits to form red wine." *ASEAN Food Journal* 14, no. 1 (2007): 15.9–17.

Csadek, Isabella, Ute Vankat, Julia Schrei, Michelle Graf, Susanne Bauer, Brigitte Pilz, Karin Schwaiger, Frans J. M. Smulders, and Peter Paulsen. "Treatment of ready-to-eat cooked meat products with cold atmospheric plasma to inactivate *Listeria* and *Escherichia coli*." *Foods* 12, no. 4 (2023): 685. doi: 10.3390/foods12040685

Cui, S. W., and Roberts K. T. "Dietary fibre fulfilling the promise of added-value formulations." In: Modern Biopolymer Science, edited by Kasapis, S., Norton, I. T., and Ubbink, J. B, Elsevier, USA (2009): 1–12.

Deepa, M. P. M., and P. S. Murthy. "Economic analysis of women empowerment through value addition of jackfruit in Bengaluru rural district." *Economic Affairs* 64, no. 2 (2019): 401–406. doi: 10.30954/0424-2513.2.2019.16

Ditudompo, Srivikorn, Saowaluk Rungchang, and Ulisa Pachekrepapol. "Quality changes and experimental accelerated shelf-life determination of ready-to-eat burgers made from striped catfish and salmon mince blend by thermal processing in retort pouches." *Journal of Food Processing and Preservation* 46, no. 8 (2022): e16801. 1–12. doi: 10.1111/jfpp.16801

Dwivedi, S. K., R. Singh, and Z. Ahmad Z. *The Seabuckthorn*, Field Research Laboratory (DRDO), Leh, Jammu and Kashmir (2006).

Ekonomou, Sotirios I., S. Bulut, K. A. G. Karatzas, and I. S. Boziaris. "Inactivation of *Listeria monocytogenes* in raw and hot smoked trout fillets by high hydrostatic pressure processing combined with liquid smoke and freezing." *Innovative Food Science and Emerging Technologies* 64 (2020): 102427. 1–11 doi: 10.1016/j.ifset.2020.102427

Esteras-Saz, Javier, Óscar de la Iglesia, Izumi Kumakiri, Cristina Peña, Ana Escudero, Carlos Téllez, and Joaquín Coronas. "Pervaporation of the low ethanol content extracting stream generated from the dealcoholization of red wine by membrane osmotic distillation." *Journal of Industrial and Engineering Chemistry* 122 (2023): 231–240. doi: 10.1016/j.jiec.2023.02.024

Ezearigo, O. E., P. O. Adeniji, and F. Ayoade. "Screening of natural spices for improving the microbiological, nutritional and organoleptic qualities of the Zobo drink." *Journal of Applied Biosciences* 76 (2014): 6397–6410. doi: 10.4314/jab.v76i1.10

Gałkowska, D., T. Fortuna, and W. Prochwicz-Zagórska. "Physicochemical quality of selected strawberry jams with fructose." *Potravinárstvo (Slovak Republic)* (2010). 22–24. doi: 10.5219/46

Gayathri, S., and L. Uthira. "Preparation and evaluation of protein enriched mango-papaya blended fruit bar." *Beverage and Food World* 49, no. 2 (2008): 17–24.

Ghafari, Sedigheh, and Sara Ansari. "Microbial viability, physico-chemical properties and sensory evaluation of pineapple juice enriched with *Lactobacillus casei*, *Lactobacillus rhamnosus* and inulin during refrigerated storage." *Journal of Food Measurement and Characterization* 12, no. 4 (2018): 2927–2935. doi: 10.1007/s11694-018-9908-z

Gómez-Guillén, María Carmen, Selene Pérez-García, Ailén Alemán, María Elvira López-Caballero, Carmen G. Sotelo, and María Pilar Montero. "Development of a ready-to-eat fish product enriched with fish oil entrapped in a κ-Carrageenan egg white fish protein hydrolysate dry powder." *Foods* 12, no. 11 (2023): 2272. 1–15. doi: 10.3390/foods12112272

Hadjicharalambous, Chrystalleni, Luca Grispoldi, Thanos Chalias, and Beniamino Cenci-Goga. "A quantitative risk assessment of *Listeria monocytogenes* from prevalence and concentration data: Application to a traditional ready to eat (RTE) meat product." *International Journal of Food Microbiology* 379 (2022): 109843. 1–13. doi: 10.1016/j.ijfoodmicro.2022.109843

Hasanuzzaman, M. D., M. Kamruzzaman, Md Mominul Islam, Sultana Anjuman Ara Khanom, Md Mashiar Rahman, Laisa Ahmed Lisa, and Dipak Kumar Paul. "A study on tomato candy prepared by dehydration technique using different sugar solutions." *Food and Nutrition Sciences* 5 (2014): 1261–1271. doi: 10.4236/fns.2014.513137

He, Jiang, Natasha Marie Evans, Huaizhi Liu, and Suqin Shao. "A review of research on plant-based meat alternatives: Driving forces, history, manufacturing, and consumer attitudes." *Comprehensive Reviews in Food Science and Food Safety* 19, no. 5 (2020): 2639–2656. doi: 10.1111/1541-4337.12610

Hossain, Mohammad A., Rana Das, Md Yasin, Humayun Kabir, and Tanvir Ahmed. "Potentials of two lactobacilli in probiotic fruit juice development and evaluation of their biochemical and organoleptic stability during refrigerated storage." *Scientific Study and Research. Chemistry and Chemical Engineering, Biotechnology, Food Industry* 23, no. 2 (2022): 131–140. www.expertmarketresearch.com/articles/top-bakery-companies-in-india

Jiang, Jiang, and Youling L. Xiong. "Technologies and mechanisms for safety control of ready-to-eat muscle foods: An updated review." *Critical Reviews in Food Science and Nutrition* 55, no. 13 (2015): 1886–1901. doi: 10.1080/10408398.2012.732624

Joshi, A., C. Kaur, and J. Prakash. "Anthocyanins rich Papaya Candy: Innovative fusion." *ICAR News Letter* 26, no. 4 (2020): 7–9pp.

Joshi, A., S. Sethi, N. Kumar, and B. Thippeswamy. "Pushpon Ke mulyavardhit prasanskrit utpaad." *Rakshya Khaadya Vigyaan Patrika* 29 (2021): 58–62, Raksha Khaadya Anusandhaan Praoyshala (DFRL), (DRDO), Sidharthnagar, Mysore, Karnatak dwara prakashit.

Joshi, A., and V. R. Sagar. "Anthocyanins rich papaya candy (tutty fruity): An innovative fusion product with enhanced functionality." *International Journal of Advanced Technology in Engineering and Science*. 4, no. 5 (2016): 417–420 (ISSN 2348-7550)

Joshi, Alka, Uma Prajapati, Shruti Sethi, Bindvi Arora, and Ram Roshan Sharma. "Fortification in fresh and fresh-cut horticultural products." *Fresh-Cut Fruits and Vegetables*, pp. 183–204. Academic Press 4, no. 5 (2020): 417–420 (ISSN 2348-7550)

Joshi, V. K., and B. L. Attri. "Importance of fruit based fermented beverage in India." *Beverage and Food World* 17, no. 4 (1990): 9–10.

Joshi, V. K., S. Sharma, and K. Kumar. "Technology for production and evaluation of strawberry wine." *Beverage and Food World* 33, no. 1 (2006): 77–78.

Kabra, Atul, Rohit Sharma, Christophe Hano, Ruchika Kabra, Natália Martins, and Uttam Singh Baghel. "Phytochemical composition, antioxidant, and antimicrobial attributes of different solvent extracts from *Myrica esculenta* buch.-ham. Ex. D. Don leaves." *Biomolecules* 9, no. 8 (2019): 357. 1–11. doi: 10.3390/biom9080357

Kale, S. J., Prerna Nath, and Charanjit Kaur. "Potentiality of freeze—thaw treatment to produce soft textured aonla (*Emblica officinalis*) candies." *Journal of Food Science and Technology* 56 (2019): 3157–3163. doi: 10.1007/s13197-019-03776-9

Kaprasob, Ratchadaporn, Orapin Kerdchoechuen, Natta Laohakunjit, Benjawan Thumthanaruk, and Kalidas Shetty. "Changes in physico-chemical, astringency, volatile compounds and antioxidant activity of fresh and concentrated cashew apple juice fermented with *Lactobacillus plantarum*." *Journal of Food Science and Technology* 55 (2018): 3979–3990. doi: 10.1007/s13197-018-3323-7

Khan, Moazzam Rafiq. "Osmotic dehydration technique for fruits preservation—A review." *Pakistan Journal of Food Sciences* 22, no. 2 (2012): 71–85.

Kitch, Courtney J., Amanda M. Tabb, Grace E. Marquis, and Rosalee S. Hellberg. "Species substitution and mislabeling of ceviche, poke, and sushi dishes sold in Orange County, California." *Food Control* 146 (2023): 109525. 1–12. doi: 10.1016/j.foodcont.2022.109525

Kumbhar, S. C., P. M. Kotecha, and S. S. Kadam. "Effect of methods of juice extraction on the quality of pomegranate wine." *Indian Food Packer* 56, no. 5 (2002): 51–53.

Leygonie, Coleen, Trevor J. Britz, and Louwrens C. Hoffman. "Impact of freezing and thawing on the quality of meat." *Meat Science* 91, no. 2 (2012): 93–98. doi: 10.1016/j.meatsci.2012.01.013

Lingappa, K., and C. Naik. "Wine preparation from carrot (*Daucus Carotia* L.)." *Indian Food Packer* 51 (1997): 10–13.

Liu, Tingting, Nazimah Hamid, Kevin Kantono, Loveena Pereira, Mustafa M. Farouk, and Scott O. Knowles. "Effects of meat addition on pasta structure, nutrition and in vitro digestibility." *Food Chemistry* 213 (2016): 108–114. doi: 10.1016/j.foodchem.2016.06.0

Makdoh, Ksanbok, Marvellous B. Lynser, and Kevin Howard M. Pala. "Marketing of indigenous fruits: A source of income among Khasi Women of Meghalaya, North East India." *Journal of Agricultural Sciences* 5, no. 1–2 (2014): 1–9.

Mantzourani, Ioanna, Chryssa Nouska, Antonia Terpou, Athanasios Alexopoulos, Eugenia Bezirtzoglou, Mihalis I. Panayiotidis, Alexis Galanis, and Stavros Plessas. "Production of a novel functional fruit beverage consisting of cornelian cherry juice and probiotic bacteria." *Antioxidants* 7, no. 11 (2018): 163.1–12. doi: 10.3390/antiox7110163

Manzoor, Arshied, Bisma Jan, Qurat Ul Eain Hyder Rizvi, Pir Mohammad Junaid, Junaid Ahmad Pandith, Ishfaq Hamid Dar, Shayeeb Ahmad Bhat, and Saghir Ahmad. "Osmotic dehydration technology for preservation of fruits and vegetables." In *Quality Control in Fruit and Vegetable Processing*, pp. 167–184, Apple Academic Press (2023).

Marani, Claudio Mauricio, Miriam Eliana Agnelli, and Rodolfo Horacio Mascheroni. "Osmo-frozen fruits: Mass transfer and quality evaluation." *Journal of Food Engineering* 79, no. 4 (2007): 1122–1130.

Mathur, Prateek, and Donald W. Schaffner. "Effect of lime juice on *Vibrio parahaemolyticus* and *Salmonella enterica* inactivation during the preparation of the raw fish dish ceviche." *Journal of Food Protection* 76, no. 6 (2013): 1027–1030. doi: 10.4315/0362-028X.JFP-12-526

Michel, Fabienne, Christina Hartmann, and Michael Siegrist. "Consumers' associations, perceptions and acceptance of meat and plant-based meat alternatives." *Food Quality and Preference* 87 (2021): 104063. 1–10. doi: 10.1016/j.foodqual.2020.104063

Mitra, R., S. V. Sangameswaran, and V. S. Govindarajan. *CFTRI Bibliographical Series* no. 23 (1979). 1974–1978.

Muhialdin, Belal J., Anis Shobirin Meor Hussin, Hana Kadum, Azizah Abdul Hamid, and Ahmad Haniff Jaafar. "Metabolomic changes and biological activities during the lacto-fermentation of jackfruit juice using *Lactobacillus casei* ATCC334." *LWT* 141 (2021): 110940. 1–8. doi: 10.1016/j.lwt.2021.110940

Muhialdin, Belal J., Hana Kadum, Mohammad Zarei, and Anis Shobirin Meor Hussin. "Effects of metabolite changes during lacto-fermentation on the biological activity and consumer acceptability for dragon fruit juice." *LWT* 121 (2020): 108992. 1–7. doi: 10.1016/j.lwt.2019.108992

Nale, R. S., P. M. Kotecha, and J. K. Chavan. "Preparation of mixed fruit toffee from tamarind, mango and papaya pulp." *Beverage Food World* 34, no. 9 (2007): 69–70.

Nath, Prerna, Ajinath Dukare, Sunil Kumar, Sakharam Kale, and Pankaj Kannaujia. "Black carrot (*Daucus carota* subsp. sativus) anthocyanin-infused potato chips: Effect on bioactive composition, color attributes, cooking quality, and microbial stability." *Journal of Food Processing and Preservation* 46, no. 1 (2022): e16180. 1–14. doi: 10.1111/jfpp.16180

Omer, Mohamed K., Avelino Álvarez-Ordoñez, Miguel Prieto, Eystein Skjerve, Tekie Asehun, and Ole Arne Alvseike. "A systematic review of bacterial foodborne outbreaks related to red meat and meat products." *Foodborne Pathogens and Disease* 15, no. 10 (2018): 598–611. doi: 10.1089/fpd.2017.2393

Poissant, Rémi, François Mariotti, Daniel Zalko, and Jeanne-Marie Membré. "Ranking food products based on estimating and combining their microbiological, chemical and nutritional risks: Method and application to Ready-To-Eat dishes sold in France." *Food Research International* 169 (2023): 112939. 1–8. doi: 10.1016/j.foodres.2023.112939

Pooja, B. K., Shruti Sethi, Rakesh Bhardwaj, Alka Joshi, Arpan Bhowmik, and Minakshi Grover. "Investigation of physicochemical quality and textural attributes of muffins incorporated with pea pod powder as a source of dietary fiber and protein." *Journal of Food Processing and Preservation* 46, no. 10 (2022): e16884. 1–11. doi: 10.1111/jfpp.16884

Poore, Joseph, and Thomas Nemecek. "Reducing food's environmental impacts through producers and consumers." *Science* 360, no. 6392 (2018): 987–992. doi: 10.1126/science.aaq0216

Ranjitha, K. "Development of fruit probiotic beverages and associated quality changes during storage". Presented paper on International Conference on Microbiology and Society—Current Trend and Future Prospects (MSCTFP-2022) from 21st to 23rd September 2022. (62nd Annual Conference of Association of Microbiologists of India) held at University of Mysore, Mysore.

Reddy, Vanitha, Asna Urooj, and Anila Kumar. "Evaluation of antioxidant activity of some plant extracts and their application in biscuits." *Food Chemistry* 90, no. 1–2 (2005): 317–321.

Reshi, Irshad Ahmad, Shabir Ahmad Dar, and Shaikh Sobiya Ansar. "An empirical study on the factors affecting consumer behavior in the fast-food industry." *Journal of Accounting Research, Utility Finance and Digital Assets* 1, no. 4 (2023): 376–381.

Ronald, Chepkwony, Joseph Wafula Matofari, and John Masani Nduko. "Antimicrobial resistance of *E. coli* strains in ready-to-eat red meat products in Nakuru County, Kenya." *The Microbe* 1 (2023): 100022. 1–12. doi: 10.1016/j.microb.2023.100022

Saurabh, V., and K. Barman. "Flowers: A potential source of human nutrition." *Journal of Postharvest Technology* 8, no. 1 (2020): 75–81.

Schmid, Eva-Maria, Asgar Farahnaky, Benu Adhikari, and Peter J. Torley. "High moisture extrusion cooking of meat analogs: A review of mechanisms of protein texturization." *Comprehensive Reviews in Food Science and Food Safety* 21, no. 6 (2022): 4573–4609. doi: 10.1111/1541-4337.13030

Sethi, S and Gupta, S. "High fibre tortilla chip" *ICAR Newsletter* 22, no. 2(2016): 4.

Sethi, Shruti, Alka Joshi, Kirti Seth, Rakesh Bhardwaj, Ajay Yadav, and Minakshi Grover. "Phytonutrient content, antioxidant potential and acceptability of muffins functionalized with soy and citrus industry waste." *Journal of Food Processing and Preservation* 46, no. 7 (2022): e16606. 1–11. doi: 10.1111/jfpp.16606

Sharma, Chetna, Param Pal Sahota, and Sarabjit Kaur. "Physicochemical and microbiological evaluation of antioxidant-rich traditional black carrot beverage: Kanji." *Bulletin of the National Research Centre* 45, no. 1 (2021): 45 (143).1–8. doi: 10.1186/s42269-021-00594-y

Shinwari, Kaunsar Jabeen, and Pavuluri Srinivasa Rao. "Stability of bioactive compounds in fruit jam and jelly during processing and storage: A review." *Trends in Food Science and Technology* 75 (2018): 181–193. doi: 10.1016/j.tifs.2018.02.002

Singh, Brahma. "Indian sea buckthorn." In: *New Age Herbals: Resource, Quality and Pharmacognosy*, edited by Singh, Brahma, and Peter, K. V. (2018): 29–54.

Singh, Om, Sanjay Pathak, Richa Singh, and Pratiksha Singh. "Changes in chemical constituents and overall acceptability of aonla mango low calorie blended beverages during storage." *Plant Archive* 14, no. 1 (2014): 343–346.

Singh, Sunita, Shruti Sethi, Sangeeta Gupta, Charanjit Kaur, and Ed Wood. "Fermentation of multigrain dough—an approach to reduce glycemic index for healthy bread." *EUREKA: Life Sciences* 5 (2019): 19–31. doi: 10.21303/2504-5695.2019.00994

Soccol, Carlos Ricardo, Maria Rosa Machado Prado, Lina Marcela Blandon Garcia, Cristine Rodrigues, Adriane Bianchi Pedroni Medeiros, and Vanete Thomaz Soccol. "Current developments in probiotics." *Journal of Microbiology and Biochemistry Technology* 7, no. 1 (2014): 011–020.

Sood, P., and R. Shri. "A review on ethnomedicinal, phytochemical and pharmacological aspects of *Myrica esculenta*." *Indian Journal of Pharmaceutical* Sciences 80, no. 1 (2018): 2–13. doi: 10.4172/pharmaceutical-sciences.1000325

Stewart, Ian M., and Ralph E. Timms. "Fats for chocolate and sugar confectionery." In: *Fats in Food Technology*, edited by Rajah, Kanes K, Sheffield Academy Press, Sheffield (2002): 169–212.

Sun, Cuixia, Jialing Fu, Yuyang Chang, Saiya Li, and Yapeng Fang. "Structure design for improving the characteristic attributes of extruded plant-based meat analogues." *Food Biophysics* (2022): 1–13. doi: 10.1007/s11483-021-09692-w

Tao, Rui, Qiyang Chen, Yi Li, Long Guo, and Zhiqin Zhou. "Physicochemical, nutritional, and phytochemical profile changes of fermented citrus puree from enzymatically hydrolyzed whole fruit under cold storage." *LWT* 169 (2022): 114009.1–13. doi: 10.1016/j.lwt.2022.114009

Thakur, M., A. Pandey, and S. C. Jain. "Good mood foods: A panacea of life." *Indian Food Industry* 31, no. 2 (2012): 45–52.

Torres, J. D., P. Talens, I. Escriche, and A. Chiralt. "Influence of process conditions on mechanical properties of osmotically dehydrated mango." *Journal of Food Engineering* 74, no. 2 (2006): 240–246.

Tran, A. M., T. B. Nguyen, V. D. Nguyen, E. Bujna, M. S. Dam, and Q. D. Nguyen. "Changes in bitterness, antioxidant activity and total phenolic content of grapefruit juice fermented by *Lactobacillus* and *Bifidobacterium* strains." *Acta Alimentaria* 49, no. 1 (2020): 103–110. doi: 10.1556/066.2020.49.1.13

Ungarov, Azizbek. "Innovative methods of fruit and vegetable processing." *International Conference on Higher Education Teaching* 1, no. 2 (2023): 107–109.

van der Vossen-Wijmenga, Wieke P., Heidy M. W. den Besten, and Marcel H. Zwietering. "Temperature status of domestic refrigerators and its effect on the risk of listeriosis from ready-to-eat (RTE) cooked meat products." *International Journal of Food Microbiology* (2023): 110516. 1–14. doi: 10.1016/j.ijfoodmicro.2023.110516

Vatansever, Serap, Mehmet C. Tulbek, and Mian N. Riaz. "Low-and high-moisture extrusion of pulse proteins as plant-based meat ingredients: A review." *Cereal Foods World* 65, no. 4 (2020): 12–14.

Verma, G., and C. S. Chopra. "Preparation and preservation of aonla-mango mixed fruit slab." *Beverage and Food World* 37, no. 1 (2010): 60–61.

Warmate, Dein, and Bukola A. Onarinde. "Food safety incidents in the red meat industry: A review of foodborne disease outbreaks linked to the consumption of red meat and its products, 1991 to 2021." *International Journal of Food Microbiology* (2023): 110240. 1–16. doi: 10.1016/j.ijfoodmicro.2023.110240

Wen, Yaxin, Chhychhy Chao, Quang Tuan Che, Hyun Woo Kim, and Hyun Jin Park. "Development of plant-based meat analogs using 3D printing: Status and opportunities." *Trends in Food Science and Technology* 132 (2023): 76–92. doi: 10.1016/j.tifs.2022.12.010

Willett, Walter, Johan Rockström, Brent Loken, Marco Springmann, Tim Lang, Sonja Vermeulen, Tara Garnett, et al. "Food in the Anthropocene: the EAT–Lancet Commission on healthy diets from sustainable food systems." *The Lancet* 393, no. 10170 (2019): 447–492. doi: 10.1016/s0140-6736(18)31788-4

Wu, Yue, Sujin Li, Yang Tao, Dandan Li, Yongbin Han, Pau Loke Show, Guangzhong Wen, and Jianzhong Zhou. "Fermentation of blueberry and blackberry juices using *Lactobacillus plantarum*, *Streptococcus thermophilus* and *Bifidobacterium bifidum*: Growth of probiotics, metabolism of phenolics, antioxidant capacity in vitro and sensory evaluation." *Food Chemistry* 348 (2021): 129083.1–16. doi: 10.1016/j.foodchem.2021.129083

Zhang, Renyu, Carolina E. Realini, Yuan H. Brad Kim, and Mustafa M. Farouk. "Challenges and processing strategies to produce high quality frozen meat." *Meat Science* 205 (2023): 109311. 1–12. doi: 10.1016/j.meatsci.2023.109311

9 Gluten-Free Aspects of RTE Foods

Latika Yadav
Assistant Professor, Department of Home Science, Vijay Singh Pathik Government Post Graduate Degree College, Kairana, Shamli, Uttar Pradesh, India

9.1 INTRODUCTION

Ready-to-eat (RTE) food products are prepackaged items that may be consumed directly without any cooking, while ready-to-cook food products require some preparation such as heating or boiling before consumption. Consumers have shown a strong preference for RTE and ready-to-eat/cook (REC) snacks in the past 5 years due to their convenience, making it the fastest-growing sector in the food industry. Consumer interest in these food products increased because of their appeal, affordability, flavor, appearance, and texture. Various technologies have been actively utilized in the manufacturing of RTE and RTC snack goods due to their user-friendly operation methods that eliminate the need for additional cooking and preparation, focusing solely on attractive packaging. RTE products encompass various sorts of food including baked, fried, sweet, extruded, dried, fast food, canned and preserved food. Traditionally cereal, millets, and pseudo-cereal consumption is primarily comprised of bakery products like cookies, biscuits, and bread; instant snacks like chips, popped and puffed mixtures, savory, etc.; extruded meals like pasta, noodles, vermicelli etc.; fast food; and cereal bars and breakfast cereals. The food compositions are easily consumed by humans and well received by customers due to their shelf stability, lightweight nature, nutritional enhancements, and ease of shopping and storage.

People who are genetically predisposed to celiac disease (CD) have a lifelong autoimmune illness that makes them permanently unable to digest gluten. The current consensus is that CD is more like an autoimmune disorder affecting multiple systems than a gastrointestinal illness. The small intestine mucosa of patients with CD showed a substantial chronic lymph epithelial inflammatory cell infiltration and large flat villi in the first reliable clinical description of the disease (Murray et al., 2003; Green et al., 2005). Subsequent research has detailed how CD patients' gut mucosal abnormalities develop in reaction to gluten. Age, duration, and the severity of extra-intestinal manifestations are major factors in the wide variety of CD clinical symptoms. Serology and biopsy-based research conducted in the Europe, Australia, South America, India, and United States (US) has recast CD's rarity as a possible prevalence of 0.33–1.06% in children and 0.18–1.2% in adults. Sahrawi has the greatest rate of occurrence among children. The significantly lower frequency of CD is a result of the rarity of CD diagnoses in other African nations. Numerous practitioners are not familiar with CD, which contributes to the high rate of undiagnosed cases, according to population surveys (Paulley, 1954; Marsh, 1992).

Mild intraepithelial lymphocytosis and complete villous atrophy are two extremes of the immune-mediated response's effects on the small intestine mucosa. In most Western countries, the clinical diagnostic rate is still generally lower, but the prevalence of CD is believed to be about 1%. There is currently no cure for gluten sensitivity (GS) other than a diet free from gluten for the rest of

DOI: 10.1201/9781032622408-9

one's life. This is because gluten causes inflammation and damage to the small intestine mucosa, which in turn causes malabsorption of micronutrients like calcium, iron, folic acid, and fat-soluble vitamins. Diets that are gluten-free (GF) by nature and GF alternatives to cereal-based diets are the hallmarks of the gluten-free diet (GFD). Ingredients in GF foods either are derived from cereals that do not contain gluten or are cereals that have undergone specific processing to remove gluten. In 2010, the worldwide market for GF products was close to $2.5 billion, driven by an increase in the percentage of people adopting a GFD compared to the anticipated number of patients clinically diagnosed with CD. This pattern is borne out by the fact that other gluten-related health issues have recently surfaced, adding to the growing list of CD-related illnesses. In certain instances of gluten reactions, neither allergic nor autoimmune mechanisms are at work; this is in addition to CD and wheat allergies. This group of people is typically thought of as having GS, a disorder in which eating gluten causes symptoms, but there are no celiac-specific antibodies, no classical celiac villous atrophy, and the human leukocyte antigen status and presence of first-generation anti-gliadin antibodies can vary (Fasano, 2005; Chand and Mihas, 2006).

There is still no agreed-upon definition of "gluten sensitivity"; therefore, we do not know how common GS is in the general population. The exact role of gluten as a GS trigger is still debatable, but many patients are making their own diagnoses and beginning GFDs without first seeing a doctor. However, fresh evidence shows that this is far from a prevalent problem. Without a CD diagnosis, 5% of the kids in one New Zealand region said they avoided gluten. In conclusion, those who have a clinical diagnosis of CD and those who suffer from a variety of diseases that necessitate a GF diet both utilize GF products. Nevertheless, even in the absence of obvious clinical signs, undiagnosed CD patients continue to eat a typical diet that includes gluten (Green and Jabri, 2003; Barker and Liu, 2008).

It is yet unclear what the long-term impact of this distribution will be on population health in terms of cost–benefit analysis. In reality, there has been a lot of study into creating GF baked goods with sensory and nutritional qualities that can replace, at least in part, the typical cereal-based products due to the ever-increasing demand for GF goods. Producing such dishes involves mechanical and sensory challenges due to GF dough's low mouthfeel, structural weakness, and difficulty in manipulating it technically. To make GF products more structurally acceptable, a wide variety of flours, starches, enzymes, proteins, and hydrocolloids have been utilized to simulate gluten's viscoelastic characteristics. Consequently, the nutritional composition and quality of GF foods differ markedly from their gluten-containing counterparts due to the varied combinations of these substances. More and more people are looking for these products, and the market for them has been expanding in recent years, because of the fact that they serve both those with medical needs and the millions of people who are looking for a GFD. Technology, taste, and nutritional value would all play a role in their creation, purchase, and consumption.

This chapter seeks to shed light on the nutritional quality of commercial RTE GF food products by addressing key problems and offering solutions to overcome them (Fasano et al., 2003; Dube et al., 2005).

9.2 READY-TO-EAT FOOD

Prepared food items are often flavored by consumers for their convenience in preparation, storage, and consumption. RTE food encompasses a variety of snack products such as cookies, breads, biscuits, rolls, chips, sandwiches, soups, fruit salad, breakfast cereals, vegetable salads, and dairy and meat products. This list can grow extensively due to the frequent addition of new products within that category on a daily basis. Fruits, vegetables, and cereals are crucial commodities for the development of RTE meals. Traditional food consumption is primarily focused on ready-made foods such as parathas, chappatis, idlis, bada, dosas, kadhi rice, rajma rice, pav bhaji, vegetable curry, mixed fruit salads, breakfast cereals, bars and biscuits, and dairy and meat products like precooked ham, sausages, kababs, and chicken products, as well as other instant food products (Catassi et al., 1999; Sood et al., 2003).

Fruits and vegetables play a crucial role in meeting consumer nutritional needs. It is a highly nutritious source of dietary fiber, vitamins, minerals, polyphenols, and other essential bioactive substances. Fresh fruits and vegetables spoil quickly because of their high moisture content and water activity, which promote the growth of microorganisms. Consumers seek fresh, healthful, easy-to-prepare, and minimally processed fruits and vegetables. Various instant food and RTE food sectors are meeting this demand by promoting items that are attractive, fresh, and economically significant. Fruits and vegetables are considered RTE food as they can be consumed directly or served in the shape of a salad. Fruit- and vegetable-based RTE products include juice, jam, jelly, marmalades, tinned goods, prepared salads, vegetable curry, and more. Strawberries and blueberries are popular due to their distinctive appearance and flavor. However, they are highly perishable and susceptible to damage during harvesting, storage, and transportation, leading to nutrient loss. Research was conducted on the mechanical properties and quality parameters of chitosan-coated edible algae on pre-packaged strawberries. This study establishes that an edible coating made from chitosan-pectin (CH-PC) can decrease respiration rate, product loss, and microbial load while enhancing nutritious value. Research was carried on the qualitative characteristics of pre-packaged blueberries with extra nutritional fiber in the edible coating. The study used sodium alginate and chitosan edible coating containing dietary fiber to assess the nutritional, microbiological, and physicochemical characteristics of preserved blueberries at 5°C for 18 days. The study showed that CH coating effectively suppressed microbial development, boosted antioxidant capabilities, and enhanced the sensory and nutritional quality of blueberries (van de Wal et al., 1999; Shan et al., 2002).

9.3 GLUTEN-FREE DIET AND RTE FOODS

9.3.1 Nutritional Status

Previous research has shown that a variety of nutritional inadequacies, including inadequate intake of calories, dietary fiber, protein, vitamins and minerals, impact 20–38% of individuals with CD. Because iron, folate, and calcium are absorbed in the proximal small bowel, malabsorption of these minerals is prevalent. Research has indicated that individuals with CD may experience iron insufficiency to a degree ranging from 12% to 69%. Although damage to the finger-like projections in the ileum, where vitamin B12 is absorbed, is very uncommon, the rate of untreated patients with vitamin B12 deficiency ranges from 8% to 41%. Inadequate intake of milk and dairy products due to malabsorption or reduced consumption may cause deficiencies in calcium, phosphorus, and vitamin D. This is often the result of trying to avoid lactose, which can develop secondary intolerance to lactose due to damaged villi producing less lactase. The degree of intestinal damage, the length of time the disease remains undiagnosed, and the degree of malabsorption all have a role in determining how severe the nutritional deficiencies described above are. Previous research has shown that rigorously adhering to a GFD can completely address the majority of these nutritional deficiencies. Prior research has demonstrated that following a GFD for 8–12 years straight is linked to particular nutrient deficits, and that adopting a GFD does not result in an appropriate intake of nutrients. Only 32%, 44%, and 46% of the women analyzed in a study of 47 American adults' eating habits obtained the necessary quantities of calcium, iron, and fiber, respectively (Dieterich et al., 1997; Ciccocioppo et al., 2005; Dewar et al., 2006).

It is most certainly the case that a large number of GF foods are made primarily using low-fiber refined flours and/or starches, which results in inadequate fiber consumption. Most of the fiber in the grain is removed during the refining process, leaving only the starchy inner layer. Past research on the nutritional makeup of processed GF food has revealed greater quantities of salt, carbohydrates, and lipids. Consuming foods heavy in fat, sugar, and calories is a common way for people with CD to make up for the limitations of a GF diet (Kemppainen et al., 1995; Kinsey et al., 2008).

As a result, people who have CD may eat unusually high levels of saturated and total fats. The diet of teenage CD patients was heavy in fat and protein and low in carbs, iron, calcium, and fiber. One possible explanation for the high prevalence of overweight patients is their overconsumption of energy, animal protein, and fats. Studies on the lipid content of GF RTE items have revealed that trans fatty acids are present. When these acids are taken in conjunction with inadequate consumption of essential fatty acids, it may cause a metabolic imbalance. It is imperative to take these components' negative effects on health into careful consideration. This is a result of the restricted food alternatives experienced by those with celiac illness, which causes them to consume a lot of packaged GF RTE products like biscuits and snacks. Consuming too much fat in the diet plays a key role in the development of diseases including obesity and coronary heart disease. In contrast, it has been suggested that people with CD may not get adequate nutrients if they carefully follow a GF diet because it may lead to their making poor dietary choices (Bode and Gudmand-Hoyer, 1996; Kemppainen, 1997).

9.3.2 The Glycemic Response of Gluten-Free Products

A wealth of evidence suggests a causal relationship between improvements in numerous pathways that are believed to be connected to disease states and the control of glucose levels after a meal. Studies have indicated that as compared to a low-glycemic diet, eating a meal with a high glycemic index (GI) causes a notable surge in inflammatory markers. Therefore, a low-GI diet is recommended by the global dietary guidelines, specifically for individuals who are overweight or have impaired glucose metabolism. Although GI is acknowledged to be important in illness situations, its usefulness for preventive measures in the general public is still up for discussion. As systemic inflammation contributes to ischemic heart disease, which is more common in people with CD, reducing inflammation may be beneficial for them (Bardella et al., 2000; Thompson, 2000). Thus, given its potential benefits for a population affected by the condition, it is imperative to evaluate the GI of food for people with CD. It is important to consider the impact on glycemic response while designing GF RTE products, even if there is a lack of research on the subject and the results are uneven. The physical makeup of food and the amount of carbohydrates it contains influence the GI. When people with CD and those in good health follow the same GFD, there are differences as well. Removing gluten from foods high in carbohydrates, such pasta, may improve their GI. This is because gluten forms a strong structure that prevents starch from expanding and breaking down too much by digestive enzymes. The phenomenon was initially observed in a scientific study whereby it was discovered that a GF bread demonstrated a noticeably elevated GI in contrast to a conventional white bread (Thompson, 1999; Thompson et al., 2005).

9.3.3 Gluten-Free Diet: A Nutritional and Safety Perspective

A variety of issues are related to food safety, such as contamination from chemical, physical, or biological causes, and emphasizing this is important. Contamination of food can be caused by purposeful application of various substances such as veterinary medications, pesticides, and other agricultural chemicals, or by the presence of environmental contaminants. Customers may be exposed to serious health hazards from these ingredients, especially if they have CD. Most goods classified as "safety" in this survey either are naturally GF or have undergone specific processing to lower their gluten content. The term "gluten-free diet" refers to a dietary plan that requires people with CD to completely avoid any food or medication that contains gluten. This is because it has been recognized that wheat is a significant factor in the development of CD in susceptible persons. As a result, researchers have established a fundamental link between the disease and the intake of wheat gluten proteins (Annibale et al., 2001; Barton et al., 2007).

Grain products containing gluten and its derivatives, such as triticale, spelt, and Khorosan wheat, must be avoided. Avoid ingesting other foods like bulgur, einkorn, semolina, as well as various wheat products like whole wheat, cracked wheat, wheat bran, and wheat germ. Furthermore, it is completely prohibited to eat any food items made from gluten-containing grains, including bread, crackers, and pasta. As malt contains a partial hydrolysate of barley prolamins, it should be avoided by those who have CD as it may be harmful to their health. Consequently, on a CD diet, eating barley malt, malt flavoring, extract, or syrup is forbidden. Although there are already a number of beer varieties with lower gluten content that may be obtained in the market, beer typically contains significant amounts of hordein. For a long time, oats were not allowed in the diet for CD because of the belief that the protein in oats, called avenin, had a risk comparable to that of CD patients. Furthermore, there is controversy around the use of oats in a GFD due to the potential for cross-contamination with gluten-containing grains, as evidenced by a number of widely accessible items in the US. Nevertheless, certain research has shown that gluten can be present in both commercially processed GF flours and naturally GF goods. It has been shown in recent research that most people with CD, adults and children alike, can handle eating uncontaminated oats—which are free of gluten-containing grains—as long as they consume them in moderation. Moreover, research indicates that people with CD who can handle uncontaminated oats might benefit from eating them. Furthermore, food flavor may enhance adherence to a GF diet and boost nutritional value (Halfdanarson et al., 2006; See and Murray, 2006).

Cereals containing gluten should be avoided, as mentioned above. However, you can eat a broad variety of unprocessed plant and animal foods, including seafood, poultry, meats, fruits, and vegetables. There has been a lot of use for rice, maize, and potatoes as GF grain substitutes. The GFD benefits considerably from the increased variety, improved flavor, and high nutritional content that results from eating nutrient-dense seeds, grains, and pulses. Teff, Fonio, job's tears, millet, teosinte, and pseudo-cereals like amaranth, quinoa, buckwheat, and Tartarian buckwheat are examples of minor cereals that are only cultivated in specific regions. You can find minor cereals and pseudo-cereals sold whole, crushed into flour, made into flakes or grits, enlarged or even mixed in with other GF RTE specialized items like pasta, cereals, crackers, and more. The selection of GF options has grown substantially during the last several years. Made with newly introduced GF ingredients and food additives, these products include breads, biscuits, and pasta. According to Jameson (2000) and Mautalen et al. (1997), people with CD need to read food labels thoroughly, looking for words like hydrolyzed wheat protein, graham flour, wheat starch, and wheat bran.

9.3.4 The Amount of Macro- and Micronutrients in Gluten-Free Foods

Although CD has historically been associated with insufficient nutrient absorption, which causes diarrhea and weight loss, there have been increased reports recently of overweight and obese people—adults as well as children—being confirmed to have CD. In addition, more and more cases are being reported in which people with CD are diagnosed later in life, even in the absence of any symptoms. Two people with CD may not realize they have an increased risk of obesity when they start a GFD because of increased absorption in the intestines and an unbalanced diet high in fats and proteins. Replacing gluten-containing meals with commercially and naturally available products that are free of gluten but high in total and saturated fats is a well-documented trend among adults and children following a GFD. Studies by Hallert et al. (2002), Farrell and Kelly (2002), Bharti et al. (2024), Ahire et al. (2024a, 2024b, 2024c), Gujarathi et al. (2024), Rai et al. (2024), and Arya et al. (2024) have all noted this pattern. Concerns over the lipid makeup of commercially available GF foods have also been raised by this. Among the GF food items that were assessed, the bread, dough/pastry/pizza, and pasta categories (comprising 25, 38, and 38 items, respectively) contained twice as much fat, which was predominantly in the form of saturated fat, as the gluten-containing alternatives. The observed occurrence could potentially be explained by the incorporation of emulsifiers and lipid-rich constituents, including animal and vegetable oils, during their production. These compounds provide advantages in bread products by facilitating the stabilization of gas bubbles, diminishing the resistance to kneading, and reducing the swelling of starch granules. In addition, emulsifiers may be utilized to optimize the structure of the bread, increase the dough's rigidity, and decrease the rate of staleness. Pasta products contain emulsifiers, which function as lubricants during the extrusion procedure and control the expansion and dispersion of starch while it is being cooked. Consequently, the surface becomes more rigid and less sticky. Hence, although emulsifiers can be omitted from pasta products, they are indispensable in baked goods (Rude and Olerich, 1996; Scott et al., 2000).

9.3.5 Legislative Considerations and Food Labeling

Removing gluten from whole foods as well as processed foods is a crucial part of ensuring the safety of the GFD. A few years ago, the term "gluten free" was added to food labels. In 1976, the Food and Agricultural Organization's Codex Alimentarius Commission and the World Health Organization formally approved the present Codex Standard for GF food. A change was made to it in 1983. The 1990s saw a thorough investigation of the word. It is important to emphasize that it is impossible to completely eradicate gluten contamination from processed GF food. Experts are still at odds on how much gluten in food a patient with CD can consume without endangering the lining of their small intestine. According to some authors, defining a precise cut-off point is required to determine the permissible degree of gluten contamination (Halfdanarson et al., 2006; Tikkakoski et al., 2007).

It is advisable to restrict gluten intake to no more than 30 mg daily, with a recommended range of 10–100 mg/day. However, it is recommended to restrict gluten consumption to no more than 50 mg daily. In the US, the Food and Drug Administration (FDA) proposes a definition for the term "gluten-free." As per the stipulations of this proposition, a GF product ought to exclude all prohibited components, including wheat, rye, barley, and their hybrid forms. Furthermore, the product must be devoid of any components derived from forbidden cereals that retain gluten, unless the gluten concentration is below 20 parts per million (ppm). European nations have recently embraced the definition of GF as specified in the Codex Alimentarius. The designation "gluten-free" shall solely apply to products that contain no more than 20 ppm of gluten. Additionally, food products such as bread, which are composed of cereals that have been subjected to specialized processing to remove the majority of gluten, will be labeled "very low gluten." This translates to a gluten content of less than 30 mg/day. Manufacturers are provided with two alternatives: an immediate implementation of the new labeling system, or an extension until January 1, 2012, during which they may acclimatize to the revised regulations. It is expected that the implementation of an innovative labeling system will aid individuals who have CD in selecting foods in a more informed and secure manner (Bode and Gudmand-Hoyer, 1996; Dahele and Ghosh, 2001).

9.3.6 Guidelines for a Gluten-Free Diet

The results consistently show that nutritional therapy is a crucial part of controlling CD, despite limitations in the length and study design of research on the nutritional status of CD. Because different countries have different lifestyles and nutritional practices, it is challenging to provide universal dietary guidelines for people with CD from diverse ethnic backgrounds (Ahmad et al., 2015a, 2015b , 2016; Jarouliya et al., 2015; Keservani et al., 2016c). To prevent malnutrition, this chapter will discuss dietary recommendations as well as the nutritional benefits of many items that can be included in a GFD. carbs. A combination of complex and simple carbohydrates should make up about 55% of total calories ingested. As previously stated, the GFD restricts the amount of carbohydrates derived from grains. Legumes are good enough, though, as are a range of grains and seeds. Unknown grains and pseudo-grains have been the subject of recent analyses of their nutritional makeup, which has revealed that they include high concentrations of dietary fiber, carbohydrates, minerals, vitamins, and phenols (Mariani et al., 1998; Ojetti et al., 2005).

9.3.7 Protein

Approximately 15% of total calories is the ideal ratio for dietary protein consumption. The primary source of protein in the GFD is animal-derived diet, which includes foods like milk, meat, dairy products, poultry, and seafood. A good source of plant-based protein is nuts, seeds, legumes, pulses, and GF cereals. Recent investigations by experts have found that the protein content of minor cereals and pseudo-cereals is substantially higher than that of wheat. In addition, these grains have noticeably higher-quality protein. The least common amino acid in grains, lysine, is present in detectable concentrations. Because arginine and histidine are important for the growth and development of newborns and toddlers, amaranth and quinoa are especially beneficial for satisfying the nutritional needs of children with CD. Moreover, a high concentration of important amino acids, such as cysteine and methionine, which are critical for human health and success, can be found in pseudo-cereals and minor cereals. The amino acid content of the protein as well as its bioavailability or digestion determine its quality. Protein nutritional quality is commonly assessed using protein digestibility, available lysine, net protein utilization, and protein efficiency ratio. Pseudo-cereals have significantly greater protein contents than cereals and are comparable to casein (Polito et al., 1992; Caponio et al., 2008).

9.3.8 Micronutrients

It is essential to improve fruit and vegetable diet in people with CD to minimize vitamin deficiencies. Eating five servings or more of fruits and vegetables a day is recommended to lower the risk of certain diseases. Vegetables and fruits are high in essential minerals and nutrients and have a low calorie density. Furthermore, they have antioxidant and phytochemical properties that provide defense against illnesses linked to oxidative damage (Judd et al., 1994; Lissner and Heitmann, 1995).

9.3.9 Dietary Fiber

Dietary fiber is a complex blend of chemicals and plant parts that are difficult to decompose. Certain dietary fibers are metabolized by bacteria in the stomach and serve significant physiological roles. Numerous studies have shown that eating a diet high in fiber effectively reduces the risk of a number of human diseases, including diabetes, coronary heart disease, and colon cancer. It should be advised that people with CD consume enough dietary fiber. A GFD has been linked to a lower consumption of dietary fiber, according to previous research (Jarouliya and Keservani, 2023; Ngan et al., 2023; Rane et al., 2023a, 2023b; Verma et al., 2023). Minor cereals and pseudo-cereals have a varying fiber content, ranging from 7 to 10 g per 100 g. In comparison to other plant-based foods, fruits, nuts, pulses, and cereals such as maize and rice contain relatively higher concentrations of these nutrients. As a result, the inclusion of these ingredients in GF RTE food products can significantly augment the intake of dietary fiber among individuals with CD (Ellis et al., 1994).

9.3.10 Lipids

It is recommended that the overall fat intake be no more than 25–30% of total calories consumed. Making unsaturated fat consumption a priority is strongly advised. It is advised that monounsaturated fatty acids account for at least 15% of total calories consumed and polyunsaturated fatty acids for at least 10%. Consuming omega-3 fatty acids and monounsaturated fats has unquestionably been associated with a lower risk of cardiovascular disease. These substances can be found in a variety of foods, such as nuts, seeds, vegetable oils, and fatty fish such as salmon, trout, and herring. Saturated fats, which are typically found in foods produced from animals and tropical oils, should be consumed in moderation. Moreover, the advancement of atherosclerosis is adversely affected by trans fatty acid ingestion. As such, it is best to limit their intake to less than 1% of total calories. The levels of saturated and trans fatty acids are higher in processed GF RTE food products compared to traditional diets, as mentioned before. Consequently, people with CD must closely monitor the presence of hydrogenated fats by reading product labels. The fat level of pseudo-cereals is higher than that of other plant-based diets. Their notable quality stems from their elevated levels of unsaturated fatty acids, especially linolenic acid, an essential omega-3 fatty acid for all animals. Eating α-linolenic acid is essential for lowering the risk of coronary heart disease's early onset and future progression. Squalene, an open-chain, highly unsaturated triterpene, is present in significant amounts in amaranth. Deep-sea fish and other marine species' livers frequently contain this chemical alone (Lohiniemi et al., 1998; Haboubi et al., 2006).

9.3.11 Vitamins

Vitamin consumption is crucial for those with CD to avoid vitamin deficits. Grains, liver, and green leafy vegetables all contain folic acid. It has been discovered that quinoa and amaranth, two GF cereals, have significantly more folic acid than wheat. Riboflavin, ascorbic acid, and tocopherols are abundant in amaranth, quinoa, and oats. Two vitamins found in buckwheat seeds are B2 and B6. Remarkably, an increasing number of specialty GF products are now being supplemented with minerals and vitamins (Janatuinen et al., 2002; Størsrud et al., 2003).

9.3.12 Mineral Salts

These are necessary nutrients. They are made up of trace elements and essential minerals. Similar to animal products, diets based primarily on plants also include a substantial amount of minerals. The mineral content of amaranth, quinoa, and oats is approximately double that of other cereals. Teff has higher iron and calcium concentrations than rice, wheat, barley, and sorghum. Because teff has a high iron content, the amount of the grain that is ingested in Ethiopia appears to be correlated with the lack of anemia. Compared to wheat, buckwheat seeds have a lower mineral content. In terms of its high quantity of nutritionally significant minerals, buckwheat performs better than several grains, including rice, sorghum, millet, and maize, except for calcium. Minerals found in the embryo indicate that the refining process is likely to have retained them to some extent (Haboubi et al., 2006; Kemppainen et al., 2007).

9.3.13 Phytochemicals

Phytochemicals, including polyphenols, are ubiquitous in plant-based foods and exert influence over their nutritional and sensory attributes. Their importance in the field of human nutrition has been acknowledged on account of the prospective benefits they offer to human health. More precisely, they possess the capacity to reduce the likelihood of developing metabolic syndrome, cardiovascular disease, ischemic stroke, gastrointestinal malignancies, and type II diabetes. Moreover, their antioxidant, antiviral, anti-allergic, antiplatelet, anti-inflammatory, and anti-cancerous properties have been documented. As previously mentioned, fruits, vegetables, wine, and teas are the primary dietary sources of these chemicals. Cereals and pseudo-cereals can be a substantial source of polyphenols, as recent studies have shown. Flavonoids are a type of polyphenolic component that can be found in fruits, vegetables, beverages, and a variety of cereals and gluten-free products

(Sharma et al., 2020; Rane et al., 2020a, 2020b, 2020c, 2023). They enhance the color of many fruits and vegetables. The pericarp of pigmented types of barley, maize, rice, rye, and wheat has also been found to contain flavonoids. This study looks at the anthocyanin content of various fruits and colored cereal grains. Cereals have been reported to contain flavonoids, which are primarily present in fruits and vegetables. Keservani et al. examined these flavonoids in multiple articles between 2014 and 2016. For instance, millet, oats, and sorghum have all been reported to contain the flavone apigenin, which is present in celery and parsley. Flavonones have been measured in oats and sorghum by research by Dillard and German (2000) and Molyneux et al. (2007).

A multitude of phytochemicals exhibit antioxidant characteristics within controlled laboratory environments. Recent research has investigated the antioxidant capacity of GF cereals as a whole. Based on the results, buckwheat and quinoa have the highest antioxidant capacity among the cereals and pseudo-cereals that were assessed. Alternative grains such as quinoa and oats may be more cost-effective in comparison to conventional alternatives for GFDs. The inclusion of quinoa and grains in one's dietary regimen may promote adherence to dietary restrictions and mitigate the financial burden commonly associated with such a regimen.

Further investigation is necessary to comprehensively assess whether the use of pseudo-cereals and minor cereals can efficiently mitigate nutritional deficiencies in individuals with treated CD while also reducing the price of GF RTE food products. Oats would increase the nutritional value and diversity of the GFD, according to a systematic review of the literature concerning their inclusion in the GFD. Further authors have corroborated the positive effect of cereals on GFD. By integrating three servings of GF alternative grains, such as quinoa and oats, into a GFD, the grain component's nutritional composition is improved. Minor cereals and pseudo-cereals constitute substantial reservoirs of proteins and carbohydrates, specifically fiber, thereby broadening the dietary alternatives accessible to individuals who have CD. In the realm of food costs, prior research has noted that commercially available GF products are significantly more expensive than comparable items (Molyneux et al., 2007; Rochfort and Panozzo, 2007; Prakash et al., 2012).

9.4 CONCLUSION

In recent years, the prevalence of CD has increased. Prior research has demonstrated that GF foods are deficient in vitamins, minerals (including iron), and fiber. The nutritional composition of GF goods is consequently an increasing source of concern. A domain ripe with promise is the application of pseudo-cereals and lesser-known grains, including teff, amaranth, buckwheat, quinoa, and sorghum. A diverse range of GF, RTE food items can be produced by incorporating obscure cereals and grain-like plants. Considerable scholarly investigation has been devoted to examining the technological and nutritional attributes of these alternative cereals, which may be utilized as a replacement for wheat. The inclusion of these cereals in the dietary regimen of individuals diagnosed with CD may result in increased protein, iron, calcium, and fiber intake.

REFERENCES

Ahire, E.D., Keservani, R.K., Balekundri, A., Rajora, A.K., Surana, K.R. Effect of nutrients on brain function. In: Nutraceutical Fruits and Foods for Neurodegenerative Disorders, Edited by Keservani, R.K., Kesharwani, R.K., Emerald, M., Sharma, A.K., Academic Press, Elsevier, 2024b, Chap 13, pp. 237–248. ISBN: 9780443189517.

Ahire, E.D., Keservani, R.K., Bharti, S.K. Role of fruit and food components in brain cognition. In: Nutraceutical Fruits and Foods for Neurodegenerative Disorders, Edited by Keservani, R.K., Kesharwani, R.K., Emerald, M., Sharma, A.K., Academic Press, Elsevier, 2024c, Chap 1, pp. 1–14. ISBN: 9780443189517.

Ahire, E.D., Surana, K.R., Khairnar, S.J., Laddha, U.D., Kshirsagar, S.J., Rajora, A.K., Keservani, R.K. Role of protein-rich diet in brain functions. In: Nutraceutical Fruits and Foods for Neurodegenerative Disorders,

Edited by Keservani, R.K., Kesharwani, R.K., Emerald, M., Sharma, A.K., Academic Press, Elsevier, 2024a, Chap 25, pp. 505–523. ISBN: 9780443189517.

Ahmad, M.F., Ahmad, S.M., Keservani, R.K., Pradhan, A. Anti-ulcer activity of tuber extracts of *Solanum tuberosum* (*Solanaceae*) in rats. Acta Fac. Pharm. Univ. Comen. LXII, 2015a (2): 32–37.

Ahmad, M.F., Ahmad, S.M., Keservani, R.K., Sharma, A.K. A study on anti-inflammatory activity of tuber extracts of *Solanum tuberosum* (*Solanaceae*) in male albino rats. Nat. Aca. Sci. Lett. India 2016, 39 (6): 421–425.

Ahmad, M.F., Keservani, R.K., Babu, D.J.M. Interaction study between *Ocimum sanctum* and glimepiride (sulfonylurea derivative) in diabetic rats. J. Chin. Pharm. Sci. 2015b, 24 (3), 156–163.

Annibale, B., Severi, C., Chistolini, A., Antonelli, G., Lahner, E., et al. Efficacy of gluten-free diet alone on recovery from iron deficiency anaemia in adult celiac patients. Am. J. Gastroenterol. 2001, 96, 132–137.

Arya, R.K.K., Kausar, M., Bisht, D., Keservani, R.K., Kumar, A. Memory-boosting fruits and foods for elderly. In: Nutraceutical Fruits and Foods for Neurodegenerative Disorders, Edited by Keservani, R.K., Kesharwani, R.K., Emerald, M., Sharma, A.K., Academic Press, Elsevier, 2024, Chap 5, pp. 77–100. ISBN: 9780443189517.

Bardella, M.T., Fredella, C., Prampolini, L., Molteni, N., Giunta, A.M., Bianchi, P.A. Body composition and dietary intakes in adult celiac disease patients consuming a strict gluten-free diet. Am. J. Clin. Nutr. 2000, 72, 937–939.

Barker, J.M., Liu, E. Celiac disease: Pathophysiology, clinical manifestations, and associated autoimmune conditions. Adv. Pediatr. 2008, 55, 349–365.

Barton, S.H., Kelly, D.G., Murray J.A. Nutritional deficiencies in celiac disease. Gastroenterol. Clin. North Am. 2007, 36, 93–108.

Bharti, S.K., Shukla, D., Mahapatra, D.K., Asati, V., Keservani, R.K. Therapeutic potential of nutraceuticals for the management of neurodegenerative disorders. In: Nutraceutical Fruits and Foods for Neurodegenerative Disorders, Edited by Keservani, R.K., Kesharwani, R.K., Emerald, M., Sharma, A.K., Academic Press, Elsevier, 2024, Chap 26, pp. 525–535. ISBN: 9780443189517.

Bode, S., Gudmand-Hoyer, E. Symptoms and haematological features in consecutive adult coeliac patients. Scand. J. Gastroenterol. 1996, 31, 54–60.

Caponio, F., Summo, C., Clodoveo, M.L., Pasqualone, A. Evaluation of the nutritional quality of the lipid fraction of gluten-free biscuits. Euro. Food Res. Technol. 2008, 223, 135–139.

Catassi, C., Rätsch, I.M., Gandolfi, L., Pratesi, R., Fabiani, E., El Asmar, R., Frijia, M., Bearzi, I., Vizzoni, L. Why is coeliac disease endemic in the people of the Sahara? Lancet 1999, 354, 647–648.

Chand, N., Mihas, A.A. Celiac disease: Current concepts in diagnosis and treatment. J. Clin. Gastroenterol. 2006, 40, 3–14.

Ciccocioppo, R., Di Sabatino, A., Corazza, G.R. The immune recognition of gluten in coeliac disease. Clin. Exp. Immunol. 2005, 140 , 408–416.

Dahele, B., Ghosh, S. Vitamin B12 deficiency in untreated celiac disease. Am. J. Gastroenterol. 2001, 96, 745–750.

Dewar, D.H., Amato, M., Ellis, H.J., Pollack, E.L., Gonzalez-Cinca, N., Wieser, H., Ciclitira, P.J. The toxicity of high molecular weight glutenin subunits of wheat to patients with coeliac disease. Eur. J. Gastroenterol. Hepatol. 2006, 18, 483–491.

Dieterich, W., Ehnis, T., Bauer, M., Donner, P., Volta, U., Riecken, E.O., Schuppan, D. Identification of tissue transglutaminase as the autoantigen of celiac disease. Nat. Med. 1997, 3, 797–801.

Dillard, C.J., German, J.B. Phytochemicals: Nutraceuticals and human health. J. Sci. Food Agric. 2000, 80 (12): 1744–1756.

Dube, C., Rostom, A., Sy, R., Cranney, A., Saloojee, N., Garritty, C., Sampson, M., Zhang, L., Yazdi, F., Mamaladze, V., Pan, I., Macneil, J., Mack, D., Patel, D., Moher, D. The prevalence of celiac disease in average-risk and at-risk Western European populations: A systematic review. Gastroenterology 2005, 128, S57–S67.

Ellis, H.J., Doyle, A.P., Day, P., Wieser, H., Ciclitira, P.J. Demonstration of the presence of coeliac-activating gliadin-like epitopes in malted barley. Int. Arch. Allergy Immunol. 1994, 104, 308–310.

Farrell, R.J., Kelly, C.P. Celiac sprue. N. Engl. J. Med. 2002, 346, 180–188.

Fasano, A. Clinical presentation of celiac disease in the pediatric population. Gastroenterology 2005, 128, S68–S73.

Fasano, A., Berti, I., Gerarduzzi, T., Not, T., Colletti, R.B., Drago, S., Elitsur, Y., Green, P.H., Guandalini, S., Hill, I.D., Pietzak, M., Ventura, A., Thorpe, M., Kryszak, D., Fornaroli, F., Wasserman, S.S., Murray, J.A., Horvath, K. Prevalence of celiac disease in at-risk and not-at-risk groups in the United States: A large multicenter study. Arch. Intern. Med. 2003, 163, 286–292.

Green, P.H., Jabri, B. Coeliac disease. Lancet 2003, 362, 383–391.

Green, P.H., Rostami, K., Marsh, M.N. Diagnosis of coeliac disease. Best Pract. Res. Clin. Gastroenterol. 2005, 19, 389–400.

Gujarathi, N.A., Aher, A.A., Sukhia, A., Patil, T.S., Goyal, Y.S., Keservani, R.K. Nutraceutical interventions in Alzheimer's disease. In: Nutraceutical Fruits and Foods for Neurodegenerative Disorders, Edited by Keservani, R.K., Kesharwani, R.K., Emerald, M., Sharma, A.K., Academic Press, Elsevier, 2024, Chap 20, pp. 379–404. ISBN: 9780443189517.

Haboubi, N.Y., Taylor, S., Jones, S. Coeliac disease and oats: A systematic review. Postgrad. Med. J. 2006, 82, 672–678.

Halfdanarson, T.R., Litzow, M.R., Murray, J.A. Hematological manifestations of celiac disease. Blood 2006, 109, 412–421.

Hallert, C., Grant, C., Grehn, S., Grännö, C., Hultén, S., Midhagen, G., Ström, M., Svensson, H., Valdimarsson, T. Evidence of poor vitamin status in coeliac patients on a gluten-free diet for 10 years. Aliment Pharmacol. Ther. 2002, 16, 1333–1339.

Jameson, S. Coeliac disease, insulin-like growth factor, bone mineral density, and zinc. Scand. J. Gastroenterol. 2000, 35, 894–896.

Janatuinen, E.K., Kemppainen, T.A., Julkunen, R.J.K., Kosma, V.M., Mäki, M., Heikkinen, M., Uusitupa, M.I.J. No harm from five-year ingestion of oats in coeliac disease. Gut 2002, 50, 332–335.

Jarouliya, U., Keservani, R.K. Developing vaccine against Covid-19 and immune response, prospects. In: COVID-19 and Immunomodulation with Special Emphasis on Nutraceutical and Herbal Formulation, Edited by Kesharwani, R.K., Keservani, R.K. Apple Academic Press, CRC Press, Taylor & Francis, 2023, Chap 5, pp. 139–160. ISBN: 9781774913178.

Jarouliya, U., Zacharia, A., Keservani, R.K., Prasad, G.B.K.S. Protective effect of *Spirulina maxima* against fructose-induced oxidative stress in Wistar rats with histopathological observations. Acta Fac. Pharm. Univ. Comen. LXII, 2015 (2): 13–19.

Judd, J.T., Clevidence, B.A., Muesing, R.A., Wittes, J., Sunkin, M.E., Podczasy, J.J. Dietary trans fatty acids: Effects on plasma lipids and lipoproteins of healthy men and women. Am. J. Clin. Nutr. 1994, 59, 861–868.

Kemppainen T. Oatmeal as a Component of a Gluten-Free Diet, Nutrient Intakes, Nutritional Status and Osteopenia in Coeliac Patients. Kuopio University Publications D, Medical Sciences, 1997, p. 188.

Kemppainen, T., Janatuinen, E., Holm, K., Kosma, V.M., Heikkinen, M., Mäki, M., Laurila, K., Uusitupa, M., Julkunen, R. No observed local immunological response at cell level after five years of oats in adults coeliac disease. Scand. J. Gastroenterol. 2007, 42, 54–59.

Kemppainen, T., Uusitupa, M., Janatuinen, E., Järvinen, R., Julkunen, R., Pikkarainen, P. Intakes of nutrients and nutritional status in coeliac patients. Scand. J. Gastroenterol. 1995, 30, 575–579.

Keservani, R.K., Kesharwani, R.K., Sharma, A.K., Gautam, S.P., Verma, S.K. Nutraceutical formulations and challenges. In: New Product Development in Nutraceuticals & Functional Foods: From Concept to Market, Edited by Bagchi, D., Nair, S., Academic Press, Elsevier, 2016c, Chap 9, pp. 161–177.

Kinsey, L., Burden, S.T., Bannerman, E. A dietary survey to determine if patients with celiac disease are meeting current healthy eating guidelines and how their diet compares to that of the British general population. Eur. J. Clin. Nutr. 2008, 62, 1333–1342.

Lissner, L., Heitmann, B.L. Dietary fat and obesity: Evidence from epidemiology. Eur. J. Clin. Nutr. 1995, 49, 79–90.

Lohiniemi, S., Collin, P., Mäki, M., Eds. Changing Features of Celiac Disease. The Finnish Coeliac Society, 1988.

Mariani, P., Viti, M.G., Montuori, M., La Vecchia, A., Cipolletta, E., Calvani, L., Bonamico, M. The gluten-free diet: A nutritional risk factor for adolescents with celiac disease? J. Pediatr. Gastroenterol. Nutr. 1998, 27, 519–523.

Marsh, M.N. Gluten, major histocompatibility complex, and the small intestine. A molecular and immunobiologic approach to the spectrum of gluten sensitivity ("celiac sprue"). Gastroenterology 1992, 102, 330–354.

Mautalen, D., Gonzalez, R., Mazure, H., Vazquez, M.P., Lorenzetti, E., Maurino, S., Niveloni, S., Pedreira, S.E., Boerr, L.A., Bai, J.C. Effect of treatment on bone mass, mineral metabolism, and body composition in untreated celiac disease patients. Am. J. Gastroenterol. 1997, 92, 313–318.

Molyneux, R.J., Lee, S.T., Gardner, D.R., Panter, K.E., James, L.F. Phytochemicals: The good, the bad and the ugly?. Phytochemistry. 2007, 68 (22–24), 2973–85.

Murray, J.A., Van Dyke, C., Plevak, M.F., Dierkhising, R.A., Zinsmeister, A.R., Melton, L. Trends in the identification and clinical features of celiac disease in a North American community, 1950–2001. Clin. Gastroenterol. Hepatol. 2003, 1, 9–27.

Ngan, D.K., Tung, B.T., Keservani, R.K., Rane, B.R. Breast cancer prevention by nutraceuticals. In: Nutraceuticals in Cancer Prevention, Management and Treatment, Edited by Keservani, R.K., Tung, B.T., Singh, S., Kesharwani, R.K., Apple Academic Press, CRC Press, Taylor & Francis, 2023, Chap 6, pp. 113–132. ISBN: 9781774913277.

Ojetti, V., Nucera, G., Migneco, A., Gabrielli, M., Lauritano, C., Danese, S., Zocco, M.A., Nista, E.C., Cammarota, G., De Lorenzo, A., Gasbarrini, G., Gasbarrini, A. High prevalence of celiac disease in patients with lactose intolerance. Digestion 2005, 71, 106–110.

Paulley, J.W. Observation on the aetiology of idiopathic steatorrhoea; jejunal and lymph-node biopsies. Br. Med. J. 1954, 2, 1318–1321.

Polito, C., Olivieri, A.C., Marchese, L., Desiderio, G., Pullano, F., Rea, F. Weight overgrowth of coeliac children on a gluten-free diet. Nutr. Res. 1992, 12, 353–358.

Prakash, D., Gupta, C., Sharma, G. Importance of phytochemicals in nutraceuticals. J. Chin. Med. Res. Dev. 2012, 1 (3), 70–78.

Rai, S., Keservani, R.K., Kumar, P., Nikam, V.K., Kachave, R.N., Kumar, Y., Kesharwani, R.K. Importance of functional foods in the management of autism. In: Nutraceutical Fruits and Foods for Neurodegenerative Disorders, Edited by Keservani, R.K., Kesharwani, R.K., Emerald, M., Sharma, A.K., Academic Press, Elsevier, 2024, Chap 8, pp. 151–171. ISBN: 9780443189517.

Rane, B.R., Ahirrao, R.A., Gujarathi, N.A., Jain, A.S., Keservani, R.K. Pathophysiology and application of herbs in the metabolic syndrome. In: The Metabolic Syndrome: Dietary Supplements and Food Ingredients, Edited by Keservani, R.K., Yadav, D., Kesharwani, R.K., Singh, S., Sandeep, K., Apple Academic Press, CRC Press, Taylor & Francis Group, 2023a, Chap 10, pp. 199–230. ISBN: 9781774911112.

Rane, B.R., Bharat, M., Keservani, R.K. COVID-19—infection and recent investigations and future prospects. In: COVID-19 and Immunomodulation with Special Emphasis on Nutraceutical and Herbal Formulation, Edited by Kesharwani, R.K., Keservani, R.K., Apple Academic Press, CRC Press, Taylor & Francis, 2023b, Chap 9, pp. 223–261. ISBN: 9781774913178.

Rane, B.R., Bharath, M.S., Patil, R.R., Keservani, R.K., Jain, A.S. Novel approaches in nutraceuticals. In: Enhancing the Therapeutic Efficacy of Herbal Formulations through Novel Drug Delivery Systems, Edited by Kesharwani, R.K., Keservani, R.K., Sharma, A.K., IGI Global International Publisher, 2020b, Chap 11, pp. 241–266. ISBN13: 9781799844532.

Rane, B.R., Patil, A.S., Keservani, R.K., Jain, A.S. Novel approaches in herbal formulation. In: Enhancing the Therapeutic Efficacy of Herbal Formulations through Novel Drug Delivery Systems, Edited by Kesharwani, R.K., Keservani, R.K., Sharma, A.K., IGI Global International Publisher, 2020a, Chap 2, pp. 43–68. ISBN13: 9781799844532.

Rane, B.R., Tadavi, S.A., Keservani, R.K. Naturopathy. In: Herbal Product Development, Edited by Sharma, A.K., Keservani, R.K., Gautam, S.P., Apple Academic Press, CRC Press, Taylor & Francis Group, 2020c, Chap 12, pp. 321–347. ISBN: 9781771888776.

Rane, B.R., Telange, T.S., Waghchaure, P.S., Keservani, R.K., Jain, A.S. Covid-19 pandemic: Role of ayurveda for the prevention and treatment. In: COVID-19 and Immunomodulation with Special Emphasis on Nutraceutical and Herbal Formulation, Edited by Kesharwani, R.K., Keservani, R.K., Apple Academic Press, CRC Press, Taylor & Francis, 2023, Chap 4, pp. 107–138. ISBN: 9781774913178.

Rochfort, S., Panozzo, J. Phytochemicals for health, the role of pulses. J. Agri. Food Chem. 2007, 55 (20): 7981–7994.

Rude, R.K., Olerich, M. Magnesium deficiency: Possible role in osteoporosis associated with gluten-sensitive enteropathy. Osteoporos Int. 1996, 6, 453–461.

Scott, E.M., Gaywood, I., Scott, B.B. Guidelines for osteoporosis in celiac disease and inflammatory bowel disease: British Society of Gastroenterology. Gut 2000, 46, 1–8.

See, J., Murray, J.A. Gluten-free diet: The medical and nutrition management of celiac disease. Nutr. Clin. Pract. 2006, 21, 1–15.

Shan, L., Molberg, O., Parrot, I., Hausch, F., Feliz, F., Gray, G.M., Sollid, L.M., Khosla, C. Structural basis for gluten intolerance in celiac sprue. Science 2002, 297, 2275–2279.

Sharma, A.K., Keservani, R.K., Gautam, S.P. Herbal Product Development, Apple Academic Press, CRC Press, Taylor & Francis Group, 2020, pp. 1–376. ISBN: 9781771888776.

Sood, A., Midha, V., Sood, N., Malhotra, V. Adult celiac disease in Northern India. Indian J. Gastroenterol. 2003, 22, 124–126.

Størsrud, S., Olsson, M., Arvidsson, R., Lenner, L., Nilsson, Å., Nilsson, O., Kilander, A. Adult celiac patients do tolerate large amounts of oats. Eur. J. Clin. Nutr. 2003, 57, 163–169.

Thompson, T. Folate, iron, and dietary fibre contents of the gluten-free diet. J. Am. Dietetic Assoc. 2000, 100, 1389–1396.

Thompson, T. Thiamin, riboflavin, and niacin contents of the gluten-free diet: Is there cause for concern? J. Am. Dietetic Assoc. 1999, 99, 858–862.

Thompson, T., Dennis, M., Higgins, L.A., Lee, A., Sharrett, M.K. Gluten-free diet survey: Are Americans with coeliac disease consuming recommended amounts of fibre, iron, calcium and grain foods? J. Hum. Nutr. Diet. 2005, 18, 163–169.

Tikkakoski, S., Savilahti, E., Kolho, K.L. Undiagnosed celiac disease and nutritional deficiencies in adults screened in primary health care. Scand. J. Gastroenterol. 2007, 42, 60–65.

van de Wal, Y., Kooy, Y.M., van Veelen, P., Vader, W., August, S.A., Drijfhout, J.W., Pena, S.A., Koning, F. Glutenin is involved in the gluten driven mucosal T cell response. Eur. J. Immunol. 1999, 10 , 3133–3139.

Verma, M., Ahire, E.D., Keservani, R.K. Colon cancer prevention by medicinal plants. In: Nutraceuticals in Cancer Prevention, Management and Treatment, Edited by Keservani, R.K., Tung, B.T., Singh, S., Kesharwani, R.K., Apple Academic Press, CRC Press, Taylor & Francis, 2023, Chap 9, pp. 163–189. ISBN: 9781774913277.

10 Ready-to-Eat Foods and Therapeutics
A Healthy Perception

Pardeep Kaur Sandhu[1] *and Gurpreet Kaur*[2*]
[1]Department of Zoology, Mata Gujri College, Fatehgarh Sahib, Punjab, India
[2]Department of Zoology, Akal University, Talwandi Sabo, Punjab, India
*Corresponding Author

10.1 INTRODUCTION

The market for ready-to-eat (RTE) foods has expanded quickly in recent years due to the fast-paced life of modern society. As per United Nation reports, the world population is predicted to be 9.2 billion by 2050, in which 70% population will be urban. The fast-paced life demands new food interventions which are convenient and need least time for preparation. RTE foods are those that are intended for direct human consumption without the need for additional cooking or processing (Elviss and Jørgensen, 2012). RTE foods are simple to handle, look fresh, are safe from a microbiological standpoint, and have a good flavor over their whole shelf life (Silberbauer and Schmid, 2017). RTE meals come in a range of forms, including fresh-cut fruits and vegetables (Ma *et al.*, 2017), cooked meat and poultry, prepared salads, and smoked or salted shellfish (Yu *et al.*, 2020). However, the consumer is also conscious about the nutrition and health status, so demands minimally processed RTE foods (Silberbauer and Schmid, 2017). Moreover the outbreak of coronavirus disease 2019 (COVID-19) pandemic put focus on consumption of food rich in micronutrients and with good antioxidant potential, resulting in increased demand of nutritious RTE meals (Suri *et al.*, 2022). Over 2 billion individuals globally have micronutrient malnutrition, also known as the "hidden hunger," which is characterized by vitamin and mineral deficiencies. Children who are malnourished struggle to grow and develop normally. It is necessary to develop a food-based, sustainable strategy with several nutritional advantages.

Ready-to-use therapeutic foods (RUTF) are quite beneficial in the rehabilitation of undernourished children. If a child exhibits hunger and is unable to ingest the daily necessary amount of calories, they can be treated with RUTF. The ability to employ the RUTF composition on-site within therapeutic programs, that is, right at home, eliminating the need to visit hospitals or nutritional treatment centers is another significant benefit of this diet. Natural disasters and war demand requirement of RUTF prepared for emergency needs. Given the raw material's trend toward price increases, this food's potential to be economically sustainable in the long run would be jeopardized. To achieve the Sustainable Development Goal 2.2 to end all forms of malnutrition, it is imperative to develop affordable RUTF for special needy groups. This can be possible by preparing formulations based on locally available fruits, vegetables, and cereals in place of specific formula RUTF (Figure 10.1). The preparation of the former by self-help groups will limit their production to a specific geographical area. So, it is important to expedite the research in preparing affordable RTE based on local ingredients at industrial level for alleviating the vulnerability to malnutrition and food insecurity (Shoge, 2022; Testa *et al.*, 2023).

 DOI: 10.1201/9781032622408-10

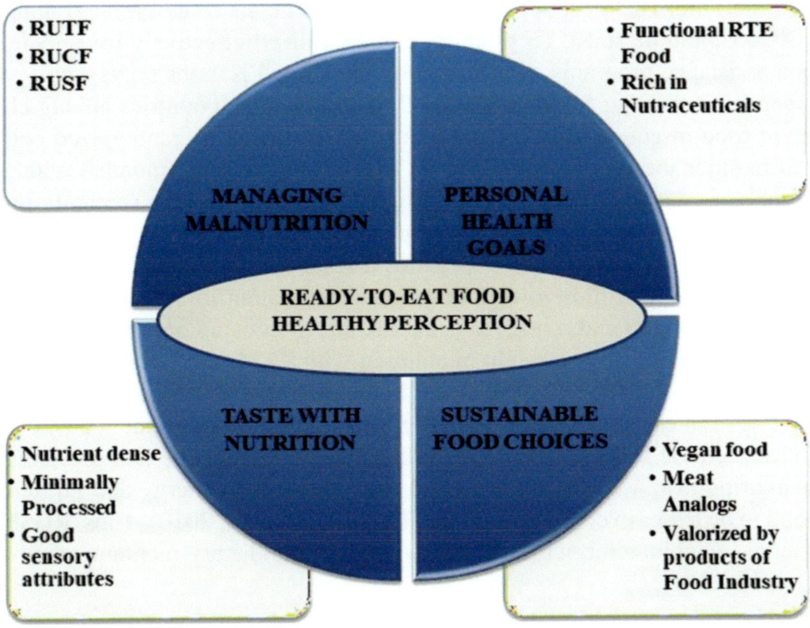

FIGURE 10.1 Ready-to-Eat Food: A Healthy Perception.

Considering the increased cost of healthcare and contribution of food in prevention of nutrition-related ailments and improvement in physical and mental health, there is significant increase in demand of functional foods (Valls *et al.*, 2013). The choice of consumers for RTE foods is not only to satiate the hunger, but also to consume it as functional foods for prevention of nutrients based noncommunicable disorders such as diabetes, obesity, and cardiovascular diseases. According to the World Health Organization (WHO), inadequate intakes of micronutrients contribute to approximately 14% of gastrointestinal cancer fatalities, 11% of ischemic heart disease deaths, and 9% of stroke-related deaths globally (Famuwagun *et al.*, 2023). Therefore, there is a need for the development of nutrient-dense, premium ready meals, which can serve as a prophylactic and therapeutic agent for the prevention of diseases. Furthermore, to cater to the needs of the persons with ailments or allergies like celiac disease and lactose intolerance, there are specific RTE formulations. This chapter explores the various types of RTE food prepared for therapeutics and for disease prevention and management.

10.2 READY-TO-EAT THERAPEUTIC FOOD FOR TREATMENT OF MALNUTRITION

Severe acute malnutrition (SAM) is a serious public health problem due to its long-term health consequences and high mortality. The foods commonly used as an approach to combat malnutrition in recent years, primarily enriched cereals and legumes, flour-based foodstuffs enhanced with legumes, and cereals with or without sugar, oil, milk, and eggs, contributed little to the success of the interventions due to their ineffectiveness in SAM treatment (Bhandari *et al.*, 2016; De Pee *et al.*, 2009; Karakochuk *et al.*, 2012; Medoua *et al.*, 2016). In recent decades RUTF, a highly fortified energy-dense paste, has provided a solution to this problem (Santini *et al.*, 2013). The majority of RUTF is made up of peanuts that have been supplemented with powdered milk, vitamins, vegetable oil, sugar, and mineral salts. RUTF has a full nutritional composition that includes amino acids, vitamins, mineral salts, and vital fatty acids, as well as a high energy density (500–540 kcal/100 g) and an extended shelf life. The most popular commercial RUTF are peanut paste (Plumpy'nut®) and solid biscuit BP100®, and these RUTF products can be utilized effectively for on-site therapeutic programs. Food assistance programs coordinated by the United Nations (UN) employed RUTF as a critical component in treating SAM in low- and middle-income countries among children aged 24–59 months in food-insecure regions (Choudhury *et al.*, 2018). A randomized controlled trial is conducted to monitor the efficacy of RUTF in children (aged 6–59 months) with SAM in the Democratic Republic of Congo. The current national standard strategy (separate protocols and products for children with severe and moderate acute malnutrition using RUTF at an increasing dose with the increment in weight in children with SAM, and ready-to-use supplementary food (RUSF) at the same concentration for children with moderate acute malnutrition) and the OptiMA strategy (a single procedure for severe and moderate acute malnutrition using only RUTF at a diminishing concentration with the increment in weight in children with SAM) were compared for 6 months in this trial. It was observed that the rate of recovery in children was observed to be noninferior in the OptiMA group (96%) in contrast to the standard group (98%) (Cazes *et al.*, 2023). Current ready-to-use food introduced the new term of food, RUSF, for the therapeutic and preventive effects of mild acute malnutrition (MAM) in children. The concentration of energy and nutrient density in RUSF was found to be lesser in contrast to the RUTF (Isanaka *et al.*, 2010). Thus, RUSF is intended to be administered in conjunction with local diets to meet the dietary recommendations of MAM youngsters (Golden, 2009).

10.2.1 Ready-to-Use Supplementary Food

Ready-to-use supplemental foods (RUSFs) are foods that may be taken without cooking or the addition of water and are fortified with micronutrients as a treatment for malnutrition. Among them are finished goods which include pastes, compacted bars, and biscuits. RUSFs are designed to provide all of the nutrients that are necessary for normal growth as well as those that are needed to sustain bodily function. The addition of child-targeted RUSF supplementation to a general food distribution led to a decrease in fever, anemia, and diarrhea (Huybregts *et al.*, 2012). Similarly, a 46-g daily administration of RUSF caused a significant boost in the levels of iron, vitamin A, folic acid, vitamin B12, and riboflavin in targeted population (Zimmermann and Hurrell, 2007). Likewise, Lopriore *et al.* (2004) also demonstrated that supplementation of nutrient-dense fat-based spread enriched with vitamins and minerals in a group of children aged 3–6 months showed a remarkable increment in hemoglobin concentration in contrast to control group. Moreover, WHO and the United Nations Children's Fund suggested the supplementation of zinc in RUSF as an ideal candidate for the cure of diarrhea (WHO, 2004). In addition to it, Sastrawan *et al.* (2021) recommend the RUSF as an adjunct therapy for the cure of MAM children during COVID-19.

10.2.2 Ready-to-Use Complementary Food

Complementary meals are essential components of early child's diet. Infants are given complementary meals in addition to breast milk when the nutrients in breast milk are no longer sufficient for their needs in terms of development and energy. WHO also favored the consumption of complementary meals high in bioactive components and antioxidants with desirable functional qualities. Complementary foods are often served as semi-liquid porridge made from tubers like sweet potatoes, cocoyams, and Irish potatoes, or from basic grains like sorghum, maize, and millet (Kehinde Alawode *et al.*, 2017). It is widely agreed upon that feeding children micronutrient-rich foods that are readily available in the area and promoting the manufacturing of inexpensive, industrially processed, fortified cereal-based supplemental meals locally are the best ways to improve the nutritional status of children in this age range (Nestel *et al.*, 2003). Various complementary food formulations based on bird meat were made: rice combined with the meat (RicQ), green banana mixed with bird meat

(BanQ), Irish potato mixed with bird meat (PotQ), and control, which included only bird meat (CtrQM). The formulations were cooked individually for 45–55 minutes at 98°C. Proximate composition analysis revealed that PotQ contained a greater moisture content (8.9 g), protein (16.7 g), crude fat (1.8 g), ash (5.4 g), and crude fiber (2.9 g) in contrast to other formulations. Furthermore, the greatest levels of beta-carotene (4.9 µg/g) and retinol (26.5 mg/kg) were found in RicQ, whereas the lowest levels were noticed in the control group (3.3 mg/kg and 1.6 µg/g, respectively) (Ngume *et al.*, 2023). In addition to it, Sogo-Temi *et al.* (2023) also screened the proximal composition of complementary food made from velvet beans, plantain, and crayfish flours. Moisture, crude protein, fat, fiber, ash, and carbohydrates in this complementary food had approximate compositions of 6.49%, 12.18%, 1.50%, 0.35%, 2.10%, and 82.93%, respectively (Table 10.1). This supplemental meal showed 64.80% *in vitro* protein digestibility, 21.80% starch digestibility, 3.29 g/100 g of *in vitro* total sugar, and 50.87% total carbohydrate. The complementary food was observed to have nutrient-dense and sensory qualities similar to commercial complementary meals, and could be used in place of more costly supplementary foods. Habte *et al.* (2023) also suggested that dried beef powder is rich in nutrients and might be used as part of a complementary food preparation recipe to lower child malnutrition. Protein, fat, fiber, ash, carbohydrate, and energy contents of meat powder were observed to be 76.12, 8.2, 0.56, 6.5, 2.8 g/100 g, and 389.30 kcal/100 g, respectively. It also consisted of a remarkable amount of minerals like potassium, sodium, phosphorus, zinc, and calcium. According to Ijeomah *et al.* (2023), the complementary food made with ricebean, fonio, crayfish and carrot showed a high amount of protein, energy, and vitamin A. Consequently, it can be utilized to treat vitamin A insufficiency and protein-energy malnutrition in newborns and preschoolers. Additionally, it was noted that the prepared complementary food contained less quantity of iron and zinc as well as fewer vitamins B, E, and C as well as minerals. Therefore, it is advised that meals that are complementary be fortified with a premix of vitamins and minerals.

It is recommended that RUTF is suitable not only for treating malnutrition in children, but they are also good for vulnerable populations (elderly people and lactating mothers) (Keyata *et al.*, 2023; Hendricks, 2010). The older adults need diet intervention to overcome age-related physical and physiological imbalance and risk of undernutrition. Payne *et al.* (2020) reported in his study that 14% of adults aged more than 65 years are at risk of nutrition-related diseases. Moreover, people recovering from severe infections and extended period of isolation at home are also vulnerable to malnutrition.

TABLE 10.1
Nutritional Composition of Ready-to-Eat Complementary Food

Product Name	Carbohydrate	Protein	Fat	Fiber
Chicken based				
Chickenrisotto and sweet corn tomato	9.4 g	2.8 g	2.1 g	1.2 g
Fruity chicken with rice	10.0 g	2.4 g	1.5 g	0.2 g
Organic vegetable and chicken hotpot	7.4 g	2.4 g	0.5 g	0.2 g
Autumn orchard chicken	10.6 g	3.1 g	2 g	1.2g
Fish based				
Spinach with salmon and parsnip	6.8 g	2.8 g	2.8 g	1.2 g
Fish pie with mash	6.2 g	5.2 g	5.6 g	1.9 g
Salmon risotto with sprinkle of cheese	8.9 g	2.9 g	1.8 g	0.7 g
Scrummy tuna penne	8.9 g	4.6 g	1.8 g	0.9 g

Source: Zand *et al.* (2012).

Undernutrition is a major factor in the disease's advancement and a common side effect of both human immunodeficiency virus (HIV) infection and acquired immunodeficiency syndrome (AIDS). Moreover, it leads to reduction in the levels of micronutrients required for both innate and acquired immunity to function properly, which increases the risk of morbidity and mortality, exacerbates the severity of the illness, and causes retardation in the recovery of HIV-infected children (Akombi *et al.*, 2017; Zegeye *et al.*, 2023). One way to improve the nutritional health of HIV/AIDS patients is through a community-based preventative food and nutrition intervention as part of an Outpatient Therapeutic Program that uses RUTF. Bahwere *et al.* (2011) monitored the effectiveness of locally made chickpea–sesame-based RUTF (CS-RUTF) in the cure of wasting in HIV-positive individuals in Malawi. They were given 500 g of CS-RUTF daily for 3–5 months. continuous weight gain (WG), static weight (SW), and continuation weight loss (WL). It has been observed that 54% of individuals detected for the continuous WG, 9.1% for the SW pattern and 37% for the continuation WL pattern. This study indicated the protective potential of CS-RUTF for the retrogression of wasting associated with HIV. In another investigation, about 59% of HIV-positive malnourished children demonstrated recovery after the intake of RUTF alone in the absence of highly active antiretroviral therapy in Dowa community-based therapeutic care program (Bahwere *et al.*, 2008).

Natural disasters not only lead to death tolls but also create post-disaster challenges of infectious diseases and poor mechanism of distribution of food. The most affected individuals are those who have compromised immunity like infants, elderly, pregnant and lactating women, and the sick and the injured people. There is need to design emergency RTE fortified foods, which can meet the nutritional requirement of this specific vulnerable group (Saggu *et al.*, 2023). Considering the cultural and traditional values, the RTE emergency meals based on local foods are prepared. Agustini *et al.* (2022) mentioned the formulation of emergency RTE formula of 10 days shelf life based on Balinese traditional recipe of glutinous rice flour and brown sugar mixed with green beans powder, cashews, peanuts, and Moringa leaf powder.

Many traditional herbs are used by lactating women to improve quality and quantity of milk. Fennel (*Foeniculum vulgare L.*) and anise (*Pimpinella anisum L.*) contain galactogogues which improve lactation. Hosseinzadeh *et al.* (2014) evaluated the effect of aqueous extracts of *P. anisum* seeds on milk production in rats and reported 68.1% increase. In a randomized controlled clinical trial on human population, a significant increase in milk volume was reported in intervention group (Khalili *et al.*, 2023) This indicates the potential of using these galactogogue herbs in RTE food. Tănase et al. (2022, 2023) prepared stem convection to prepare RTE purees of sweet potato and zucchini fortified with these herbs and reported improved nutritional and rheological properties.

10.2.3 Alternative Sources of RUTF

Despite RUTFs' potential as a therapy for SAM, 80% of SAM-affected children lacked access to them (Choudhury *et al.*, 2018). According to UNICEF (2020) if RUTF intervention reaches the affected group by year 2025, the number of children with stunted development will be reduced by 65 million, saving 3.7 million lives. RUTFs were failing to appropriately handle SAM for two main reasons: availability and accessibility. The primary barriers to expanding the availability of RUTFs are the high costs associated with production, shipping, and ingredients (Clark, 2020). The commonly used RUTF is peanut-based and costs USD 41–51 per kid. It would be prudent to create new alternative formulations that exclude milk and peanut butter.

Based on many investigations, RUTF products obtained with locally sourced ingredients can be an ideal option to resolve the cost and acceptability issues commonly observed with typical RUTF (Violette *et al.*, 2013). Four main ingredients are used in alternative RUTF formulations (Collins *et al.*, 2006; Collins and Henry, 2004). These include a cereal as the main ingredient, a protein source that can be derived from either an animal or vegetable source, an amino acid and vitamin supplement (derived from fruits, vegetables, or a combination of both), and an energetic supplement (e.g., lipids, oil, sugar, etc.). Customers have responded more favorably to RUTF made from locally sourced raw materials, because they are accustomed to the texture, flavor, and aroma of the raw materials (Wakhu-Wamunga and Wamunga, 2017). Crops that are readily accessible locally are used to create RUTF worldwide. Hassan *et al.* (2016) created a new RUTF based on chickpeas and evaluated its effectiveness and chemical makeup using an animal model. While the commercial RUTF "Plumpy'nut", which is based on peanuts, showed different levels of moisture (2.02%), ash (3.63%), crude protein (12.56%), crude fat (32.33%), fiber (2.15%), and nitrogen-free extract (49.33%), the developed pea-based RUTF contained 2.07% moisture, 3.51% ash, 12.42% crude protein, 31.33% crude fat, 3.17% crude fiber, and 50.63% nitrogen-free extract. It was observed that RUTF made of chickpeas were found to be similar with Plumpy'nut in terms of their protein quality and nutritional content. In Bangladesh, rice–lentil and chickpea-based RUTF are made with locally sourced raw ingredients. This RUTF mixture contained 5.1–6.0% protein, 14.8–15.9% total fat, 1.0–1.2% moisture, 267.6–24.9% carbohydrates, and 1.9–2.5% ash (Ahmad *et al.*, 2014). UNICEF created a fish-based wafer snack locally in Colombia to help SAM. Its production costs are 20% lower than those of the usual RUTF formulation, and children with SAM who use it report acceptable taste acceptance (Sigh *et al.*, 2018).

Apart from it, insects are rich in proteins and there is high scope of using insect-based RUTF for treating malnutrition. Akande *et al.* (2023) formulated a RUTF where milk powder was replaced by migratory locust powder (LP). The protein content was observed to be19.58–26.38%, and it also contained a remarkable amount of thiamine, niacin, vitamin D and E. This RUTF did meet the same quantity of amino acids which is given by Food and Agriculture Organization (FAO) 2011 for children aged 6 months to 3 years. Moreover, it also demonstrated antioxidant potential. Apart from this, *Zonocerus variegatus* (variegated grasshopper) is widely consumed by people in Northern Nigeria and is a rich source of protein and vitamin B. Olagunju *et al.* (2023) prepared an RUSF based on local ingredients: millets, soybean, and variegated grasshopper in a ratio of 70:20:10. The insects were boiled in water to remove flavors and then oven-dried at 60°C for 12 hours. The milled insect powder was added to millets and soybean flour and the prepared formulation in the form of pellets was fed to weaning albino rats in a controlled experimental design. The acclimatized albino rats were induced with acute malnutrition and after feeding with prepared therapeutic pellets, weight, tail length, and hematological parameters were recorded. The study showed efficacy of the said formulation in alleviating malnutrition.

10.3 READY-TO-EAT FUNCTIONAL FOOD

There is increasing demand of nutritious functional food rich in bioactive components that have the ability to promote health and lower the risk of diseases. So the focus of RTE food industry is substantially inclined toward development of novel functional RTE products (Figure 10.1).

10.3.1 Ready-to-Drink Beverages

Tea is one of the most consumed beverages worldwide. Besides having attractive color and aroma, it is rich in antioxidants. Ready-to-drink tea (RDT) is a growing nonalcoholic beverage industry based on black or green tea and is consumed either hot or cold. Considering the increasing trend of consuming tea as a healthy beverage, the RDT tea market size is expected to reach 38.96 billion USD by 2027 (Kale and Deshmukh, 2020; Dubey *et al.*, 2020). There is wide exploration of technological advances to prepare novel products based on healthier options of RDT tea. The RDT formulation containing 2% (w/v) rosella leaves and 0.3% (w/v) stevia leaves was considered safe for diabetic people as rosella reduces blood glucose and stevia provides sweetness with zero calories (Saraswati and Taniwiryono, 2020). RDT with riceberry rice tea and mulberry leaf green tea infusion is rich in anthocyanin and gets acceptable sensory acceptance (Ngamdee, 2022). Faustina *et al.* (2022)

prepared an RDT by incubating tea leaves with exogenous enzymes: tyrosinase and beta glucosidase which enhances the color and aroma of the product. Ready-to-drink (RTD) iced tea prepared from microencapsulated extracts of spearmint and peppermint are rich in phenolic compounds rosmarinic and eriocitrin, respectively (Alaşalvar and Cam, 2020). Kombucha tea is a fermented beverage prepared from tea, sugar syrup, and bacteria and yeast starter. Tanner *et al.* (2022) prepared an RTD kambucha tea with the addition of sea grapes. The formulation was rich in ascorbic acid and was able to increase liver superoxide dismutase enzyme in mice fed on cholesterol and a fat-rich diet. An antioxidant-rich ready-to-drink beverage was prepared from snake fruit to which flower extracts of roselle flower and butterfly pea flower were added (Purnawijayanti and Nai, 2020). The fermented milk-based beverages prepared with pearl millets have high nutritional value and longer shelf life. Beverage prepared with a combination of pearl millets with whey milk or buttermilk or skimmed milk sweetened with sugar date powder and fermented with yogurt starter culture has low glycemic index (GI) and glycemic load (Alwohaibi *et al.*, 2023).

10.3.2 Ready-to-Eat Grain-Based Food

The intake of refined form of commonly used cereals like wheat and rice leads to deficiency of micronutrients and dietary fiber, so alternative food grains are used in preparation of RTE cereal-based food. Minor grains like millets are rich in bioactive compounds, dietary fiber, micronutrients, more resistant starch, and low GI (Tripathi *et al.*, 2021). So, they are used in the preparation of a wide range of RTE products like cakes, pasta, crackers, and snacks. Akeem *et al.* (2023) reported the crackers made from the sesame seed (defatted), germinated pearl millet, and tigernut composite flour (defatted), and compared them with the wheat crackers. The developed crackers exhibited higher concentrations of protein, ash, fiber, calcium, and energy (376–381 kcal/100 g) in contrast to the wheat crackers. In addition, *in vitro* protein digestibility was found to be more in developed crackers than wheat crackers. Thus, composite flours made from germinated pearl millet, defatted sesame seed, and defatted tigernut flours can be used as an alternative option for the preparation of nutritious and healthful crackers. Similarly, RTE breakfast cereal made from popped pearl millet showed a marked amount of proteins, dietary fibers, folic acid, and minerals such as calcium, phosphorus, and iron (Kumari *et al.*, 2019). Likewise Shaivya and Sunita (2016) also prepared the pearl millet upma mixes which exhibited 13% protein, 53% carbohydrate, 23% fat, calcium (24 mg), and ascorbic acid (20 mg). They also developed quinoa upma mixes and proximate composition analysis indicated 13% protein, 53.49% carbohydrate, 23.17% fat, calcium (27 mg), and ascorbic acid (15 mg). These RTE products are highly recommended for preventing the neurodegenerative and cardiac diseases. In addition to it, Kahlon *et al.* (2016) reported the gluten-free snacks made from Quinoa which can be a healthy alternative for the people who are allergic to gluten.

Apart from this, grain amaranth is a reservoir of bioactive components, and it possesses several medicinal properties. This underutilized crop can be used for the production of value-added RTE products and it has huge potential to improve the food security for millions of rural and impoverished urban households in developing nations, where malnutrition has serious health effects (Aderibigbe *et al.*, 2022). Singh *et al.* (2022) reported the functional snack bar which are prepared from the different combination of oats, amaranth grain, and banana peel powder. Out of all combinations, amaranth grain (60%), oats (25%), and banana peel powder (15%) were observed to be healthier in comparison to the other combinations and exhibited better organoleptic and nutritional properties. These snacks contained a significant amount of protein, essential amino acid, minerals, dietary fiber, phenols, and demonstrated marked antioxidant activity. Thermogravimetric analysis showed the presence of active constituents even at higher temperature which is an added property to its functional value. These snacks showed higher acceptability by the consumers and microbial analysis also showed that these snacks can be stored for 60 days without any microbial infection. Similarly, RTE breakfast cereal was prepared from popped pearl millets and amaranth grains, and groundnut powder. This RTE product is highly nutritious and contained a significant amount of protein. Moreover, it is a healthy option for the elderly and celiac patients. This RTE breakfast cereal contained significant quantities of zinc and iron, which makes it an ideal food product for lactating mothers and anemic patients (Mishra, 2022). In another investigation, snacks were prepared from amaranth, soycake, and shallot. These snacks exhibited low GI and contained significant amounts of minerals and bioactive components. They also showed remarkable antioxidant property and can be used to alleviate malnutrition, particularly in children and hyperglycemia in adults (Olagunju *et al.*, 2022). Nkesiga *et al.* (2022) prepared baby food of 4–6 months of shelf life for children under 5 years from orange fleshed sweet potato, amaranth grains, and soybean flour through extrusion cooking and analyze it for its nutritional efficacy. The nutritional analysis indicates high energy value and enrichment with vitamin A and micronutrients with reduced value of antinutrients. Similarly, intake of RTE oat cereal in obese adults showed significant reduction in low-density lipoprotein (LDL) cholesterol and total cholesterol in comparison to the adults who take low fiber food (Maki *et al.*, 2010). Furthermore, Yaman *et al.* (2019) demonstrated that GI in RTE breakfast with oats were found to be lower in contrast to those comprising rice and corn. It has been seen that more glycemic food taken in morning can result in insulin insufficiency, higher blood glucose levels, and augmented fattiness. Thus, oats can be utilized for the production of value added RTE food products with low GI.

10.3.3 Plant-Based Dairy Products

People now prioritize plant-based diet over animal-based diet including dairy products as the former is considered as healthier and an environment sustainable option. The market for dairy alternatives is expected to rise to 25.1 billion USD in 2026, which is almost double in comparison to that in 2018. Plant-based beverages include nuts (almonds, cashews, hazelnuts), coconut, grains (rice, millets) seeds (hemp, chia, flaxseed), and legume (pea, soy). To make them comparable to milk in nutrients, the products are fortified with calcium, vitamin D, and vitamin B12 (Craig *et al.*, 2023). Plant-based milks are lactose-free substitutes to dairy milk. Łopusiewicz et al. (2020) monitored the effectiveness of flaxseed oil cake plant milk to form a yogurt-like beverage with the use of yogurt lactic acid bacteria (LAB). As a prebiotic, inulin was introduced, which caused increment in the levels of total polyphenols, and enhanced the viability of LAB. Therefore, these beverages serve as an ideal option as a novel, nondairy product for the intake of LAB consumption for vegans and lactose-intolerant people. Similarly, a ready-to-drink soy milk beverage with tea aroma showed a remarkable amount of total flavonoids, total phenols, and antioxidant property (Table 10.2). Moreover, it also contained significant quantities of ash, fibers, proteins, and carbohydrates (Susanna *et al.*, 2023). The GI of several plant-based dairy alternatives is lesser than that of cow's milk. It has been detected that almond milk and unsweetened soy milk exhibited reduced GI than cow's milk; thus it is useful for diabetic people for regulating the concentration of sugar in their blood (Bouchenak and Lamri-Senhadji, 2013; Khaliq *et al.*, 2023). Dairy alternatives made from soy such as soy milk and tofu contain isoflavones and soy protein and these can substitute animal protein in the diet and can cause remarkable decline in the concentration of LDL cholesterol. A meta-analysis of 46 studies that the Food and Drug Administration (FDA) deemed to be relevant demonstrated that intake of soy protein through plant-based dairy analogs caused a significant reduction in LDL and total cholesterol in adults (Jenkins *et al.*, 2019; Khaliq *et al.*, 2023). Moreover, RTE products made from oats like oatmeal and oat milk also revealed a marked decline in the concentration of LDL cholesterol in hypercholesterolemic adults (Thongoun *et al.*, 2013). Several plant-based dairy substitutes like fortified soy and almond milk contain significant amounts of calcium and vitamin D, which assist in osteogenesis in the developing fetus in pregnant women (Kovacs, 2011). Additionally, Sandhosh *et al.* (2023) reported the formulation of ice cream made from coconut milk and coconut cream as a

TABLE 10.2
Various Plant-Based Dairy Substitutes with Their Active Constituents and Health Benefits

S. No.	Plant-Based Dairy Substitutes	Active Constituents	Health Benefits	References
1	Flaxseed	Secoisolariciresinol diglucoside Omega-3 fatty acid	Neuroprotective Cardioprotective Enhances osteogenesis and immunity	Mueed *et al.*, 2022
2	Walnut	Chlorogenic acid Catechins	Antioxidant property, decreases glycemic index, activity, neuroprotective	Lee *et al.*, 2016 Sánchez-González *et al.*, 2017
3	Hazelnut	Tocopherols Flavonols	Cardioprotective and antioxidant property	Di Nunzio, 2019
4	Peanut	Niacin Phenolics	Immunomodulatory Cardioprotective Antioxidant	Arya *et al.*, 2016
5	Sunflower	Luteolin Quercetin	Anti-inflammatory and maintains blood pressure and cholesterol.	Kulczyk *et al.*, 2023
6	Hemp	γ-tocopherol Linolenic acid	Decreases clotting in blood Neuroprotective	Crescente *et al.*, 2018
7	Soybean	Daidzein Genistein	Antitumor activity	Wang *et al.*, 2023
8	Almond	Tocopherol Arabinose	Preserves gut microbiota and exhibits antioxidant properties	Zahrani and Shori, 2023
9	Oat	Avenanthramides β-glucans	Decreases cholesterol, antioxidant and anti-inflammatory activity	Sang and Chu, 2017
10	Coconut	Gallic acid Phenolics	Neuroprotective Anti-fatigue	Halim *et al.*, 2023
11	Sesame	Sesamin Sesamolin	Decreases cholesterol, antitumor and antiviral activity	Dossou *et al.*, 2023
12	Chickpea	Carotenoids Flavones	Antidiabetic, cardioprotective anti-inflammatory, and antitumor activity	Kaur and Prasad, 2021

substitute for milk. Proximate composition analysis indicated carbohydrate (14%), moisture (76%), fiber (0.16%), fat (8%), and protein (1.8%). These nondairy products are superior for the people who are suffering from lactose intolerance.

Plant-based formulations have the limitation of digestibility of proteins and absorption of minerals due to the presence of antinutrients. The raffinose family of oligosaccharide present in legume can induce abdominal flatulence as they are digested in colon rather than small intestine, leading to poor tolerance of legume-based RUTF. However, extrusion cooking and germination reduce the flatulence factors by 47–60% and 72–80%, respectively (Enn, 2016).

10.3.4 RTE Meat and Meat Analogue Products

The demand for animal protein is increasing along with growing world population. However, intake of red meat in ultra-processed meat products increases the risk of cancer and cardiovascular ailments. It also generates negative impact on environment and creates concern for animal welfare. The preparation of meat formulations with plant-based ingredients can lower the cost and can make it functional food rich in plant-based bioactive compounds. The consumers who wish to adopt vegan diet are interested in meat analogs with flavor, texture, and taste similar to meat products and also fortified with nutraceuticals with health-promoting features. Gluten and soya are commonly used in meat analogs; however, Szpicer *et al.* (2022) prepared formulation of gluten and soy-free, antioxidant-rich meat analogous with microencapsulated anthocyanin-derived from black chokeberry.

Mazumder *et al.* (2023) use gray oyster mushroom and chickpea flour as an alternative to soy protein to prepare emulsion-type sausages. The novel innovative ideas of preparing meat analogs prepared from seitan and soya enriched with insect flour derived from tenebrio molitor is reported by Ardila *et al.* (2023). The bright red color, taste, and texture of meat products get affected by oxidation of lipids and proteins. To prevent this, the addition of natural antioxidants leads to healthier options in RTE meat products free from toxic chemicals. The addition of 1% *Solanum lycopersicum* in preparation of pork and beef meatballs stabilizes the color and flavor (Wojtasik-Kalinowska *et al.*, 2022). The addition of 250 mg of olive extract per kilogram of meat prevents the lipid oxidation in beef patties by 15 days (Gómez *et al.*, 2020).

10.3.5 RTE-Based on Fruits and Vegetables

There are several RTE items made from fruits and vegetables, including juices, jams, jellies, marmalades, canned goods, prepared salads, etc. Because of their distinctive flavor and texture, some fruits, like blueberries and strawberries, are often consumed; however, due to their perishable nature which results in nutritional loss, they sustain severe harm throughout the processes of harvesting, storage, and transportation. Several studies reported the edible coatings prepared with chitosan and algae (*Palmaria palmata*) on RTE strawberries prevent the nutrient loss in strawberries and enhance their shelf life (Rico *et al.*, 2019). Moreover, this coating also enhanced the levels of anthocyanin and vitamin C in the product. In another investigation, formulation of RTE fruit flakes was done using dried apple, papaya, and banana. Nutritional profile of fruit flakes revealed that these flakes contained a remarkable amount of carbohydrates (73.74%) and proteins (4.40%). It was observed that fat content in flakes were found to be 1.98%. Moreover, the proportion of calcium and iron is detected to be 23–24%. In addition, RTE fruit flakes were also observed to be microbiologically safe (Philip, 2021). Apart from this, the hepatoprotective effect of phenolic-rich extract of *Rhodomyrtus tomentosa* fruit was determined on mice with nonalcoholic fatty liver disease. RTE fruit extract exhibited reduction in liver function markers, oxidative stress indices, levels of lipid profile markers, and inflammatory cytokines (Wang *et al.*, 2022). The addition of indigenous vegetables in the form of dry powder or aqueous slurry makes bread a polyphenol-rich baked product. The textural property of bread gives it an organoleptic adaptability and its fortification makes it nutritious. Appropriate intake of vegetables in diet checks the deficiency of micronutrients. As per the FAO and WHO guidelines, the minimum per day intake of vegetables required is 400 g (Famuwagun *et al.*, 2023).

Jams are consumed as RTE food throughout the world. They are prepared from pulp of variety of fruits boiled with sugar and pectin to form a firm consistency. Pectin also has cholesterol and blood sugar-lowering properties (Baker, 1994). Fiber-rich jam with low GI is of consumer interest who is prediabetic. Younis *et al.* (2015) prepared papaya jam with the addition of salt and sodium bicarbonate-treated mosambi peels. Papaya is rich in vitamin A and mosambi peels are rich in polyphenol, which have antioxidant properties. The mosambi peel being rich in pectin has high water holding capacity, which prevents syneresis of jam and provides it firmness. The untreated mosambi peel imparts bitterness to jam, but the treated mosambi peel in concentration up to 5% was accepted in sensory evaluation. This formulation is not only nutritious but also a sustainable way of avoiding food loss as peels are the major by-products during extraction of juice from citrus fruits (Licandro and Odio, 2002).

Famuwagun *et al.* (2023) demonstrated that MicroVeg initiative devised novel systems of production, use, and value addition possibilities to improve nutrition and economic empowerment of the resource-poor rural people in Africa, particularly women. The value addition and food fortification studies were specifically focused on three vegetables—local amaranth, fluted pumpkin, and African eggplant—which are widely consumed in the West Africa subregion. They are paying close attention to the recent surge in the production of enriched bakery goods and pastries, notably in the optimization of nutritional qualities. Thus, bread was started to fortify or enrich bread with vegetable-based ingredients to increase its nutritional value. Each of the dry leafy vegetables was added to the bread in amounts of 1%, 3%, and 5%. Maize ogi was additionally fortified at 1%, 3%, and 5% with the wet equivalent of dried pumpkin, while pastry was fortified at 1%, 3%, and 5% with the only extract derived from amaranth leafy vegetable. It was observed that bread and pastry formulation with 3% addition of dry vegetable leaf powders contained more quantity of nutrients and polyphenols with marked antioxidant activity and can improve mother and child nutrition. Consumer acceptance of the flavor and color of pastry formulations with 3% dry vegetable leaf powders was somewhat lower, but because of the possible health advantages, people were nevertheless prepared to purchase the vegetable-fortified items in place of the standard ones.

10.3.6 Functional Foods Based on Food Industry by Products

Fish skin is the by-product of fish food processing industry and being rich in collagen, and is beneficial for skin, blood circulation, and bones. Wang *et al.* (2022) prepared puffed snack from skin of tilapia which when seasoned with seaweeds and tomato had good sensory acceptance. Avocado seeds are the waste product generated from avocado during production of avocado oil. They have low lipid but high starch content. They contained 69.1% carbohydrates and 19.5% dietary fiber, which makes it suitable raw material for preparation of extruded snacks (Table 10.3). The challenge in using avocado seeds is safety issue due to presence of toxin persin in oil cells of fruit and leaves. Moreover, seeds contain amygdalin, a cyanogenic glycoside which when metabolized by plant enzymes produce cyanohydrins. The cyanohydrins decompose to toxic hydrogen cyanide (Bolarinwa *et al.*, 2015; Permal *et al.*, 2023). RTE snacks were prepared from avocado seeds by extrusion cooking in friction cooker and found high antioxidant capacity and total phenolic content than that prepared from brown rice or malted barley. Moreover, the prepared snacks have amygdalin and persin at nontoxic levels (Permal *et al.*, 2023). A formulation of bovine hamburger enriched with 3% pumpkin peel flour prepared by Hartmann *et al.* (2020) had improved nutrition and showed increase in cooking yield, retention of moisture, and reduction of shrinkage and with acceptability index more than 70%. Varghese *et al.* (2022) prepared high-energy cookies with a unique formulation of jack fruit seed powder, finger millet powder, Moringa powder, Xanthum gum, vitamin and mineral premix, and tested it on Sprague–Dawley rats. It was found that this product has good protein quality with better digestibility and can be used for treating malnutrition in adolescents.

10.3.7 Utilization of Natural Compounds as Preservative and Flavoring Agent in RTE

Food preservation and safety is another challenge in the production of RTE. Various preservatives and flavoring agents have been utilized for the enhancement of shelf life and sensory quality of RTE food products. Consumers today seek for food that is fresher, safer, and healthier stuff and is free of artificial preservatives. Thus, phytoconstituents or edible coatings from natural resources constitute a promising substitute for the enhancement of sensory attributes in RTE food without any mixing of chemical preservatives. Vanillin is the active component of vanilla beans and is utilized as a

TABLE 10.3
Functional Foods Based on Food Industry By-Products

By-Products From Food Industry	RTE Products	Summary of Findings	References
Mango peel	Biscuits	Dietary fiber content augmented (21%) More polyphenols (4.50 mg/g) and carotenoid (247 μg/g) More antioxidant activity	Lesmana *et al.*, 2021
Mango peel	Sponge cake	High dietary fiber with low fat, calorie	Noor Aziah *et al.*, 2011
Prickly pear peel	Biscuits	More concentration of fiber (20.70 g/100 g), and phenols (2776 mg/100 g)	Bouazizi *et al.*, 2020
Pineapple peels	Jam, and canned juice	Protein content (14.9%)	Aruna, 2019
Peel from Nendran	Cookies	More total dietary fiber content Phenolic content was observed to be 5.28 mg GAE in cookie made from Nendran peel Higher antioxidant activity	Arun *et al.*, 2015
Almond skins	Biscuits	Enhancement in phenolic content and antioxidant activity	Pasqualone *et al.*, 2020
Mango peel and decocted roselle calyces	Tortilla chips	Lower *in vivo* glycemic index Higher phenolic content and ABTS (2,2′-azino-bis-(3-ethylbenzothiazoline-6-sulfonic) acid) radical scavenging activity	Mayo-Mayo *et al.*, 2020
Mango peel	Bread	Less *in vivo* glycemic index	Chen *et al.*, 2019
Potato peel	Fries, Chips	Antiobesity effect	Elkahoui *et al.*, 2018
Yellow passion fruit peel and flour	Cookies	More content of fiber, protein, and ash	Andrade *et al.*, 2012
Avocado seeds	Brown rice and malted barley RTE snacks	Displays antioxidant property due to the presence of phenols	Permal *et al.*, 2023 Siol and Sadowska, 2023

flavoring agent in food industry. It also exhibited antimycotic and bacteriostatic properties (Cassani *et al.*, 2016). The sensory qualities of fruit purees with varying vanillin concentrations were assessed by Cerrutti and Alzamora (1996), who revealed that the purees retained the characteristic fruit flavor after the administration of vanillin. Additionally, Yousaf *et al.* (2010) reported that supplementation of banana juice supplemented with inulin and oligofructose remained acceptable and were ranked the best for all sensory metrics.

10.4 CONCLUSION

Change in lifestyle of consumers demands for convenient RTE meals of longer shelf life. The aware consumer demands for nutrient-dense healthier options of RTE in place of ultra-processed products and that too with good sensory attributes. RTE food with functional ingredients catering to the specific needs of the consumers is of interest which can help them to achieve their personal health goals. Moreover, there is increased concern for sustainable food system as well to minimize the adverse impact of food industry preparing RTE food products. There is need to explore technologies for preparation of minimally processed RTE food formulations based on both locally available, environmentally sustainable cheaper raw ingredients and waste generated from food processing industry. The production of novel RUTF foods is a significant step that may serve as a viable stand-in for an ideal approach to solve the global problem of hunger and malnutrition. Attempts to develop novel

RTE foods with the potential of disease prevention and management will cater to the health needs of specific consumer target.

REFERENCES

Aderibigbe, O. R., O. O. Ezekiel, S. O. Owolade, J. K. Korese, B. Sturm, and O. Hensel. "Exploring the potentials of underutilized grain amaranth (*Amaranthus* spp.) along the value chain for food and nutrition security: A review." Critical Reviews in Food Science and Nutrition 62, no. 3 (2022): 656–669.

Agustini, Ni Putu, I. Gusti Putu Sudita Puryana, and I. Komang Agusjaya Mataram. "Modification of traditional balinese food as disaster emergency food." Jurnal Nutrisia 24, no. 1 (2022): 40–47.

Ahmad, Hafiz Bashir, Muhammad Amir Amin, I. Hussian, C. M. Rafique, M. Naveed, M. A. Awais, M. Shafiq, and M. Aqeel. "Effect of different sowing dates on yield contributing traits of Mash bean (*Vigna mungo* L.)." International Journal of Agronomy and Agriculture Research 5, no. 6 (2014): 42–48.

Akande, Olamide Abigael, Matthew Oluwamukomi, Oluwatooyin Faramade Osundahunsi, Oluwole Steve Ijarotimi, and Ivan Muzira Mukisa. "Evaluating the potential for utilising migratory locust powder (*Locusta migratoria*) as an alternative protein source in peanut-based ready-to-use therapeutic foods." Food Science and Technology International 29, no. 3 (2023): 204–216.

Akeem, Sarafa Adeyemi, Balikis Oluwakemi Mustapha, Ridwan Opeyemi Ayinla, Oluwaseun Ajibola, Winnifred Oluwaseyi Johnson, and Olaide Akinwunmi Akintayo. "Physical characteristics, nutritional composition and acceptability of gluten-free crackers produced from germinated pearl millet (*Pennisetum glaucum*), defatted-sesame seed (*Sesamum indicum*) and defatted-tigernut (*Cyperus esculentus*) composite flours." Discover Food 3, no. 1 (2023): 22.

Akombi, Blessing J., Kingsley E. Agho, Dafna Merom, Andre M. Renzaho, and John J. Hall. "Child malnutrition in sub-Saharan Africa: A meta-analysis of demographic and health surveys (2006–2016)." PLoS One 12, no. 5 (2017): e0177338.

Alaşalvar, Hamza, and Mustafa Çam. "Ready to drink iced teas from microencapsulated spearmint (*Mentha spicata* L.) and peppermint (*Mentha piperita* L.) extracts: physicochemical, bioactive and sensory characterization." Journal of Food Measurement and Characterization 14, no. 3 (2020): 1366–1375.

Alwohaibi, Azzah A. A., Asmahan A. Ali, Sally S. Sakr, Isam A. Mohamed Ahmed, Raghad M. Alhomaid, Khalid A. Alsaleem, Mohammed Aladhadh, Hassan Barakat, and Mohamed F. Y. Hassan. "Valorization of different dairy by-products to produce a functional Dairy–Millet Beverage Fermented with *Lactobacillus paracasei* as an adjunct culture." Fermentation 9, no. 11 (2023): 927.

Andrade, J. K. S., L. C. de O. Barretto, M. Denadai, N. Narain, and J. A. B. dos Santos. "Harnessing passion fruit peel flour (*Passiflora edulis* f. Flavicarpa) for the preparation of stuffed cookies." In ISHS Acta Horticulturae 1198: III International Symposium on Medicinal and Nutraceutical Plants and III Conference of National Institute of Science and Technology for Tropical Fruits (2012): 205–212.

Ardila, Paula, Adrián Honrado, Pedro Marquina, José Antonio Beltrán, and Juan B. Calanche. "Innovative plant-based burger enriched with Tenebrio molitor meal: Characterization and Shelf-Life." Foods 12, no. 18 (2023): 3460.

Arun, K. B., Florence Persia, P. S. Aswathy, Janu Chandran, M. S. Sajeev, P. Jayamurthy, and P. Nisha. "Plantain peel—a potential source of antioxidant dietary fibre for developing functional cookies." Journal of Food Science and Technology 52 (2015): 6355–6364.

Aruna, T. E. "Production of value-added product from pineapple peels using solid state fermentation." Innovative Food Science & Emerging Technologies 57 (2019): 102193.

Arya, Shalini S., Akshata R. Salve, and Salve Chauhan. "Peanuts as functional food: A review." Journal of Food Science and Technology 53 (2016): 31–41.

Bahwere, Paluku, Ellen Piwoz, Marthias C. Joshua, Kate Sadler, Caroline H. Grobler-Tanner, Saul Guerrero, and Steve Collins. "Uptake of HIV testing and outcomes within a Community-based Therapeutic Care (CTC) programme to treat severe acute malnutrition in Malawi: A descriptive study." BMC Infectious Diseases 8, no. 1 (2008): 1–13.

Bahwere, Paluku, Hedwig Deconinck, Theresa Banda, and Steve Collins. "Effective therapeutic feeding with chickpea sesame based ready-to-use therapeutic food (CS-RUTF) in wasted adults with confirmed or suspected AIDS." World Journal of AIDS 1, no. 4 (2011): 169–181.

Baker, Robert A. "Potential dietary benefits of citrus pectin and fiber." Food Technology (Chicago) 48, no. 11 (1994): 133–139.

Bhandari, Nita, Sanjana Brahmawar Mohan, Anuradha Bose, Sharad D. Iyengar, Sunita Taneja, Sarmila Mazumder, Ruby Angeline Pricilla et al. "Efficacy of three feeding regimens for home-based management of children with uncomplicated severe acute malnutrition: A randomised trial in India." BMJ Global Health 1, no. 4 (2016): e000144.

Bolarinwa, Islamiyat F., Caroline Orfila, and Michael R. A. Morgan. "Determination of amygdalin in apple seeds, fresh apples and processed apple juices." Food Chemistry 170 (2015): 437–442.

Bouazizi, Souhir, Giuseppe Montevecchi, Andrea Antonelli, and Moktar Hamdi. "Effects of prickly pear (*Opuntia ficus-indica* L.) peel flour as an innovative ingredient in biscuits formulation." LWT 124 (2020): 109155.

Bouchenak, Malika, and Myriem Lamri-Senhadji. "Nutritional quality of legumes, and their role in cardiometabolic risk prevention: a review." Journal of Medicinal Food 16, no. 3 (2013): 185–198.

Cassani, L., B. Tomadoni, G. Viacava, A. Ponce, and Maria del Rosario Moreira. "Enhancing quality attributes of fiber-enriched strawberry juice by application of vanillin or geraniol." LWT-Food Science and Technology 72 (2016): 90–98.

Cazes, Cécile, Kevin Phelan, Victoire Hubert, Harouna Boubacar, Liévin Izie Bozama, Gilbert Tshibangu Sakubu, Bruno Bindamba Senge et al. "Optimising the dosage of ready-to-use therapeutic food in children with uncomplicated severe acute malnutrition in the Democratic Republic of the Congo: A non-inferiority, randomised controlled trial." EClinicalMedicine 58 (2023): 1–15.

Cerrutti, Patricia, and Stella M. Alzamora. "Inhibitory effects of vanillin on some food spoilage yeasts in laboratory media and fruit purees." International Journal of Food Microbiology 29, no. 2–3 (1996): 379–386.

Chen, Yan, Lei Zhao, Ting He, Zhuoshen Ou, Zhuoyan Hu, and Kai Wang. "Effects of mango peel powder on starch digestion and quality characteristics of bread." International Journal of Biological Macromolecules 140 (2019): 647–652.

Choudhury, Nuzhat, Tahmeed Ahmed, Md Iqbal Hossain, M. Munirul Islam, Shafiqul A. Sarker, Mamane Zeilani, and John David Clemens. "Ready-to-use therapeutic food made from locally available food ingredients is well accepted by children having severe acute malnutrition in Bangladesh." Food and Nutrition Bulletin 39, no. 1 (2018): 116–126.

Clark, Lisa F. "Are innovative ready to use therapeutic foods more effective, accessible and cost-efficient than conventional formulations? A review." Outlook on Agriculture 49, no. 4 (2020): 267–277.

Collins, S., and J. Henry. "Alternative RUTF formulations." Community-based Therapeutic Care (CTC). Emergency Nutrition Network Supplement 2 (2004): 35–37.

Collins, Steve, Nicky Dent, Paul Binns, Paluku Bahwere, Kate Sadler, and Alistair Hallam. "Management of severe acute malnutrition in children." The Lancet 368, no. 9551 (2006): 1992–2000.

Craig, Winston J., Virginia Messina, Ian Rowland, Angelina Frankowska, Jane Bradbury, Sergiy Smetana, and Elphee Medici. "Plant-based dairy alternatives contribute to a healthy and sustainable diet." Nutrients 15, no. 15 (2023): 3393.

Crescente, Giuseppina, Simona Piccolella, Assunta Esposito, Monica Scognamiglio, Antonio Fiorentino, and Severina Pacifico. "Chemical composition and nutraceutical properties of hempseed: An ancient food with actual functional value." Phytochemistry Reviews 17 (2018): 733–749.

De Pee, Saskia, and Martin W. Bloem. "Current and potential role of specially formulated foods and food supplements for preventing malnutrition among 6-to 23-month-old children and for treating moderate malnutrition among 6-to 59-month-old children." Food and Nutrition Bulletin 30, no. 3_suppl3 (2009): S434–S463.

Di Nunzio, Mattia. "Hazelnuts as source of bioactive compounds and health value underestimated food." Current Research in Nutrition and Food Science Journal 7, no. 1 (2019): 17–28.

Dossou, Senouwa Segla Koffi, Fang-tao Xu, Komivi Dossa, Z. H. O. U. Rong, Ying-zhong Zhao, and Lin-hai Wang. "Antioxidant lignans sesamin and sesamolin in sesame (*Sesamum indicum* L.): A comprehensive review and future prospects." Journal of Integrative Agriculture 22, no. 1 (2023): 14–30.

Dubey, Kriti Kumari, Madhura Janve, Aratrika Ray, and Rekha S. Singhal. "Ready-to-drink tea." Trends in Non-alcoholic Beverages (2020): 101–140.

Elkahoui, Salem, Glenn E. Bartley, Wallace H. Yokoyama, and Mendel Friedman. "Dietary supplementation of potato peel powders prepared from conventional and organic russet and non-organic gold and red potatoes reduces weight gain in mice on a high-fat diet." Journal of Agricultural and Food Chemistry 66, no. 24 (2018): 6064–6072.

Elviss, N. C., and F. Jørgensen. "Investigations prompted by the discovery of high levels of Salmonella in fresh herbs in the UK." In Case Studies in Food Safety and Authenticity, pp. 116–122. Woodhead Publishing, 2012.

Enn, Emergency Nutrition Network. "Developing food supplements for moderately malnourished children: Lessons learned from RUTF." Field Exchange 50 (2016): 25.

Famuwagun, Albert A., Odunayo C. Adebooye, Tunji V. Odunlade, Kehinde A. Taiwo, Durodoluwa J. Oyedele, and Rotimi E. Aluko. "Potential of indigenous vegetable-fortified food products for improved human nutrition and health in West Africa." In *Food Security Challenges and Approaches*. IntechOpen, 2023.

Faustina, Dea Risfika, Rachmad Gunadi, Aprilia Fitriani, and Supriyadi Supriyadi. "Alteration of phenolic and volatile compounds of tea leaf extract by Tyrosinase and β-Glucosidase during preparation of ready-to-drink tea on farm." International Journal of Food Science 2022, (2022): 1–8.

Golden, Michael H. "Proposed recommended nutrient densities for moderately malnourished children." Food and Nutrition Bulletin 30, no. 3_suppl3 (2009): S267–S342.

Gómez, Inmaculada, Celia García-Varona, María Curiel-Fernández, and Miriam Ortega-Heras. "Effects of an extract from olive fruits on the physicochemical properties, lipid oxidation and volatile compounds of beef patties." Foods 9, no. 12 (2020): 1728.

Habte, Kifle, Meseret Azene, Yohannes Chanyalew, Samson Girma, Chala Bashea, Asrat Yehualshet, Getamesay Behailu, Abiy Abebe, and Masresha Tessema. "Nutrient density and microbial safety of open-air-dried beef meat and its biochemical and organ histopathology effects in albino rats: A promising ingredient for complementary food formulation." International Journal of Food Science 2023, (2023): 1–11.

Halim, Hani Hafeeza, Mohd Sabri Pak Dek, Azizah Abdul Hamid, Nazamid Saari, Mohd Izwan Mohd Lazim, Faridah Abas, Ahmad Ngalim, Amin Ismail, and Ahmad Haniff Jaafar. "Novel sources of bioactive compounds in coconut (*Cocos nucifera* L.) water from different maturity levels and varieties as potent skin anti-aging strategies and anti-fatigue agents." Food Bioscience 2023, 51: 102326.

Hartmann, Giovanna Luiza, Allana Mariny Marconato, M. M. Romeiro Santos, L. A. Do Amaral, E. F. Dos Santos, and D. Novello. "Addition of pumpkin peel flour affect physicochemical and sensory characteristics of bovine burger." International Journal of Regulation and Governance 8, no. 2 (2020): 254–263.

Hassan, J. U., S. Iqbal, N. Naeem, Frasat Nasir, and M. Nasir. "Development, efficacy and comparative analysis of novel chickpea based ready-to-use therapeutic food." Journal of Animal & Plant Sciences 26, no. 6 (2016): 1843.

Hendricks, Kristy M. "Ready-to-use therapeutic food for prevention of childhood undernutrition." Nutrition Reviews 68, no. 7 (2010): 429–435.

Hosseinzadeh, Hossein, Mohsen Tafaghodi, Shirin Abedzadeh, and Elahe Taghiabadi. "Effect of aqueous and ethanolic extracts of *Pimpinella anisum* L. seeds on milk production in rats." Journal of Acupuncture and Meridian Studies 7, no. 4 (2014): 211–216.

Huybregts, Lieven, Freddy Houngbé, Cécile Salpéteur, Rebecca Brown, Dominique Roberfroid, Myriam Ait-Aissa, and Patrick Kolsteren. "The effect of adding ready-to-use supplementary food to a general food distribution on child nutritional status and morbidity: a cluster-randomized controlled trial." PLOS Medicine 9 (2012): e1001313.

Ijeomah, O. C., E. S. Ukpong, and C. Azubuike. "Chemical Composition of Fonio (Digitaria-Exilis)/Ricebean (Vigna Umbellata) Based Complementary Food Fortified With Carrot and Crayfish." International Journal of Agriculture and Biological Sciences (2023): 47–53.

Isanaka, Sheila, Thomas Roederer, Ali Djibo, Francisco J. Luquero, Nohelly Nombela, Philippe J. Guerin, and Rebecca F. Grais. "Reducing wasting in young children with preventive supplementation: a cohort study in Niger." Pediatrics 126, no. 2 (2010): e442–e450.

Jenkins, David J. A., Sonia Blanco Mejia, Laura Chiavaroli, Effie Viguiliouk, Siying S. Li, Cyril W. C. Kendall, Vladmir Vuksan, and John L. Sievenpiper. "Cumulative meta-analysis of the soy effect over time." Journal of the American Heart Association 8, no. 13 (2019): e012458.

Kahlon, Talwinder S., Roberto J. Avena-Bustillos, and Mei-Chen M. Chiu. "Sensory evaluation of gluten-free quinoa whole grain snacks." Heliyon 2, no. 12 (2016): 1–12.

Kale, R., and R. Deshmukh. Ready-to-Drink (RTD) Tea Market. Portland, OR, USA: Allied Market Research (2020): 299.

Karakochuk, Crystal, Tina van den Briel, Derek Stephens, and Stanley Zlotkin. "Treatment of moderate acute malnutrition with ready-to-use supplementary food results in higher overall recovery rates compared with a corn-soya blend in children in southern Ethiopia: An operations research trial." The American Journal of Clinical Nutrition 96, no. 4 (2012): 911–916.

Kaur, Ravneet, and Kamlesh Prasad. "Technological, processing and nutritional aspects of chickpea (*Cicer arietinum*)—A review." Trends in Food Science & Technology 109 (2021): 448–463.

Kehinde Alawode, Ebunoluwa, M. A. Idowu, A. A. Adeola, E. K. Oke, and S. A. Omoniyi. "Some quality attributes of complementary food produced from flour blends of orange flesh sweet potato, sorghum, and soybean." Croatian Journal of Food Science and Technology 9, no. 2 (2017): 122–129.

Keyata, Ebisa Olika, Yetenayet B. Tola, Geremew Bultosa, and Sirawdink Fikreyesus Forsido. "Bioactive compounds, antioxidant capacity, functional and sensory properties of optimized complementary weaning flour processed from sorghum, soybean, and karkade (*Hibiscus sabdariffa* L.) seeds." Scientific African 19 (2023): e01457.

Khalili, Sona, Leila Amiri-Farahani, Shima Haghani, Arash Bordbar, Asie Shojaii, and Sally Pezaro. "The effect of *Pimpinella anisum* herbal tea on human milk volume and weight gain in the preterm infant: A randomized controlled clinical trial." BMC Complementary Medicine and Therapies 23, no. 1 (2023): 19.

Khaliq, Ayman, Akhilesh Kumar Mishra, Muhammad Nouman Shaukat, and Ahmad Rabbani. "Plant-based dairy substitutes: Current scenario and future prospects." Innovations in Agriculture 6 (2023): 01–10.

Kovacs, Christopher S. "Calcium and bone metabolism disorders during pregnancy and lactation." Endocrinology and Metabolism Clinics 40, no. 4 (2011): 795–826.

Kulczyk, Ewa, Emilia Drozłowska-Sobieraj, and Artur Bartkowiak. "Novel milk substitute based on pea, bean and sunflower seeds with natural bioactive stabilisers." Plants 12, no. 12 (2023): 2303.

Kumari, Ritu, Karuna Singh, Rashmi Singh, Neelam Bhatia, and M. S. Nain. "Development of healthy ready-to-eat (RTE) breakfast cereal from popped pearl millet." Indian Journal of Agricultural Sciences 89, no. 5 (2019): 877–881.

Lee, Jisoo, Yoo-Sun Kim, Jae Hwan Lee, Seung Chul Heo, Kook Lae Lee, Sang-Woon Choi, and Yuri Kim. "Walnut phenolic extract and its bioactive compounds suppress colon cancer cell growth by regulating colon cancer stemness." Nutrients 8, no. 7 (2016): 439.

Lesmana, David, Yoanes M. Vianney, Yohanes A. Goenawan, Karina Natalie, Johan Sukweenadhi, Gisela Buschle-Diller, Yayon P. Mukti, Christina M. Erawati, and Maria G. M. Purwanto. "Valorization of peel-based agro-waste flour for food products: A systematic review on proximate composition and functional properties." ACS Food Science & Technology 2, no. 1 (2021): 3–20.

Licandro, G., and C. E. Odio. "Citrus by-products." In *Citrus: The Genus Citrus*, Giovanni Dugo, Angelo Di Giacomo. CRC Press (2002): 159–178.

Lopriore, C., Guidoum, Y., Briend, A., & Branca, F. "Spread fortified with vitamins and minerals induces catch-up growth and eradicates severe anemia in stunted refugee children aged 3–6 years." The American Journal of Clinical Nutrition 80, no. 4 (2004), 973–981.

Łopusiewicz, Łukasz, Emilia Drozłowska, Paulina Siedlecka, Monika Mężyńska, and Artur Bartkowiak. "Preparation and characterization of novel flaxseed oil cake yogurt-like plant milk fortified with inulin." Journal of Food & Nutrition Research 59, no. 1 (2020): 61.

Ma, Liang, Min Zhang, Bhesh Bhandari, and Zhongxue Gao. "Recent developments in novel shelf life extension technologies of fresh-cut fruits and vegetables." Trends in Food Science & Technology 64 (2017): 23–38.

Maki, Kevin C., Jeannemarie M. Beiseigel, Satya S. Jonnalagadda, Carolyn K. Gugger, Matthew S. Reeves, Mildred V. Farmer, Valerie N. Kaden, and Tia M. Rains. "Whole-grain ready-to-eat oat cereal, as part of a dietary program for weight loss, reduces low-density lipoprotein cholesterol in adults with overweight and obesity more than a dietary program including low-fiber control foods." Journal of the American Dietetic Association 110, no. 2 (2010): 205–214.

Mayo-Mayo, Guadalupe, Abril Navarrete-García, Yanik I. Maldonado-Astudillo, Javier Jiménez-Hernández, David Santiago-Ramos, Gerónimo Arámbula-Villa, Patricia Álvarez-Fitz, Mónica Ramirez, and Ricardo Salazar. "Addition of roselle and mango peel powder in tortilla chips: A strategy for increasing their functionality." Journal of Food Measurement and Characterization 14 (2020): 1511–1519.

Mazumder, Md Anisur Rahman, Naphat Sujintonniti, Pranchalee Chaum, Sunantha Ketnawa, and Saroat Rawdkuen. "Developments of plant-based emulsion-type sausage by using grey oyster mushrooms and chickpeas." Foods 12, no. 8 (2023): 1564.

Medoua, Gabriel Nama, Patricia M. Ntsama, Anne Christine A. Ndzana, Véronique J. Essa'a, Julie Judith T. Tsafack, and Henriette T. Dimodi. "Recovery rate of children with moderate acute malnutrition treated with ready-to-use supplementary food (RUSF) or improved corn–soya blend (CSB+): A randomized controlled trial." Public Health Nutrition 19, no. 2 (2016): 363–370.

Mishra, Sunita. "Physicochemical properties and microstructure of ready-to-eat (RTE) breakfast cereal made from popped pearl millet, amaranth grains and groundnut powder." International Journal of Research in Engineering, Science and Management 5, no. 6 (2022): 340–342.

Mueed, A., S. Shibli, S. A. Korma, P. Madjirebaye, T. Esatbeyoglu, and Z. Deng. Flaxseed Bioactive Compounds: Chemical composition, functional properties, food applications and health benefits-related gut microbes. Foods 11, no. 20 (2022): 3307.

Nestel, P., A. Briend, B. De Benoist, E. Decker, E. Ferguson, O. Fontaine, A. Micardi, and R. Nalubola. "Complementary food supplements to achieve micronutrient adequacy for infants and young children." Journal of Pediatric Gastroenterology and Nutrition 36, no. 3 (2003): 316–328.

Ngamdee, Paradorn. "Evaluation of physical, health promoting properties and storage stability of ready-to-drink mixed tea from Riceberry rice and mulberry leaf." RMUTI Journal Science and Technology 15, no. 3 (2022): 1–14.

Ngume, L. S., L. K. Katalambula, M. J. Munyogwa, R. J. Mongi, and H. Lyeme. "Formulation and nutritional properties of qualea-bird-meat-based complementary foods for children (6–23 months) in Tanzania using a linear programming technique." NFS Journal 30 (2023): 1–7.

Nkesiga, Jackson, Joseph O. Anyango, and Peninah N. Ngoda. "Nutritional and sensory qualities of extruded ready-to-eat baby foods from orange-fleshed sweet potato enriched with amaranth seeds, and soybean flour." Research Journal of Food Science and Nutrition 7, no. 5 (2022): 120–140.

Noor Aziah, A. A., W. Lee Min, and Rajeev Bhat. "Nutritional and sensory quality evaluation of sponge cake prepared by incorporation of high dietary fiber containing mango (*Mangifera indica* var. Chokanan) pulp and peel flours." International Journal of Food Sciences and Nutrition 62, no. 6 (2011): 559–567.

Olagunju, Aderonke Ibidunni, Olufunmilayo Sade Omoba, Timilehin David Oluwajuyitan, and Olalekan Francis Akinrinlola. "Physicochemical composition, antioxidant properties and glycemic index of optimized extruded snacks from blends of amaranth, soycake and shallot flours." SSRN Electronic Journal (2022): 1–19.

Olagunju, Mobolaji Timothy, Olunike Rebecca Abodunrin, Elizabeth Oluwafolakemi Aleru, Ifeoluwa Eunice Adewole, and Folahanmi Tomiwa Akinsolu. "Effects of complementary food formulated from millet, soybean, and grasshopper on hematological parameters of malnourished weanling albino rats." The North African Journal of Food and Nutrition Research 7, no. 15 (2023): 108–116.

Pasqualone, Antonella, Barbara Laddomada, Fatma Boukid, Davide De Angelis, and Carmine Summo. "Use of almond skins to improve nutritional and functional properties of biscuits: An example of upcycling." Foods 9, no. 11 (2020): 1705.

Payne, Liz, P. Harris, Daniela Ghio, Joanna Slodkowska-Barabasz, Michelle Sutcliffe, J. Kelly, M. Stroud, Paul Little, Lucy Yardley, and Leanne Morrison. "Beliefs about inevitable decline among home-living older adults at risk of malnutrition: A qualitative study." Journal of Human Nutrition and Dietetics 33, no. 6 (2020): 841–851.

Permal, Rahul, Tim Chia, Giada Arena, Cassandra Fleming, Jack Chen, Tony Chen, Wee Leong Chang, Brent Seale, Nazimah Hamid, and Rothman Kam. "Converting avocado seeds into a ready to eat snack and analysing for persin and amygdalin." Food Chemistry 399 (2023): 134011.

Philip, Akhil. "Development of ready to eat fruit flakes." Food and Agriculture Spectrum Journal 2, no. 1 (2021): 25–30.

Purnawijayanti, Hiasinta A., and Hildagardis Meliyani Erista Nai. "Antioxidant activity in ready-to-drink beverage made from snake fruit (*Salacca edulis* Reinw) enriched with butterfly pea (*Clitoria ternatea*) and Roselle (*Hibiscus sabdariffa*) flower extracts." In Proceedings of the 1st IPB International Conference on Nutrition and Food, Malaysian Journal of Medicine and Health Sciences, 16, (2020): 109–110.

Rico, Daniel, Beatriz Barcenilla, Agate Meabe, Cristina González, and Ana B. Martín-Diana. "Mechanical properties and quality parameters of chitosan-edible algae (*Palmaria palmata*) on ready-to-eat strawberries." Journal of the Science of Food and Agriculture 99, no. 6 (2019): 2910–2921.

Saggu, Aashmeen Kaur, Vidisha Tomer, Ashwani Kumar, and Pragya Pandey. "Consideration of phytonutrients, probiotics and prebiotics for enhanced immunity during disaster relief situation—A review." Clinical Nutrition Open Science 47, (2023): 131–146.

Sánchez-González, Claudia, Carlos J. Ciudad, Veronique Noe, and Maria Izquierdo-Pulido. "Health benefits of walnut polyphenols: An exploration beyond their lipid profile." Critical Reviews in Food Science and Nutrition 57, no. 16 (2017): 3373–3383.

Sandhosh, K., D. Vidhya, Raj Abitha, M. Indumathi, and S. Indu. "Formulation and development of ice apple (*Borassus flabellifer*) flavored ice cream utilizing coconut milk." Journal of Technical Education 65 (2023): 65.

Sang, Shengmin, and YiFang Chu. "Whole grain oats, more than just a fiber: Role of unique phytochemicals." Molecular Nutrition & Food Research 61, no. 7 (2017): 1600715.

Santini, Antonello, Ettore Novellino, Vincenzo Armini, and Alberto Ritieni. "State of the art of ready-to-use therapeutic food: A tool for nutraceuticals addition to foodstuff." Food Chemistry 140, no. 4 (2013): 843–849.

Saraswati, A. R., and D. Taniwiryono. "Formulation of ready to drink (RTD) made from roselle (*Hibiscus sabdariffa*. L) tea and stevia (*Stevia rebaudiana*) leaf safe for diabetics." Indonesian Journal of Applied Research (IJAR) 1, no. 1 (2020): 1–9.

Sastrawan, S., M. Menap, L. Sulaiman, and H. Hendrayani. "Development of home-based ready-to-use supplementary food (RUSF-HB) to overcome nutrition-related problems among children under five during the covid-19 pandemic." In IOP Conference Series: Earth and Environmental Science 883, no. 1 (2021): 012074. IOP Publishing.

Shaivya, Mishra, and Mishra Sunita. "Nutritional analysis of value-added product by using pearl millet, quinoa and prepare ready-to-use upma mixes." International Journal of Food Science and Nutrition 1, no. 5 (2016): 41–43.

Shoge, M. O. "Bambara groundnut as ready to use therapeutic food (RUTF) for children under 5 years of age in North-West Nigeria." Book of Proceedings of the 1st Faculty of Science International Conference FSIC, pp. 318–330. Air Force Institute of Technology, Kaduna, Nigeria, 2022. ISBN: 9789787946589.

Sigh, Sanne, Nanna Roos, Daream Sok, Bindi Borg, Chhoun Chamnan, Arnaud Laillou, Marjoleine A. Dijkhuizen, and Frank T. Wieringa. "Development and acceptability of locally made fish-based, ready-to-use products for the prevention and treatment of malnutrition in Cambodia." Food and Nutrition Bulletin 39, no. 3 (2018): 420–434.

Silberbauer, Alina, and Markus Schmid. "Packaging concepts for ready-to-eat food: Recent progress." Journal of Packaging Technology and Research 1 (2017): 113–126.

Singh, Arshya, Aparna Kumari, and Anil Kumar Chauhan. "Formulation and evaluation of novel functional snack bar with amaranth, rolled oat, and unripened banana peel powder." Journal of Food Science and Technology 59, no. 9 (2022): 3511–3521.

Siol, Marta, and Anna Sadowska. "Chemical composition, physicochemical and bioactive properties of avocado (*Persea americana*) seed and its potential use in functional food design." Agriculture 13, no. 2 (2023): 316.

Sogo-Temi, C. M., O. A. Idowu, and A. M. Lateef. "Nutritional quality assessment of complementary food from plantain, velvet bean and crayfish." Valley International Journal Digital Library (2023): 143–148.

Suri, Shweta, Anuradha Dutta, Navin Chandra Shahi, Anupama Singh, Rita Singh Raghuvanshi, and Chandra Shekhar Chopra. "Formulation, process optimization, and quality evaluation of barnyard millet (*Echinochloa frumentacea*)-based ready-to-eat (RTE) extruded snack." Journal of Food Processing and Preservation 46, no. 11 (2022): e16936.

Susanna, N. C., T. A. K. A. Thennakoon, J. W. A. Sajiwanie, and L. Kanapathyraj. "Development of ready-to-drink soymilk beverage incorporated with broken mixed fanning (BMF) liquid and powdered tea aroma." In *International Conference on Applied Sciences (ICAPS_ SUSL 2023)*, Faculty of Applied Sciences, Sabaragamuwa University of Sri Lanka (2023): 60.

Szpicer, Arkadiusz, Anna Onopiuk, Mateusz Barczak, and Marcin Kurek. "The optimization of a gluten-free and soy-free plant-based meat analogue recipe enriched with anthocyanins microcapsules." LWT 168 (2022): 113849.

Tănase, Luiza-Andreea, Doina-Georgeta Andronoiu, Oana-Viorela Nistor, Gabriel-Dănuţ Mocanu, Elisabeta Botez, and Bogdan Ioan Ştefănescu. "Sweet potatoes puree mixed with herbal aqueous extracts: A novel ready-to-eat product for lactating mothers." Processes 11, no. 7 (2023): 2219.

Tănase, Luiza-Andreea, Oana-Viorela Nistor, Gabriel-Dănuţ Mocanu, Doina-Georgeta Andronoiu, Adrian Cîrciumaru, and Elisabeta Botez. "Effects of heat treatments on various characteristics of ready-to-eat zucchini purees enriched with anise or fennel." Molecules 27, no. 22 (2022): 7964.

Tanner, Melvin J., Fahrul Nurkolis, Nelly Mayulu, Nurpduji Astuti Taslim, Endang Achadi, Happy Kurnia Permatasari, Defny Silvia Wewengkang, and Hardinsyah Hardinsyah. "Sea grapes kombucha tea improves liver-superoxide dismutase (SOD) serum in mice fed on cholesterol-and fat-enriched diet: A novel probiotic ready-to-drink rich in ascorbic acid." Proceedings of the Nutrition Society 81, no. OCE2 (2022): e66.

Testa, Mario, Francesco Polese, and Sergio Barile. "Breaking the cycle of malnutrition through sustainable business models: The case of ready-to-use therapeutic foods." In Case Studies on the Business of Nutraceuticals, Functional and Super Foods, pp. 121–144. Woodhead Publishing, 2023.

Thongoun, Pimonphan, Patcharanee Pavadhgul, Akkarach Bumrungpert, Pratana Satitvipawee, Yashna Harjani, and Anne Kurilich. "Effect of oat consumption on lipid profiles in hypercholesterolemic adults." Journal of the Medical Association of Thailand = Chotmaihet thangphaet 96 (2013): S25–S32.

Tripathi, Manoj Kumar, Debabandya Mohapatra, Rajpal S. Jadam, Sharad Pandey, Vaishali Singh, Vishnu Kumar, and Anil Kumar. "Nutritional Composition of Millets." Millets and Millet Technology (2021): 101–119.

UNICEF. "Levels and trends in child malnutrition." Report (2020): 21–25.

Valls, J., N. Pasamontes, A. Pantaleón, S. Vinaixa, M. Vaqué, A. Soler, S. Millán, and X. Gómez. "Prospects of functional foods/nutraceuticals and markets." Natural Products (2013): 2491–2525.

Varghese, Cinu, Prem Prakash Srivastav, and M. S. Roopesh. "High-energy cookies for undernourished adolescents: In vivo rat assay of protein quality and evaluation of storage conditions on cookies shelf-life." Future Foods 6 (2022): 100154.

Violette, William J., Aurélie P. Harou, Joanna B. Upton, Samuel D. Bell, Christopher B. Barrett, Miguel I. Gómez, and Erin C. Lentz. "Recipients' satisfaction with locally procured food aid rations: Comparative evidence from a three country matched survey." World Development 49 (2013): 30–43.

Wakhu-Wamunga, Florence, and Brian J. Wamunga. "Locally developed ready-to-use-therapeutic-food (RUTF) for management of malnutrition using animal models." Journal of Clinical Nutrition and Dietetics 3, no. 1 (2017): 10.

Wang, Chunfang, Jinpeng Chen, Wenguo Tian, Yanqi Han, Xu Xu, Tao Ren, Chengwang Tian, and Changqing Chen. "Natto: A medicinal and edible food with health function." Chinese Herbal Medicines 15, no. 3 (2023): 349–359.

Wang, Lin, Wenyan Chen, Ruoyu Zhou, Yuhan Ren, Kang Wang, Ning Jiang, and Ruichang Gao. "Physicochemical and sensory properties of a tilapia skin-based ready-to-eat snack prepared by infrared drying and air frying." Applied Food Research 2, no. 2 (2022): 100155.

Wang, Ruimin, Linling Yao, Xue Lin, Xiaoping Hu, and Lu Wang. "Exploring the potential mechanism of *Rhodomyrtus tomentosa* (Ait.) Hassk fruit phenolic rich extract on ameliorating nonalcoholic fatty liver disease by integration of transcriptomics and metabolomics profiling." Food Research International 151 (2022): 110824.

Wojtasik-Kalinowska, Iwona, Monika Marcinkowska-Lesiak, Anna Onopiuk, Arkadiusz Szpicer, Agnieszka Wierzbicka, and Andrzej Poltorak. "The effect of *Solanum lycopresicum* L. extract on quality and flavor stability of ready to eat meat products." Animal Science Papers & Reports 40, no. 4 (2022): 423.

World Health Organization, United Nations Children's Fund. Clinical management of acute diarrhoea: WHO/UNICEF joint statement (2004). Available: http://whqlibdoc.who.int/hq/2004/WHO_FCH_CAH_04.7.pdf (accessed 10 May 2011).

Yaman, Mustafa, Hafsa Sena Sargın, and Ömer Faruk Mızrak. "Free sugar content, in vitro starch digestibility and predicted glycemic index of ready-to-eat breakfast cereals commonly consumed in Turkey: An evaluation of nutritional quality." International Journal of Biological Macromolecules 135 (2019): 1082–1087.

Younis, Kaiser, Rayees Ul Islam, Kausar Jahan, Basharat Yousuf, and Aradhita Ray. "Effect of addition of mosambi (*Citrus limetta*) peel powder on textural and sensory properties of papaya jam." Cogent Food & Agriculture 1, no. 1 (2015): 1023675.

Yousaf, Muhammad Sohail, Salmah Yusof, Mohd Yazid Bin Abdul Manap, and S. U. R. A. I. N. I. Abd-Aziz. "Storage stability of clarified banana juice fortified with inulin and oligofructose." Journal of Food Processing and Preservation 34 (2010): 599–610.

Yu, Shubo, Pengfei Yu, Juan Wang, Chun Li, Hui Guo, Chengcheng Liu, Li Kong et al. "A study on prevalence and characterization of Bacillus cereus in ready-to-eat foods in China." Frontiers in Microbiology 10 (2020): 3043.

Zahrani, Ashwag Jaman A. L., and Amal Bakr Shori. "Viability of probiotics and antioxidant activity of soy and almond milk fermented with selected strains of probiotic *Lactobacillus* spp." LWT 176 (2023): 114531.

Zand, Nazanin, Babur Z. Chowdhry, David S. Wray, Frank S. Pullen, and Martin J. Snowden. "Elemental content of commercial 'ready to-feed' poultry and fish based infant foods in the UK." Food Chemistry 135, no. 4 (2012): 2796–2801.

Zegeye, Martha Kassahun, Aysheshim Kassahun Belew, Addisalem Damtie Aserese, and Derese Bekele Daba. "Time to recovery from malnutrition and its predictors among human immunodeficiency virus positive children treated with ready-to-use therapeutic food in low resource setting area: A retrospective follow-up study." Health Science Reports 6, no. 1 (2023): e959.

Zimmermann, Michael B., and Richard F. Hurrell. "Nutritional iron deficiency." The Lancet 370, no. 9586 (2007): 511–520.

11 Nutrition and Other Associated Benefits of Ready-to-Eat Foods

Rekha Rani,[1*] *Payal Karmakar,*[2] *Simrat Kaur,*[3] *and Chinju Davis*[4]
[1]Department of Dairy Chemistry, College of Dairy and Food Technology, AU, Jodhpur, Rajasthan, India
[2]Food Science and Nutrition Department, University of Minnesota, Minneapolis, Minnesota, MN, USA
[3]Department of Chemistry, Mata Gujri College, Fatehgarh Sahib, Punjab, India
[4]Quality Assurance and Management Division, Central Institute of Fisheries Technology, Cochi, Kerala
*Corresponding Author

11.1 INTRODUCTION

Food that is ready to eat (RTE) does not require heating before consumption. RTE and ready-to-cook foods have gained significant market share in recent years because they are the closest alternatives to our everyday diet. Due to social modification, urbanization, and cultural changes, there have been significant changes in food preferences and cooking methods. Due to hectic lifestyles, people prefer quick and simple cooking methods and time-consuming cooking techniques (Temgire et al., 2021). Modern people have altered their lifestyles and behaviors toward RTE and RTE foods because of so many fast foods available in the market and products are fresh in taste and easy to consume (Alvarez et al., 2018; Temgire et al., 2021). In contrast to ready-to-cook food products, which require preparations like heating or boiling before consumption, RTE food products are packaged and intended for direct consumption without any cooking. Because they are simple to prepare, store, and enjoy, RTE food products are generally accepted and well-liked by consumers. RTE food includes a wide range of snack products, such as biscuits, rolls, sandwiches, soups, chips, fruit and vegetable salads, breakfast cereals, dairy products, and so on. Because there are new products falling under that category almost every day, this list can get very lengthy. Cereals, fruits, and vegetables are the main ingredients used in the production of RTE meals. A variety of ready-made meals and snacks, such as idlis, dosas, pav bhaji, mixed fruit salads, vegetable curry, cereal biscuits and bars, chapattis, and meat items, such as precooked sausage, ham, and chicken goods, predominate in traditional food intake (Lucera et al., 2017).

In the latest research "India Ready-to-eat Food Market Forecast and Opportunities, 2019" from TechSci Research, the country's RTE food industry was expected to develop at a compound annual growth rate (CAGR) of approximately 22% from 2014 to 2019. In 2017, the Japanese market for RTE food and meals reached a value of over JPY 10 trillion. The main items of these ready-to-eat food/meal vendors are cooked rice (50%) and everyday dishes (34%), particularly lunch boxes, rice balls, sandwiches, croquettes, and vegetable salads in the Tokyo metropolitan region. The primary

 DOI: 10.1201/9781032622408-11

sales channels for these products are convenience stores (32%) specialized stores (29%) and food supermarkets around 26% (White Paper of Japan Ready-Made Meal Association (2018). In India, the market for prepared foods was valued at 261 million in 2017; by 2023, it was expected to expand to 647 million, with a compound annual growth rate of more than 16% CAGR rate (Research and Market India Ready-To-Eat Food (RTE) Market Study, 2013–2023). An estimated US$ 182 billion is spent on food in India, which makes up over two-thirds of the country's whole retail industry. As per McKinsey & Co., a consulting firm, projects that India's retail food market will increase from approximately US$ 70 billion in 2008 to US$ 150 billion by 2025, making up a sizable portion of the global food sector, which is projected to increase from US$ 175 billion to US$ 400 billion by 2025. Sometimes people believe that RTE food is not very healthy, yet some healthier RTE items do not have very good nutrition (Remnant & Adams, 2015). According to the survey on prepared meals in Asia, RTE cuisine is enjoyed in numerous nations. The Asian market for RTE food is currently competitive. According to the data, 49% of respondents in Asia Pacific said they occasionally bought ready meals, while 25% said they did so regularly (Nielsen *et al.*, 2017). RTE and processed food are rapidly expanding sectors, particularly in Thailand, Vietnam, and Taiwan, as a result of busy lifestyles, the availability of ready meals, and the growth of retail outlets. Chinese consumers also enjoy Western RTE foods such as frozen pizza and chilled salad. China now has more convenience stores than ever before, which will contribute to the future expansion of ready meals. Because prepared meals are less expensive at retail stores than at restaurants, youths in Indonesia typically visit there to hang out with their friends. Customers may also get hot water and microwaves from the stores, which makes it convenient.

To finish their jobs, people in Hong Kong typically put in longer hours at the office. Customers typically select the ready meal as it is quicker and less expensive. Customers in Taiwan are demanding better RTE food, so one company decided to work with an organic farm to create healthier chilled RTE meals. Even though they are becoming increasingly popular, quick meals are not yet common in all nations such as Malaysia and Singapore. Because the products have a long shelf life, Malaysian customers believe that RTE food is fortified with food coloring and preservatives. As a result, consumers who cook from scratch and use more fresh ingredients are likely to find satisfaction in ready-to-cook products. In a similar way, Singapore people stopped regularly consuming RTE meals as health consciousness increased. It is interesting to note that consumers in Vietnam frequently purchase vegetarian-ready meals because of their religious convictions and nutritional advantages, but they have no interest in ready meals that are less salty and organic. As a result, RTE products vary among Asian nations based on consumer choices and attitudes (Thienhirun and Chung, 2017). Additionally, consumers' desires for items like snacks that can be consumed on the go are expanding. The "snacking and grazing" consumption pattern that is prevalent in more industrialized nations is facilitated by urban living. Hot snacks are quickly being replaced by readily transportable and hygienic snack meals as Indian consumers adopt a more global lifestyle. Sharma (2009) has noted that traditional Indian meals continue to be underrepresented in the RTE market in terms of their variety and regional preferences. Certain RTE Indian meal dishes, like lentils (dal makhani), chhole, palak paneer, rajma, etc., appear to be particularly popular across all regions (north, east, south, and west), among both vegetarian and non-vegetarian consumers, based on the category's visible trends. As a result, these dishes are included in the brand portfolios of most players. Reaching and making RTE foods easily accessible to potential customers is another aspect of marketing them. Additionally, they discuss how crucial retail commerce is to promoting RTE meals. Contemporary retail markets can contribute significantly to the growth of demand for products like RTE, just as contemporary trade does so by increasing these products' visibility and awareness.

RTE meals, often known as ready meals, are elaborate combinations of precooked dishes bundled together and sold through chilled retail chains to provide customers with quick meal solutions. These could consist of a wide range of foods, such as rice-based dinners with sauce, veggies, and/or meats; pasta and sauce combos with or without meat; etc. The manufacture of ready meals presents significant operational hurdles due to the broad range of possible ingredients. RTE foods include canned goods, convenience foods, frozen foods, fast food, dry foods, preserved foods, and more (Patel D. and Rathod R. (2017) for consumption without further heating or processing. This definition covers both open and pre-wrapped RTE products and is meant to apply whether the RTE meal may be enjoyed hot or cold. The term "further heating or processing" does not apply to food preparation tasks that the consumer chooses to do over otherwise RTE food items, such as light washing, slicing, chopping, portioning, marinating, or preservation (Food Standards Agency, 2011).

11.2 TYPES OF RTE FOODS

According to this definition, a variety of processed foods, such as fruit, prepared salads and vegetables, dairy products (milk, cheese, and spreads), cookies, crisps, breads, pies, sandwiches, and rolls, might be considered RTE items. The list can be very lengthy, and it is growing longer and longer as new goods are introduced to the food market almost daily. Costa et al. (2001) categorized several kinds of convenience foods, such as chilled goods and take-out meals, under RTE food. Types of RTE foods are presented in Table 11.1 and Figure 11.1.

11.3 WHY IS RTE FOOD PRODUCTS' MARKET INCREASING?

The attributes of food, including flavor, texture, and appearance; socioeconomic elements like accessibility, cost, and culture; biological elements like the need for energy and nutrients; and psychological elements, such as conduct, emotions, and eating attitudes, affect consumers' choices of foods (Patel and Rathod 2017). Figure 11.2 illustrates the reason to buy RTE foods.

11.3.1 Change in Lifestyle

The demand for RTE food products in the nation is anticipated to increase throughout the forecast period due to changing consumer lifestyles and expanding employment opportunities. A smaller family size and a busy lifestyle are additional reasons to purchase RTE food, which is available at a variety of retail stores and on the street easily. The number of single people and dual-income households has increased as a result of the busy lifestyle. Around the world, people's lives are changing and transitioning more quickly in tandem with changes in food consumption. Time constraints have become important factors in determining food purchase and consumption patterns. Over the past two decades, there has been a sharp decline in physical activity in all sections of the Indian population due to the increase in families with both parents working outside the home. The food business increased the availability of prepared meals and convenience foods in response to these

TABLE 11.1
Types and Examples of Different RTE Foods

Sl. no.	Types	Examples
1.	RTE based on fruits and vegetables	Canned products, juice, jelly, jam, marmalades, prepared salad, vegetables curry, etc.
2.	RTE based on cereals and pulses	Instant snacks, fast food, breakfast cereals, cereal biscuit and bar, extruded snacks, bakery products, nutria bars, cereal-based chocolate bars
3.	RTE based on meat, fish, and poultry	Chicken nuggets
4.	RTE based dairy products	Ice cream, flavored milk, whey-based beverages, milk-based smoothies, milk sweets, milk chocolates

FIGURE 11.1 Different types of RTE foods.

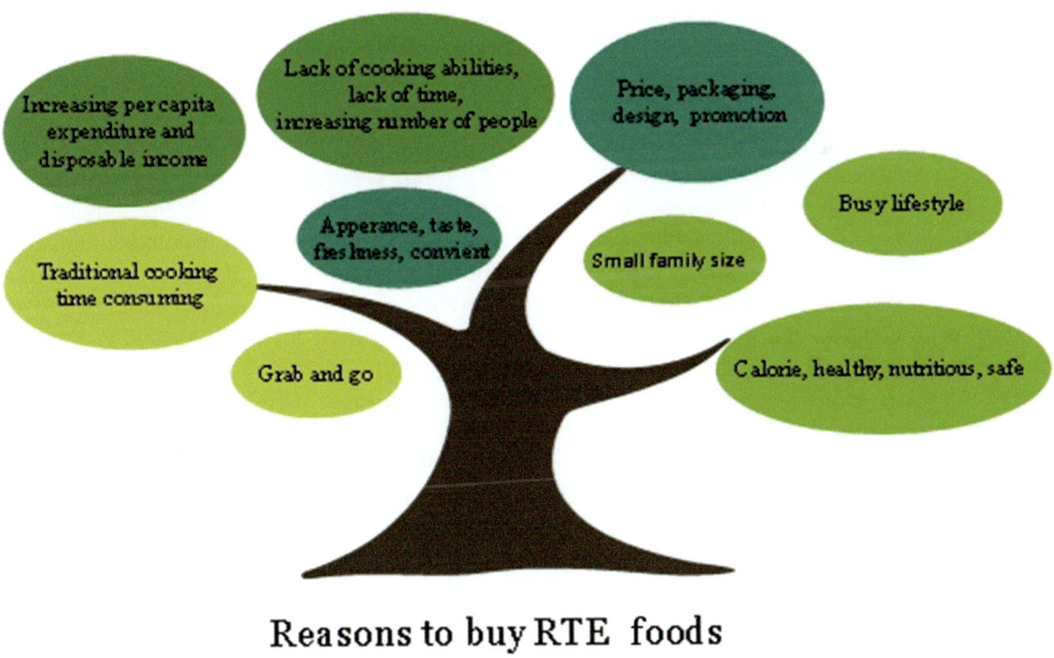

FIGURE 11.2 Pictorial representation of reasons to buy RTE foods.

new family concerns. Unfortunately, the food industry has also been pushing meals that are high in sugar and fat, which has led to an excessive amount of energy-dense mechanization in the home, workplace, and transportation sectors in both rural and urban places. One significant risk factor for a number of non-communicable diseases has been identified as physical inactivity. Therefore, the primary reason for buying RTE meals is convenience. Young people make up the majority of the workforce in India, where there is a large number of consumer base. Because of their hectic schedules, these young consumers hardly ever have time for traditional cooking, which further increases the market potential for prepared foods (Patel and Rathod, 2017).

It is projected that the market would expand due to a number of factors, including a growing middle class and wealthy customer base, rising per capita disposable income, rising per capita expenditure on prepared meals, and an increasing number of working people. The notion and implementation of healthy eating in the RTE food and meal sector in Japan were evaluated by Working Group 2 of the International Life Sciences Institute Japan's Healthy Diet Research Committee (WG2). RTE food/meal suppliers mainly referred to Dietary Reference Intakes for Japanese, the Japanese Food Guide. They decreased energy, carbohydrates, and salt in "healthy" food while increasing dietary fiber, various vegetables, whole-grain cereals, millet rice, and soybean products. Several industries decreased the salt amount without drawing attention to the practice. As the primary driver of healthy food sales, they are always working to enhance flavor. In Western countries, when assessing nutrients, it is important to place more emphasis on energy and nutrients that are related to long-term health and contribute to the burden of disease (e.g. sodium, fiber) compared to other nutrients that are not of concern, being consumed in sufficient quantities by all adults (e.g. protein). The nutritional risk assessment was relatively wide and comprehensive, quantitative, and fast to perform based on the regulatory information provided on the RTE dish pack (Poissant et al., 2023). In France, the "Nutriscore," which is a logo with the letters A through E on the front of the packaging, was created to help consumers make more informed decisions by making nutritional information easier for them to understand (Sante Publique, 2023).

From previous studies, convenience was found to be the primary consideration for customers when making meal choices. When buying meals, the primary considerations are taste, price, freshness, and ease of consumption, besides from convenience. Appearance also plays a key role in consumer expectations before they purchase and eat the food. The consumers will decide whether the food is good or bad from their experiences and visual appearances of the food. Not only are the qualities of the product essential, but the packaging design also influences the consumer's tendency to purchase. As the customers cannot fully evaluate a product inside, design plays a significant role in buying a product in consumers' minds. Locations with a wide selection of food have an impact on consumers' choices, and promotions also have an impact on their intention to buy. Promotion was also another key attribute that induced consumers to purchase RTE food in convenience stores if promotion was applied, it would be another key factor that made a rush in consumers' minds to purchase RTE food. That is the reason that many retail stores typically offer different types of promotions to attract consumers. Additionally, consuming RTE food is also motivated by a lack of cooking abilities and a lack of desire to spend time preparing meals. Calories may play a significant role in the decision of consumers who are health-conscious to buy RTE meals. Cardello and Wolfson (2013) mentioned that, when compared to high-calorie products, low-calorie products produced an 82% increase in sales and accounted for 75% of new product sales

Geographically dense areas are typically the sites of ready-meal production to maximize logistical efficiency during delivery. The high intrinsic value and high preparation cost of ready meals contribute to this focus, but the need to optimize shelf-life benefits from reducing the chill transport and storage chain is just as significant. The key drivers for growth in the sector are : (1) A 20% increase in per capita food consumption spending during the same period of time due to an 8% increase in per capita income over the previous five years, from 2005 to 2010. (2) The size of the middle class to the very rich class is expected to grow by more than 300% between 2005 and 2015.

Because of this, there will be a growing need for food products to satisfy consumers' needs for convenience, variety, health, and evolving tastes. (3) The emergence of tier 1 and tier 2 cities will be a major factor in future growth because of the limited availability of items in these markets, rising incomes, and growing awareness. Significant lifestyle changes will cause a huge shift in consumers' spending patterns. (4) An increasing number of nuclear working families are motivated by a strong need for convenience and have little time for cooking. (5) An increase in the prevalence of chronic illnesses including diabetes, asthma, and obesity is driving up demand for health-conscious goods. (6) A rise in consumer awareness brought about by media coverage and the promotion of new product categories and cuisines by celebrity chefs. The fact that rural Indians are frequently exposed to these processed goods through their urban family members means that this problem is not exclusive to urban areas (Patel and Rathod, 2017).

In Thailand, Thai customers mostly buy prepared meals as they are fresh, hygienic, and convenient (Nuntasriwiwat, 2012). In addition, while purchasing frozen food goods, consumers are more focused on product features than on location and price (Chatthipmongkol & Jangphanish, 2016). Toda (2015) mentioned three primary factors contributing to the popularity of quick meals among Japanese consumers. First, as the proportion of employed women in Japan rose, they tended to prioritize saving time for cooking. Second, because dining out is expensive , many customers have chosen to dine at home since the 1990s bubble economy broke. Third, customers who prioritized their health and safety would prefer to order food that was higher in nutrients. In trains, many stations in Japan serve RTE meals, it is known as the station box lunch (Ekiben) which has been available around the train station for more than 100 years. Like Western fast food, as Ekiben is primarily cold, customers can eat it on trains without having to reheat it. It is low in price, quick, and easy to use. In response to the needs of the modern world, a variety of RTE foods, such as tuna-mayonnaise onigiri, are available in the market. These foods mix both traditional and modern cultures. Furthermore, one factor that contributed to an RTE food product's popularity among Japanese young women was its taste (Thienhirun and Chung, 2017).

11.3.2 Transition in Food Consumption Patterns

Convenience food has replaced home-cooked meals in the modern consumer's diet (Dublin, 2014; Thienhirun and Chung, 2017). Many consumers buy RTE items from retail stores as they do not have time to make meals due to their hectic schedules and overtime work (Gatley et al., 2014; Thienhirun and Chung, 2017). RTE food is prepared food that can be eaten right away or requires minor preparation, like heating, before consumption (FDA, 2009). It requires very little preparation and no additional ingredients. RTE foods are typically purchased by city dwellers to consume in an automobile during rush hour or at an office. Due to many environmental conditions, women's work, changing lifestyles, and an increase in per capita income, there have been changes in the patterns of food consumption and purchases. In emerging nations like India, the amount of processed food and non-cereal-based items consumed per person is rising quickly. Over the past 20 years, the total energy intake in industrialized nations like the United States has increased for all age groups. There has been a shift from eating meals to snacks and from consuming energy at home to using energy outside the home (Patel and Rathod, 2017). Worldwide, similar transitions are being noticed. Dietary choices have changed because of food availability, modernization of the food business, and urbanization, which has increased money. Supermarket usage has expanded throughout nations, from cities to villages, and from middle-class and upper-class households to the working class. Urban consumers now spend a higher percentage of their income on processed meals. Processed food increasingly accounts for a larger portion of the income of urban consumers. Africa, Eastern Europe, and Asia are all experiencing the same thing, albeit at varying speeds and phases. However, over time, the amount of energy obtained from vegetables, animals, and cereals has decreased. However, there is a rise in the consumption of fats in various nations. According to reports, North America and Europe consumed the most fat, whereas Africa consumed the least. The countries going through a transition saw a 9-g decrease in per capita fat intake, while developed countries saw the lowest increase (4 g per capita). The largest increase in fat consumption per capita was seen in emerging countries (14 g). In the United States, children's overall energy intake has not changed, but their diet's macronutrient intake (Patel and Rathod, 2017). Intake of foods high in carbohydrates and protein has increased, whereas intake of total energy derived from fats has decreased. Intake of foods high in carbs and protein has increased, whereas intake of total energy derived from fats has decreased. Changes in eating habits could explain why children's levels of obesity are rising. For example, there has been a rise in the quantity of meals consumed outside the house, food availability, portion sizes, snacking, and meal skipping (Rani et al., 2021). The consumption of high-fat foods, soft drinks, and snacks is also steadily rising.

11.4 READY-TO-USE THERAPEUTIC FOOD (RUTF)

Therapeutic foods that are ready to use are fortified, high-energy foods that may be consumed right away and are meant to treat children suffering from severe acute malnutrition. Many times, ready-to-use therapeutic meals (RUTF) are used to manage severe acute malnutrition (SAM) at the patient's home (Bhutta et al., 2013). Malnutrition has been treated with imported Ready to Use Therapeutic Foods (RUTF) in several developing (and undeveloped) nations, such as Nigeria. Soy Kunu, a locally derived and cooked food blend made of peanuts, soybeans, and millet, is also used in Northern Nigeria. A subset of therapeutic foods known as "ready-to-use therapeutic food" (RUTF) is made primarily of peanuts, sugar, oil, and milk powder. It is an energy-dense, micronutrient-enriched paste with a nutritional profile akin to that of the traditional F-100 milk-based diet and food used in inpatient therapeutic feeding programs (Isanaka et al., 2009; Godswill et al., 2020). Nutritional composition of typical RUTF food should consist of energy: 520 to 550 kcal/100 g; moisture content: 2.5% maximum; lipids: 45% to 60% total energy; proteins: 10% to 12% total energy; calcium: 300 to 600 mg/100 g; iodine: 70 to 140 μg/100 g; copper: .4 to 1.8 mg/100 g; magnesium: 80 to 140 mg/100 g; iron: 10 to 14 mg/100 g; potassium: 1100 to 1400 mg/100 g; phosphorus (excluding phytate): 300 to 600 mg/100 g; sodium: 290 mg/100 g maximum; zinc: 11 to 14 mg/100 g; selenium: 20 to 40 μg; vitamins A: 0.8 to 1.1 mg/100 g, vitamin D: 15 to 20 μg/100 g; vitamin K: 15 to 30 μg/100 g; vitamin E: 20 mg/100 g minimum; B1: 0.5 mg/100 g minimum; B6: 0.6 mg/100 g minimum; B2: 1.6 mg/100 g minimum; B12: 1.6 μg/100 g minimum; C: 50 mg/100 g minimum; niacin: 5 mg/100 g minimum; folic acid: 200 μg/100 g minimum; pantothenic acid: 3 mg/100 g minimum; biotin: 60 μg/100 g minimum; omega-6 fatty acids: 3% to 10% total energy; omega-3 fatty acids: 0.3% to 2.5% of the total energy (Godswill et al., 2020). The F-100 formula was used to create the RUTFs, which contain the same ingredients plus peanut butter (World Health Organization, 1999). Instead of leaving food as a powder, peanut butter transforms it into a viscous liquid product with altered physical and functional qualities. A standard recipe for Therapeutic Food Ready to Use: full-fat milk: 30% weight; sugar: 28% weight; peanut butter: 25% weight; mineral vitamin mix: 1.6% weight; vegetable oil: 15% weight.

As the components must be suspended in a liquid, lipid or fat-based liquid is required for the creation of RUTFs. Dairy products and milk should provide half of the protein in therapeutic foods that are ready to use.

Uncertainty surrounds the ideal iron dosage in ready-to-use therapeutic meals (RUTF) used to treat simple SAM in community settings. The aim of this systematic analysis was to determine whether or not children aged six months or older benefited from a higher iron dose in RUTF when compared to the standard iron dose in the peanut-based RUTF recommended by the World Health Organization (WHO).

TABLE 11.2
RTE Foods in Different Countries

Country	Variants
Thailand	Thai ready meals typically contained only one or two main dishes (e.g., curry and spicy salads) and rice. Sandwiches and rice balls. Pad Thai, green curry, Tom yum, and Pad Ka-Prao (Thienhirun and Chung, 2015)
Japan	Chicken Teriyaki rice. Onigiri, sandwiches, and salad. Main dishes (e.g., grilled meats and raw fish) and side dishes (e.g., pickles and beans, sandwiches, spaghetti, and hamburgers). Bento lunch boxes with different kinds of high fiber rice such as brown and red rice with grains (Thienhirun and Chung, 2015)
Africa	Samosas, pizza, porotah, veggie patties, shingara, cereals
India	Milk desserts, Beverages, Snacks, Flavored milk, Namkeen, Biscuits, Breads

11.5 RTE FOODS IN DIFFERENT COUNTRIES

Different countries have their own RTE-based foods which are presented in Table 11.2.

11.5.1 Thailand

The cost has switched from eating out to eating RTE food, which made up 65% of all food expenses in Bangkok (National Statistical Office, 2013a). Moreover, according to the survey, approximately 47.8% of the 26,520 homes consumed prepared meals on a weekly basis (National Statistical Office, 2013b). The popular Thai ready meals usually contained rice and one or two main dishes (such as curries and spicy salads). The primary dish varieties may differ depending on the country's various regions. However, international cuisine is also offered in train stations and street markets; examples include rice balls and sandwiches. Because the cost of raw materials has decreased, the price of RTE food in Thailand is rather consistent (Euromonitor, 2015b), which approximately ranges from 1 to 5 dollars. Ready meal sales at retail saw a 13% increase in volume in 2015, totaling 163.5 million dollars (Euromonitor, 2015b). In addition, it is anticipated that the RTE food industry will make more money in the upcoming years. RTE food held about a 30% market share in retail establishments (Marketing History Association, 2014). Most Thai participants who bought RTE meals typically do so when they do not have time to cook and eat them by themselves.

11.5.2 Japan

In 2014, the growth rate of prepared meals in Japan was 4%, generating $14.28 billion in revenue (Euromonitor, 2015c). The best performers, making up almost half of all sales channels, are delicatessens (32.3%) and convenience stores (28.1%), where consumers may purchase a range of RTE meals (Yano Research Institute, 2014). For example, the RTE food category at Family Mart represented almost 32% of net sales (Thienhirun & Chung, 2018).

11.5.3 Africa

There has been a steep rise in the shift of the dietary preferences of consumers toward RTE foods. The whole world shows this trend and Africa is not an exception. Rising urbanization in middle-income households is one of the primary factors (Barrett et al., 2020). The increase in the purchasing power of the urban population influences the rising demand for ready-to-cook and RTE foods that help save food preparation time for consumers in this fast-running world (Barrett et al., 2019). However, these diversified diets can also associate themselves with some safety challenges in African middle-income countries (Waage et al., 2022). Informal food markets are more prominent in African cities which include mostly street-vended RTE foods (Abrahale et al., 2019). A survey of street food vendors in Johannesburg, South Africa, showed that about 30% of the urban population depends on street foods for their daily dietary needs (Steyn et al., 2011) which calls for African countries to develop domestic standards and regulations for the protection of consumers.

Due to their convenience, RTE and REC snacks have become increasingly popular among consumers in the last five years, leading to the fastest-growing food sector. These food products attracted more attention from consumers because of their appealing qualities, affordable pricing, flavor, texture, appearance, etc. (Patel and Rathod, 2017; Temgire et al., 2021). The manufacture of RTE and RTC snack products has made widespread use of a variety of technologies because of their simple operating procedures, which obviates the need for extra cooking and preparation and instead concentrates on beautiful packaging. Because of their simple operating procedures that obviate the need for extra heating and preparation and instead concentrate only on appealing packaging, a variety of technologies have been widely employed in the manufacturing of RTE and RTC snack goods. Foods that are sweet, salted, fried, canned, fast food, baked, dried or preserved, extruded, etc. are all included in the category of RTE items. A variety of items, including bakery goods, extruded meals, quick snacks, fast food, breakfast cereals, biscuits, bars, etc., predominate in the traditional cereal intake. Because these food formulations are shelf-stable, lightweight, nutritionally enriched, and simple to shop for and store, they are easy for humans to consume and are highly accepted by customers. These days, RTE and ready-to-cook products are widely available in the food industry as they are the closest alternatives to our everyday meals. Most young consumers prefer to spend their money on things that are ready to cook and eat. Young customers, who are easy targets for RTE manufacturing companies, are driving up demand for these foods because of their convenience, pleasing taste, and long shelf life. Consumer interest in ready meals and snacks was primarily piqued by its widespread availability, convenience, and propensity for nibbling (Patheeparambi, 2020; Temgire et al., 2021). As a result, RTE and ready-to-cook food products are becoming increasingly popular among Indian and international consumers in place of conventional cooking methods. Cereals, pulses, and millets are used to make a variety of convenience foods, such as pasta, noodles, baked goods, extruded goods, fermented food, and weaning food (Jaybhaye, 2014; Temgire et al., 2021). The main categories of food that are referred to as RTE meals include bakery food, extruded food, fermented food, frozen food, and weaning food. Occasionally, these foods require brief preparation before consumption (Borah et al., 2019; Temgire et al., 2021).

When a food's primary function is to partially satiate hunger, it might provide quick energy and nutrients. Its convenient size and small packaging make it simple to handle and convert (Kumar et al., 2019). All set to cook and eat foods including fruits, vegetables, salad, processed meat, half-boiled meat, processed fish, samosas, pizza, porotah, veggie patties, shingara, etc. are primarily made as snacks or in the form of frozen food (Hassan et al., 2020, Temgire et al., 2021). Sant Anna et al., (2020) bring an overview on microbial safety of minimally processed vegetables. According to their statement, minimally processed veggies that are RTE must retain their nutritional and sensory qualities to lower the microbial load, inhibit the growth of dormant pathogens, and avoid cross-contamination. They raised public, business, and governmental awareness of foodborne illnesses linked to RTE consumption. Thus, to provide consumers with high-quality products, effective sanitization and safety throughout the production of RTE items are essential. An increasing trend is the consumption of RTE foods like pastries, French fries, and packed pancakes (samosas, patties), which are replacing traditional lunch or dinner fare like roti, vegetable curries, naan, and non-vegetarian products. RTE foods are said to provide empty calories instead of a diet that is well-balanced. It is stated that RTE foods offer empty calories rather than a well-balanced diet. The antioxidative and anticarcinogenic benefits of beta-carotene, vitamin C, and the B complex vitamins have become increasingly apparent in recent times. These vitamins are abundant in fruits and vegetables. Additionally, there is growing agreement that fruits and vegetables, particularly their micronutrients and other active ingredients, can help prevent the onset of chronic illnesses like cancer and heart disease.Moreover, there is growing agreement that fruits and vegetables, particularly their micronutrients and other active ingredients, can help prevent the onset of chronic illnesses like cancer and heart disease. For every element under investigation, the consumption of RTEC was substantially linked to a higher coverage of daily dietary needs. Elevated RTE levels consuming cereal were linked to both higher dairy product consumption and higher-quality breakfasts.

These days, iron deficiency anemia (IDA) is one of the major global health issues that affects both wealthy and poor nations equally. The metabolic syndrome is extremely prevalent in low-income women and their children (Reshi et al., 2020). It is estimated to affect 2 billion people worldwide, which puts a significant strain on the healthcare system and impedes economic expansion. The World Health Organization (WHO) identified India as one of the most seriously afflicted nations, with a significant frequency of anemia in women and preschoolers. Due to a diet low in iron, IDA is more common in children, adolescents, and pregnant women. It also affects 40% to 80% of mothers (Ezzati et al., 2002; Reshi et al., 2020).

11.6 DAIRY-BASED RTE FOODS

11.6.1 Acidophilus Milk

Due to increasing consumer demand, many novel products are coming in the scenario. In the sector of RTE foods, dairy products play a leading role. Particularly in the last few decades, the fermented varieties have gained much popularity on health-promoting grounds. They are even labeled as functional foods or nutraceuticals (Granato et al. 2010). The single or multiple bacterial species in these products are attributed to the beneficial qualities of these products. The rate of usage of fermented milk products has increased significantly in the last few decades (Pinto et al. 2017).

11.6.1.1 Health Benefits

Prebiotic food helps in stimulating growth of good bacteria, which increases the viable cell count during fermentation and storage, thereby enriching nutritional value besides enhancing sensory characteristics (Li et al. 2016). Although acidophilus milk has all the macronutrients like protein, vitamins, lipids, minerals, and microelements, the fermentation process is mainly responsible for the enrichment of macronutrients and the production of novel compounds. This process helps to improve the nutritional value of milk (Borresen et al. 2012) as found in the case of cheese, yogurt, and kefir. In addition to the added flavor, digestive promotion and creation of new functional compounds, fermentation with lactic acid bacteria lead to the production of essential vitamins such as B1, B2, B6, B12, A, and E (Borresen et al. 2012). Moreover, the different enzymatic reactions, i.e., glycolysis, lipolysis, and proteolysis result in the formation of many compounds with several biological benefits (Beermann and Hartung 2013).

Due to the rising health awareness among consumers, there has been tremendous demand for the use of solely natural ingredients to replace synthetic ingredients. To date, numerous studies have been conducted on the added value of acidophilus milk by fortifying it with many natural additives. Value addition to these probiotic food items has gained much popularity because of their enhanced nutritional value. There are several additives that can be used for fortification that lead to the production of functional foods. The most commonly used prebiotics are oligosaccharides, inulin, fructooligosaccharides (FOS), and galactooligosaccharides (GOS), which stimulate the growth of bifidobacteria and lactobacilli. These carbohydrates are not digestible by human enzymes but rather fermented by the gut microbiome. They showed beneficial effects in reducing the clinical symptoms of diarrhea by increasing the Bifidobacterium count in the human gut (Rigo Adrover et al. 2019). They also show beneficial effects on skin health by mode of reduction of phenol production by gut microbiome (Miyazaki et al. 2014; Kano et al. 2013). Fruit matrices such as black mulberry, red grapes, and cornelian cherry could act as probiotic vehicles by increasing the pH of the final product, thereby enhancing the stability and viability of the probiotic culture, especially for longer periods as lowering pH decreases the viable number of probiotic cultures. (Barat and Ozcan 2018). Moreover, fruits act as a storehouse of phenolic compounds and dietary fibers that can successfully improve gut health and increase the product's overall nutritional value (Barat and Ozcan 2018). The addition of dates and honey to acidophilus milk showed enrichment in carbohydrates, proteins, fibers, and unsaturated fatty acids such as linoleic and a-linolenic acids which are considered essential components as they are not synthesized inside our body (Hamad, Ismail, and Elraghy 2018). These are associated with an increase in antioxidant activity and provide protective effects against cardiovascular diseases, rheumatoid arthritis, and diabetes (Ismail, Hamad, and Elraghy 2018). Green tea is highly rich in catechins content, which has several health benefits related to oxidative stress, atherosclerosis, and obesity management (Najgebauer-Lejko 2014), hence it can be considered as an additive to fermented milk. Thus, it can be summed up that fortification of acidophilus milk with natural additives is a widely used approach to increase its nutritional value and improve its sensory characteristics and boost consumer satisfaction. However, detailed research involving these additives and their nutritional value has yet to be investigated.

11.6.1.2 Antioxidant

Acidophilus milk is an antioxidant-rich food which makes it less susceptible to oxidation during storage and is thus less prone to change in color and flavor (Chan et al. 2018). Moreover, ingesting such food reduces the risks of oxidative stress-related diseases, i.e., cancer and cardiovascular and inflammatory diseases (Pisoschi and Pop, 2015), growth and/or increased shelf-life stability upon storage (Li et al. 2016).

11.6.2 Yogurt

There has been a fast-growing consumer demand for yogurt and yogurt-related products which led to it becoming the fastest growing dairy category in the global market. Yogurt is considered to be one of the most nutrition-dense dairy foods. It is a very popular fermented dairy product obtained worldwide having high consumer acceptability due to its health benefits beyond just basic nutrition. It is a rich source of calcium and the calcium provided is in bio-available form and that too in significant amounts. It is a storehouse of almost all the essential amino acids and also a good source of milk proteins with a higher biological value. Yogurt is considered a probiotic carrier food that can deliver significant amounts of probiotic bacteria into the body which can claim specific health benefits once ingested, thereby promoting human health. Moreover, yogurt is reported to claim improved lactose tolerance, immune enhancement, and prevention of gastrointestinal disorders.

11.6.2.1 Health Benefits

Yogurt can help in achieving the overall nutritional recommendations such as Recommended Dietary Allowances of each nutrient on a daily basis. To exemplify, it provides calcium in bio-available form which is reported to provide 41% of the recommended daily requirement of calcium for a 5-year-old through a serving of 50 g of yogurt. Lactose is the main carbohydrate found in milk. Many individuals have lactose intolerance. It has been reported that lactose intolerance is associated with low calcium intake and bone mineral density. Yogurt is effective for individuals with lactose intolerance to attain all the benefits of milk products without causing associated discomfort. Proper wholesome nutrition and health depend upon the optimum balance of the intestinal microflora and yogurt acts as a probiotic carrier food and an ideal source for the delivery of viable probiotic strains, L. acidophilus and Bifidobacterium bifidum which are the most common probiotics used in the dairy industry (Parodi 2016). Probiotics are reported to have therapeutic effects such as prevention of urogenital infections, alleviation of constipation, protection against diarrhea, prevention of infant diarrhea, prevention of hypercholesterolemia, protection against colon/bladder cancer, and prevention of osteoporosis. However, probiotics are claimed to have other beneficial effects such as maintenance of normal intestinal flora, enhancement of the immune system, reduction of lactose intolerance and serum cholesterol levels, treatment of inflammatory bowel disease (IBD), and enhance anticarcinogenic activity, thereby resulting in an overall immunological enhancement.

Yogurt is a nutrient-dense food. Its easy availability and convenience in including in diets can help people to eat yogurt as part of a balanced meal that can contribute to improved physical health (Derrien et al., 2015). Yogurts are now being manufactured in numerous styles and varieties with different fat contents, flavors, and textures suitable for different meal occasions and plates as a snack, dessert, sweet, or savory food. Further studies need to be conducted to determine the role of yogurt in overweight and obesity, which can give a better idea regarding the recommendation of regular fat and/or fat-free yogurt in dietary guidelines.

11.6.3 Ice Cream

Ice cream can be defined as a frozen dairy product that is sweetened. It is consumed mainly as an after-meal dessert, or sometimes even as a snack. It is usually sweetened with sugar or sugar alternatives, or naturally by fruits. Flavorings and colorings are added in combination with stabilizers and emulsifiers. All the ingredients used for ice cream preparation like sugars, fats, dairy, stabilizer, water, and all the other required substances need to be correctly blended and emulsified together so there is nothing left behind that may reduce the quality of the final product. The standard ice cream serving temperature ranges from −11 to −18 °C.

11.6.3.1 Health Benefits

Ice cream has often been considered "junk" food, indicating its unhealthy nature. However, in reality, it is just the opposite. Ice cream is a relatively well-balanced, wholesome, easily digestible, and delicious food. Ice cream is often considered to be a nutritious part of the diet. Ice cream contains about four times as much carbohydrates as milk. In developing a formulation of the satisfactory composition, some of the factors that need to be considered are a personal preference of company management or customer demands for flavor, body and texture, and color characteristics of the finished product, that is, natural or fortified flavor with artificial flavoring; chewy to heavy, higher overrun; or more cooling body and texture characteristics. Even though ice cream is considered a nutritious food, it is high in fat and sugar which may be its limitation for consumption (Jadhav et al., 2014). The type of oil used in developing ice cream may be a risk for ice cream consumers especially in developing countries

In recent studies, it has been observed that ice cream finds utility in being used as a probiotic carrier because of its storage stability at low temperatures. However, the high fat and sugar content of ice cream makes it unsuitable for many individuals. Ice cream is high in sugar (20.7/100 g) and fat (16.0) which leads to health problems in public (Tripathi and Giri, 2014). This can also lead to microbial growth due to its high nutritional and water content. These are the two main issues with ice cream that need to be addressed to make it a better product. Research has been going on for the substitution of sugar with other similar but low-calorie ingredients which can be a solution to these issues.

Ice cream can be classified on the basis of its ingredients or the type of flavors used. Ice cream can be rightly termed as a nutritious food containing high amounts of carbohydrates, fats, protein, vitamins, and also important minerals. In addition to nutrition advantages, ice cream can be used as a probiotic carrier. Ice cream has the unique property that helps people suffering from stomach or throat ailments when other foods cannot be taken. Although ice cream is stored at a low (−18°C) storage temperature, contamination can occur at different stages which may also affect public health. A critical factor that needs to be addressed is to identify the strains of probiotics microorganisms that can survive in low storage temperatures.

11.6.4 Sweets

11.6.4.1 Vermicelli

MTR vermicelli is made with a special process that ensures it is nonsticky and does not form lumps. There are different varieties like short cut, long cut, or roasted vermicelli. One can use MTR vermicelli in so many marvelous ways by making creamy sweet kheer, savory upma, or tossing it with boiled vegetables for a wholesome pulao.

11.6.4.2 Instant Sweet Mixes

MTR Instant Sweet Mixes are a delightful and easy way to make the most mouthwatering delicacies at home. Whether it is a special occasion or just a regular meal, following the simple step-by-step instructions on the pack can easily help in preparing gulab jamun, badam feast, vermicelli payasam. A few research findings of value-added RTE foods are presented in Tables 11.3 and 11.4 RTE foods with enhanced nutritional benefits are provided.

11.7 EFFECT ON HEALTH

These food products undergo extreme processing, which either increases or diminishes their nutritional value and microbiological quality (Nayma et al. 2020). While these foods are readily available, convenient, and healthful, consuming too much of them might be detrimental to our well-being. TechnoPak (2009) in its annual report identifies the different trends in the Indian food market. It has been noted that 99% of consumers take health and nutrition into account while making purchases. An increasing number of people are moving toward the convenience and change that RTE provides: 355 to 40% of housewives regularly shop for packaged and convenience foods items like noodles, pasta, and soup powders, and "ready to cook" offerings from companies like MTR and Aashirvaad, regardless of whether they are shopping from traditional format or modern format grocery store.

11.7.1 Nutritional Benefits

Agte et al. (2002) studied folic acid, ascorbic acid, riboflavin, thiamine, and beta-carotene in 263 cooked food samples and 260 meals representing dietary patterns of Asia, Africa, Europe, the USA, and Latin America by spectrophotometry and photofluorometry. A broad range of beta-carotene (84–2038 mcg%), riboflavin (0.01–0.48 mg%), thiamine (0.04–0.36 mg%), vitamin C (1–28 mg%), and folate (26–111 mcg%) was observed in individual foods. Nutritional based RTE foods are given in Table 11.5. Bakery products and sweets were found to be poor sources and green leafy vegetables and fruits were good sources of these five vitamins. The differences between RTE foods and meals consumed during lunch or dinner were prominent for beta-carotene, ascorbic acid, riboflavin, and folic acid.

11.7.2 Adverse Health Effects

Certain fast-food items have elevated levels of fat, salt, sugar, and other chemicals. Overindulgence has a negative impact on health and can lead to unwanted health concerns like obesity, heart difficulties, and other issues (Hassan et al., 2020). Junk food items such as pizza, burgers, sandwiches, and seared chicken are high in fat content and contain ingredients like cheese and vegetable oil that contribute to weight gain. Foods high in sugar and salt, such as cakes, biscuits, bread, chips,

TABLE 11.3
Value-Added RTE Foods

Product	Value Addition	References
Ready-to-eat fruit flakes	Formulated from cereal mix with addition of dried banana, papaya, and apple flakes. Based on a number of factors, consumers find the sensory quality of fruit flakes to be somewhat satisfactory. Once hot milk and sugar are added, this food maintains its original flavor and is ready to be consumed.	Patheeparambil (2020)
Ready-to-eat extruded snack	Utilizing a composite combination of soybean, sorghum, and rice flour to create a protein-rich product	Omwamba and Mahungu (2014)
Fortified extruded product	Wherein 20% (w/w) of corn grit, defatted soybean meal, germinated brown rice meal, and mango peel fiber were added to create fortified extruded snacks. The product's taste and nutritional value are enhanced by the abundant protein and fiber content of the byproduct. It is also enhanced with antioxidant activity and phenolic chemicals, which are successfully utilized as dietary supplements.	Korkerd et al. (2016)
Breakfast cereal using low-amylose rice flour, seeded banana (Musabalbisiana, ABB) and carambola (*Averrhoa carambola* L.) pomace	Fiber and mineral rich product. Pomace from the carambola juice industry can be used to make products rich in minerals and useful ingredients. The primary goal of the research is to reduce agricultural waste and make use of naturally occurring phytochemicals using the extruded technique.	Borah et al. (2019)
Finger millets-based RTE extruded snacks	Enhanced with a variety of grains, including cornmeal, rice flour, cheese, and banana powder, high-quality extruded product.	Akhil et al. (2018)
Ready-to-eat breakfast cereal products using extrusion technology	Compared to other samples that included 12 morning cereals made of rice and corn, this one had a lower glycemic index. When made with oats, the extruded sample is sugar-free.	Yaman et al. (2019)
Buckwheat and bread waste RTE snack	This product lowers fat and is high in fiber, protein, and minerals. These nutritious treats are created from buckwheat and leftover bread. Consuming fruit, milk, and ready-to-eat cereals at breakfast is beneficial for a balanced diet.	Temgire et al. (2021)
Baobab-based ready-to-eat sorghum cowpea blend snack bars (Kenya)	This cereal-based snack bar can be used to boost nutrient intake from homemade snacks by diversifying your diet.	Momanyi et al. (2020)
	Treatment/processing	
Yao meat products	A unique alternative that checks the quality and shelf life of ready-to-eat meat products is nano emulsion based active coating (NEAC), which combines a blend of natural antimicrobials.	Liu et al. (2020)
Semi-dried ready-to-eat tomato	The combined effects of reduced atmospheric pressure and an antibacterial agent (potassium sorbate) extend the shelf life of semi-dried tomatoes	Lucera et al. (2017)
Ready-to-eat strawberries.	Edible algae (Palmaria palmata) coated in chitosan. In addition to lowering product loss and respiration rate, CH-PC can lower microbial burden and increase nutrient value.	Rico et al. (2019)
Ready-to-eat blueberries	The qualitative characteristics of edible coatings with additional dietary fiber. They used dietary fiber, chitosan edible coating, and sodium alginate, and they kept the mixture at 5°C for 18 days. It greatly reduces the growth of microorganisms, boosts antioxidant qualities, and improves the nutritional value and overall sensory appeal of blueberries.	Alvarez et al. (2018)

TABLE 11.4
RTE Foods with Nutritional Benefits

Product	Nutritional Benefits	References
Barley-based extruded ready-to-eat snacks	Low glycemic index (GI), low in iron. Barley has a low glycemic load and GI, but a high β-glucan content (GL)	Reshi et al. (2020)
Healthy RTE breakfast cereal from popped pearl millet	Breakfast cereal's moisture level was 3.16%, while its protein, fat, and crude fiber concentrations were 9.34%, 8.26%, and 0.1 g per 100 g, respectively. The estimated amounts of calcium, phosphorus, iron, zinc, folic acid, and other minerals and vitamins per 100 g were 62.79%, 239.53%, 5.02%, 103.21%, and 3.24%, respectively. Additionally, the investigation showed that RTE breakfast cereals had a respectable level of dietary fiber overall (6.13%). The designed morning cereal met the RDA for energy, protein, fat, calcium, and iron in each serving with 22.8%, 12.80%, and 13%, respectively.	Kumari et al. (2019)
Ready-to-eat iron-rich extruded snacks for anemic population (barnyard millet, defatted soy flour, amla (Indian gooseberry) and rice flour)	Iron (15.71 mg/100 g) and protein (18.91 g/100 g).	Suri et al. (2020)
Ready-to-eat-breakfast cereals (RTEBCs) in Irish adults and its impact on adequacy and safety of micronutrient intakes and compliance with dietary recommendations	Found daily intake of carbohydrate (8.1%), dietary fiber (9.8%), starch (10.8%), and non-starch polysaccharides (NSP) (10.8%) in consumers. Increased consumption was associated with a more fiber-dense diet and with greater compliance with dietary recommendations for fat, carbohydrate, and NSP. Fortified RTEBCs contributed significantly to mean daily intakes of iron (18%), thiamin (14%), riboflavin (17%), niacin (15%), vitamin B6 (13%), total folate (18%), and vitamin D (10%), and most of the contribution was from micronutrients added to RTEBCs is associated with improved compliance with dietary recommendations for fat, carbohydrate and fiber, with a more micronutrient-dense diet and a reduced risk of dietary inadequacy for calcium, iron, riboflavin, and folate, without increasing the risk of excessive intakes of micronutrients.	Galvin et al. (2002)
Finger millet-based extruded products with composite flour (finger millet, maize, rice, soybean, Bengal gram, and skimmed milk powder)	Composite flour (finger millet:maize:rice:soybean) in the ratios of 20:50:20:10. Product is rich in minerals and fiber	Sawant et al. (2013)

Kurkure, lollipops, and so on, make it tastier. However, when consumed in excess, these foods have a negative impact on health.

Berthold-Pluta et al. (2017) determined the microbiological quality of RTE food products derived from plants. Twenty samples each of vegetables, sprouts, and unpasteurized fruit and vegetable juice were among the 60 RTE samples that were examined. According to this investigation, cronobacter spp. was found in about 21 samples, with sprout samples having the highest level of microbial contamination. Garibaldi et al. (2019), after inoculating RTE swordfish with the probiotic strain Lactobacillus paracasei for three months at 40°C, assess the quality of the fish.

TABLE 11.5
RTE Products' Population Studies

Population Studies	Outcomes	Reference
A random sample of 3534 subjects aged 2 to 24 years in Spain was studied. Food and nutrient intakes were determined by a 24-hour recall. RTEC consumption was assessed by a quantitative food frequency questionnaire. Cereal consumption was classified into non-consumers and daily intakes between 1 and 20 g, 21 and 40 g, and more than 40 g. After excluding the under reporters, the final sample consisted of 2852 individuals.	About half of the population (49.8%) reported eating RTEC. The profile of macronutrients improved when cereal consumption increased. While the intake of niacin and folate improved in practically all groups and that of calcium, iron, and vitamin D in at least half of the groups, the intake of thiamine, riboflavin, and vitamin B6 rose significantly with increasing consumption of RTEC in all age–sex groups.	van den Boom et al. (2006)
Dietary characteristics of breakfast cereals in European adolescents due to its popularity among all the breakfast options. Milk and fruit intake at breakfast gives higher glucose and fructose which is better for the consumer than bread and other breakfast.	Cereals that are ready to eat are not the best option for a complete meal; milk and fruits	Michels et al. (2016)

11.8 PACKAGING AND SAFETY OF RTE FOODS

To efficiently contain the food and expose it to the consumer in an appealing way, a variety of packaging styles must be used. Although ready meals are more expensive to produce than their component ingredients, they also fetch a greater price at the register. They also spoil far more easily than their component ingredients alone because of their complexity and the high degree of manipulation required in their creation. Concerns about microbiological development and the relatively short shelf life of these items are critical to both the end user's safety and the successful launch and sales of these products. Because of its widespread effect on buyers, its presence at the pivotal point when a purchase decision is made, and the high degree of consumer involvement when they actively scan packages to inform their decisions, package design is a very important medium. Different reactions are elicited by the package design. Particularly in the Indian setting, the importance of packaging has greatly increased while purchasing RTE meals. It has been noted that various RTE food varieties call for various packaging options. Food spoiling in packages is a result of microbial growth, particularly in environments with high oxygen content. Remaining oxygen levels in RTE meals can be effectively reduced with the use of modified atmosphere packaging, or MAP for many years; food oxidation has been controlled and the organoleptically determined shelf life of foods has been extended by using nitrogen, typically in combination with CO_2, as the preferred MAP gas. By preventing the formation of microorganisms, carbon dioxide gives the MAP atmosphere antibacterial activity and increases the shelf life of food as evaluated by microbes. Because of argon's distinct physical characteristics, particularly its increased density, it is significantly more effective than nitrogen at eliminating and/or excluding oxygen from packages. In practical applications, argon can yield substantially lower quantities of residual oxygen. Microbes' respiration is successfully suppressed by argon-based MAP mixes because, in contrast to nitrogen, argon suppresses oxidase activity. Because argon MAP requires less carbon dioxide (CO_2) than nitrogen MAP to limit microbial growth, utilizing argon in combination with CO_2 yields a higher level of antimicrobial action than nitrogen MAP. Because carbon dioxide degrades food products, using less of it will result in improved food quality preservation. Therefore, even though argon MAP is far better at preventing oxidation and microbiological growth and extends the shelf life of a variety of ready meals, nitrogen

MAP is still a reasonably priced method of food preservation. Additionally, argon MAP has been successfully designed, tested, and financially marketed for the preservation of many products, including ready meals.

As prepared and RTE meals are more likely to be contaminated by consumers and to be exposed to pathogens during the post-process phase, food safety procedures are unquestionably essential. For processors nowadays, preventing cross-contamination prior to product packaging is the most important concern. Either the internal environment—such as microbial populations on food contact surfaces—or the exterior environment—such as airborne pollutants and heated temperatures—can induce contamination in food processing.

11.9 PRECAUTIONARY MEASURES

Enlisted are a few of the precautionary measures that need to be followed with respect to RTE foods.

- Understanding which category of RTE foods suit which safest format: pre-packed or foods on request.
- Storage conditions should be read and followed properly to ensure their safety.
- Strictly follow the expiration date of pre-packed and foods on request RTE dairy and meat products.
- Maintaining a good food safety attitude toward the maintenance of the food cold chain, by keeping the RTE dairy and meat products in the refrigerator.
- Storing the RTE meat and dairy food products on separate shelves from fresh fruits/vegetables in the refrigerator.

It was analyzed from the consumers purchasing habits that they usually purchase RTE dairy or meat food products mostly in the supermarket, and in pre-packed format and these products are more frequently consumed as ingredients for salads or sandwiches. These RTE products, after opening, are usually stored in the refrigerator for a period of about 5 days but vary with the food products. Globally, the results reveal that consumers are conscious and follow assertive attitudes toward food safety. Most importantly, appropriate hygienic practices should be followed by the workers in food preparing places. Not only that, consumers also play an important role in terms of food safety, to decrease the risk of food contamination.

11.10 FUTURE STUDIES

However, further work needs to be done to assess the bioavailability of these fermented milk products and also to analyze the changes that occur in the human gut microbiome after ingestion. Very limited studies have been conducted on the interaction between different strains within the milk or with the starter culture during the process of fermentation and storage. More research should be performed on these interactions to optimize the fermentation conditions and to simultaneously investigate the influence of different variables on product quality using statistical design.

11.11 GAP BETWEEN DEVELOPMENT AND FUTURE RECOMMENDATIONS

SAM affects millions of individuals. Lack of nutrition has a detrimental effect on all body functions, leading to major pathological problems like edema and even death (Vijay, 2018). Microbiological (M), chemical (C), and nutritional (N) risks are assessed separately by expert groups in food safety agencies such as the European Food Safety Authority (EFSA) in Europe. The Food and Agriculture Organization (FAO) has recommended Multi-Criteria Decision Analysis (MCDA) to rank public health impacts and include factors such as economic losses, food security, consumer perception, and sociocultural concerns (FAO, 2017).

11.12 CONCLUSION

According to the research on consumers' perceptions, preferences, and choices for RTE foods, consumers primarily choose brands based on factors like safety and trust. These two factors foster enduring brand associations. Trust and safety have emerged as the most important criteria when buying RTE food goods, despite the fact that consumers still take other considerations into account. Purchase intention is not much affected by brand qualities, brand endorsements, or brand ethical issues. Thus, it can be inferred that a variety of factors influence our food preferences, including price, functional aspects, health, sensory aspects, and sociocultural aspects, as well as the consumer's sociodemographic characteristics, which include age, income level, gender, and location and size of home. The next stage could be to create a rule that combines the safety and nutritional dimensions into an on-pack food scoring logo, or it could involve making this information available to the general population through mobile apps that aggregate numerous hazards associated with microbiological, chemical, and nutritional factors.

REFERENCES

Abrahale, K., Sousa, S., Albuquerque, G., Padrãao, P., Lunet, N. (2019). Street food research worldwide: a scoping review. *Journal of Human Nutrition and Dietetics*, 32, 152–174.

Agte, V., Tarwadi, K., Mengale, S., Hinge, A., Chiplonkar, S. (2002). Vitamin profile of cooked foods: how healthy is the practice of ready-to-eat foods? *International Journal of Food Sciences and Nutrition*, 53, 197–208.

Akhil, K. G., Gopika, P., Baiju, S., & Sudheer, K. P. (2018). *Development and quality evaluation of hot extruded RTE products from speciality rice* (Doctoral dissertation, Department of Post-Harvest Technology and Agricultural Processing).

Alvarez, M.V., Ponce, A.G., Moreira, M.R. (2018). Influence of polysaccharide-based edible coatings as carriers of prebiotic fibres on quality attributes of ready-to-eat fresh blueberries. *Journal of the Science of Food and Agriculture*, 98(7), 2587–2597.

van den Boom, A., Serra-Majem, L., Ribas, L., Joy Ngo, R.D., Pérez-Rodrigo, C., Aranceta, J., Fletcher, R. (2006). The contribution of ready-to-eat cereals to daily nutrient intake and breakfast quality in a Mediterranean setting. *Journal of the American College of Nutrition*, 25(2), 135–143.

Barat, A., Ozcan, T. (2018). Growth of probiotic bacteria and characteristics of fermented milk containing fruit matrices. *International Journal of Dairy Technology*, 71(March), 120–129.

Barrett, C.B., Reardon, T., Swinnen, J., Zilberman, D. (2019). Structural transformation and economic development: insights from the agri-food value chain revolution, pp. 1–56. Ithaca, NY: Dyson School of Applied Economics and Management, Cornell University.

Barrett, C.B., Reardon, T., Swinnen, J., Zilberman, D. (2020). Agri-food value chain revolutions in low- and middle-income countries. *Journal of Economic Literature*, 60, 1316–1377.

Beermann, C., Hartung, J. (2013). Physiological proper ties of milk ingredients released by fermentation. *Food Functions*, 4(2), 185–199.

Berthold-Pluta, A., Garbowska, M., Stefańska, I., et al. (2017). Microbiological quality of selected ready-to eat leaf vegetables, sprouts and non-pasteurized fresh fruit–vegetable juices including the presence of *Cronobacter* spp. *Food Microbiology*, 65, 221–230.

Bhutta, Z.A., Das, K.J., Rizvi, A., Gaffey, F.M., Walker, N., Horton, S., Webb, P., Lartey, A., Black, E.R., Lancet Nutrition Interventions Review Group, Maternal and Child Nutrition Study, Group. (2013). Evidence-based intervention for the improvement of maternal and child nutrition: what can be done and at what cost? *Lancet*. 382(9890), 452–477. doi: 10.1016/s0140-6736(13)60996-4.

Borah, A., Chutia, H., Balasubramanian, S., Mahanta, C. (2019). Role of food extrusion in development of healthy food products: a review. *Acta Scientific Nutrition Health*, (4), 39–44.

Borresen, E.C., Henderson, A.J., Kumar, A., Weir, T.L., Ryan, E.P. (2012). Fermented foods: patented approaches and formulations for nutritional supplementation and health promotion. *Recent Patents on Food, Nutrition & Agriculture*, 4(2), 134–140.

Cardello, H., Wolfson, J. (2013). Lower-calorie foods and beverages drive healthy weight commitment foundation companies' sales growth: Interim report. Retrieved from the Healthy Weight Commit Foundation website. www.hudson.org/sites/default/files/researchattachments/attachment/1107/lowercalhealthyweightcommitment--may2013.pdf.

Chan, C.-L., Gan, R.-y., Shah, N.P., Corke, H. (2018). Enhancing antioxidant capacity of *Lactobacillus acidophilus* fermented milk fortified with pomegranate peel extracts. *Food Bioscience*, 26(December), 185–192.

Chatthipmongkol, M., Jangphanish, K. (2016). Factors influencing consumer decision making process of Thai frozen food products. *International Business Management*, 10, 166–175.

Chung, S. (2015). Localization of FamilyMart, a Japanese convenience store, in Taiwan (in Japanese language). *Industrial Engineering & Management*, 12, 133–155. DOI: 10.4172/2169-0316.1000169.

Ciliberto, M. A., Sandige, H., Ndekha, M. J., Ashorn, P., Briend, A., Ciliberto, H. M., & Manary, M. J. (2005). Comparison of home-based therapy with ready-to-use therapeutic food with standard therapy in the treatment of malnourished Malawian children: a controlled, clinical effectiveness trial. *The American journal of clinical nutrition*, *81*(4), 864-870.

Costa, A.I.A., Schoolmeester, D., Dekker, M., Jongen, W.M.F. (2007). To cook or not to cook: a means-end study of motives for choice of meal solutions. *Food Quality and Preference*, 18, 77–88.

Derrien, M., van Hylckama Vlieg, J.E. (2015). Fate, activity, and impact of ingested bacteria within the human gut microbiota. *Trends in Microbiology*. 23(6), 354–366.

Dublin. (2014). India ready-to-eat food market forecast & opportunities, 2019—leading companies are ITC, Kohinoor, Haldiram's, McCain, MTR. Retrieved from the Research and Markets website. www.businesswire.com/news/home/20140815005337/en/Research-and-Markets-India-Ready-to-Eat-Food-Market-Forecast-Opportunities-2019---Leading-Companies-are-ITC-Kohinoor-Haldirams-McCain-MTR

Euromonitor. (2015a). Market research on ready meals. Retrieved from the Euromonitor website.

Euromonitor. (2015b). Ready meals in Thailand. Retrieved from the Euromonitor website www.euromonitor.com/ready-meals-in-thailand/report

Euromonitor. (2015c). Ready meals in Japan. Retrieved from the Euromonitor website www.euromonitor.com/ready-meals-in-japan/report

Ezzati, M., Lopus, A.D., Dogers, A., Vanders, H.S., Murray, C. (2002). Selected major risk factors and global and regional burden of disease. *Lancet*, 360, 34–60. DOI: 10.1016/S0140-6736(02)11403-6

FAO. (2017). Food safety risk management evidence-informed policies and decisions, considering multiple factors. FAO Guidance Materials Food Safety and Quality Series 4. Retrieved from www.fao.org/3/i8240en/I8240EN.pdf accessed April 25, 2023.

Food and Drug Administration (FDA). (2009). FDA food code 2009: Chapter1—purpose & definitions. Retrieved from the FDA website. www.fda.gov/food/fda-food-code/food-code-2009

Food Standards Agency. (2011). *E. coli* O157 control of cross-contamination guidance for food business operators and enforcement authorities. www.food.gov.uk/sites/default/files/media/document/ecoli-cross-contamination-guidance.pdf

Galvin, M.A., Kiely, M., Flynn, A. (2002). Impact of ready-to-eat breakfast cereal (RTEBC) consumption on adequacy of micronutrient intakes and compliance with dietary recommendations in Irish adults. *Public Health Nutrition* 6(4), 351–363.

Gatley, A., Caraher, M., Lang, T. (2014). A qualitative, cross-cultural examination of attitudes and behaviour in relation to cooking habits in France and Britain. *Appetite*, 75, 71–81.

Giribaldi, M., Gai, F., Peiretti, P.G., et al. (2018). Quality of ready-to-eat swordfish fillets inoculated by *Lactobacillus paracasei* IMPC 2.1: quality of probiotic ready-to-eat swordfish fillets. *Journal of the Science of Food and Agriculture*, 99 (1), 199–209. doi: 10.1002/jsfa.9161.

Godswill, A.C., Somtochukwu, I.V., Amagwula, I.O. (2020). Ready-to-use therapeutic foods (RUTFs) for remedying malnutrition and preventable nutritional diseases. *International Journal of Advanced Academic Research | Sciences, Technology and Engineering*, 6(1), 47–81.

Granato, D., Branco, G. F., Nazzaro, F., Cruz, A. G., Faria, J. A. (2010). Functional foods and nondairy probiotic food development: trends, concepts, and products. *Comprehensive Reviews in Food Science and Food Safety*, 9(3), 292–302.

Hamad, M.N.F., Ismail, M.M., Elraghy, E.M. (2018). Effect of goat's milk supplementation with Tamr on free fatty acids, antioxidant activity, and probiotic bacteria of Bio-Rayeb milk. *Journal of Dietary Supplements*, 15(1), 108–123.

Hassan, S.A., Bhateja, S., Arora, G., Prathyusha, F. (2020). Impact of junk food on health. *Journal of Management Research and Analysis*, 7(2), 57–59.

Isanaka, S., Nombela, N., Djibo, A., Poupard, M., Van Beckhoven, D., Gaboulaud, V., Guerin, P.J., Grais, F.R. (2009). Effect of preventive supplementation with the ready-to-use therapeutic food (RUTF) on the nutritional status, mortality & morbidity of children 6–60 months in Niger: a cluster randomized trial. *JAMA*. 301(3): 277–285.

Ismail, M., Hamad, M., Elraghy, E.M. (2018). Quality of Rayeb milk fortified with Tamr and honey. *British Food Journal*, 120 (2):499–514.

Jadhav, A.S., Raut, P.D. (2014). evaluation of microbiological quality of ice creams. Marketed in Kolhapur City, Maharashtra, India *International Journal of Current Microbiology and Applied Sciences* 3(9): 78–84.

Jaybhaye, R.V., Pardeshi, I.L., Vengaiah, P.C., Srivastav, P.P. (2014). Processing and technology for millet-based food products: a review. *Journal of Ready to Eat Food*, 1(2), 32–48.

Kano, M., Masuoka, N., Kaga, C., Sugimoto, S., Iizuka, R., Manabe, K., Sone, T., Oeda, K., Nonaka, C., Miyazaki, K., et al. (2013). Consecutive intake of fermented milk containing bifidobacterium breve strain yakult and gal acto-oligosaccharides benefits skin condition in healthy adult women. *Bioscience of Microbiota, Food and Health*, 32(1), 33–39.

Korkerd, S., Wanlapa, S., Puttanlek, C., Uttapap, D., & Rungsardthong, V. (2016). Expansion and functional properties of extruded snacks enriched with nutrition sources from food processing by-products. *Journal of Food Science and Technology*, 53, 561–570.

Kumar, P., Verma, A.K., Kumar, D., Umaraw, P., Mehta, N., Malav, O.P. (2019). Meat snacks: a novel technological perspective. In *Innovations in Traditional Foods*. Woodhead Publishing, pp. 293–321.

Kumari, R., Singh, K., Singh, R., Bhatia, N., Nain, M.S. (2019). Development of healthy ready-to-eat (RTE) breakfast cereal from popped pearl millet. *Indian Journal of Agricultural Sciences*, 89(5), 877–881.

Li, S., Ma, C., Gong, G., Liu, Z., Chang, C., Xu, Z. (2016). The impact of onion juice on milk fermentation by *Lactobacillus acidophilus*. *LWT—Food Science and Technology* 65(January), 543–548.

Lievore, Paolla, Deise R. S. Sim~ oes, Karolline M. Silva, Northon L. Drunkler, Ana C. Barana, Alessandro Nogueira, and Ivo M. Demiate. 2015. Chemical characterisation and application of acid whey in fermented milk. Journal of Food Science and Technology 52 (4):2083–2092. doi: 10.1007/s13197-013-1244-z.

Liu, Q., Zhang, M., Bhandari, B., Xu, J., & Yang, C. (2020). Effects of nanoemulsion-based active coatings with composite mixture of star anise essential oil, polylysine, and nisin on the quality and shelf life of ready-to-eat Yao meat products. *Food Control*, 107, 106771.

Lucera, A., Conte, A., Gammariello, D., Del Nobile, M.A. (2017). Ready-to-eat semi-dried tomatoes: study to preserve the product quality. *Journal of Food Processing and Preservation*, 41(5), e13175.

Malik, T., Saxena, M., Sonu, K.S. (2018). Development of a ready to cook curry. *International Journal of Innovative Science and Research Technology*, 3(3), 354–357.

Manary, J.M. (2005). Comparison of the home-based therapy with the ready-to-use therapeutic food (RUTF) with standard therapy in treatment of malnourished Malawian children: controlled, clinical effectiveness trial. *American Journal of Clinical Nutrition*, 81(4), 864–870. doi: 10.1093/ajcn/81.4.864.

Michels, N., De Henauw, S., Beghin, L., Cuenca-García, M., Gonzalez-Gross, M., Hallstrom, L., ... & Claessens, M. (2016). Ready-to-eat cereals improve nutrient, milk and fruit intake at breakfast in European adolescents. *European Journal of Nutrition*, 55, 771–779.

Miyazaki, K., Masuoka, N., Kano, M., Iizuka, R. (2014). Bifidobacterium fermented milk and galacto-oligosaccharides lead to improved skin health by decreasing phenols production by gut microbiota. *Beneficial Microbes*, 5(2), 121–128.

Momanyi, D., Owino, W., & Makokha, A. (2020). Formulation, nutritional and sensory evaluation of baobab based ready-to-eat sorghum and cowpea blend snack bars. *Scientific African*, 7, e00215.

Najgebauer-Lejko, D. (2014). Effect of green tea supplementation on the microbiological, antioxidant, and sensory properties of pro biotic milks. *Dairy Science & Technology*, 94(March), 327–339.

National Statistical Office. (2013a). Report of the 2013 household socio-economic survey whole kingdom.

National Statistical Office. (2013b). Summary of the consumer behavior survey 2013 (in Thai language).

Nayma, K., Das, K.C., Alice, E.J., Mehbub, M.F., Islam, M.T. (2020). Extension of shelf-life of ready-to-cook (RTC) pangas fish (*Pangasianodon hypophthalmus*) curry by modified atmosphere packaging at chilled storage. In *IOP Conference Series: Earth and Environmental Science*. IOP Publishing, 414(1), 012015.

Nielsen, E. M., Björkman, J. T., Kiil, K., Grant, K., Dallman, T., Painset, A., ... & Strachan, N. (2017). Closing gaps for performing a risk assessment on Listeria monocytogenes in ready-to-eat (RTE) foods: activity 3, the comparison of isolates from different compartments along the food chain, and from humans using whole genome sequencing (WGS) analysis. *EFSA Supporting Publications*, 14(2).

Nuntasriwiwat, E. (2012). Consumer behavior towards frozen food in Paragon shopping mall, Bangkok. Retrieved from The Chiang Mai University library

Omwamba, M., & Mahungu, S. M. (2014). Development of a protein-rich ready-to-eat extruded snack from a composite blend of rice, sorghum and soybean flour. *Food and Nutrition Sciences, 2014*.

Parodi, P.W. (2016). Cooperative action of bioactive components in milk fat with PPARs may explain its anti-diabetogenic properties. *Medical Hypotheses*, 89, 1–7.

Patel, D., Rathod, R. (2017). Ready to eat food perception, food preferences and food choice: a theoretical discussion. *World Wide Journal of Multidisciplinary Research and Development*, 3(8), 198–205.

Patheeparambil, A.P. (2020). Development of ready to eat fruit flakes. *Food and Agriculture Spectrum Journal*, 1(4).

Pinto, Jorge M. S., Sergio Sousa, Dina M. Rodrigues, Francisco X. Malcata, Armando C. Duarte, Teresa A. P. Rocha-Santos, Ana C. Freitas, and Ana M. Gomes. 2017. Effect of probiotic Co-cultures on physico-chemical and biochemical properties of small ruminants' fermented milk. International Dairy Journal 72 (September):29–35. doi: 10.1016/j.idairyj.2017.04.005.

Pisoschi, A.M., Pop, A. (2015). The role of antioxidants in the chemistry of oxidative stress: a review. *European Journal of Medicinal Chemistry*, 97(June), 55–74.

Poissant, R., Mariotti, F., Zalko, D., Membre, J.M. (2023). Ranking food products based on estimating and combining their microbiological, chemical and nutritional risks: method and application to ready-to-eat dishes sold in France. *Food Research International*, 169, 112939. https://doi.org/10.1016/j.foodres.2023.112939

Rani, R., Dharaiya, C. N., Singh, B. (2021). Importance of not skipping breakfast: a review. *International Journal of Food Science and Technology*, 56, 28–38.

Remnant, J., Adams, J. (2015). The nutritional content and cost of supermarket ready meals. Cross-sectional analysis. *Appetite*, 92, 36–42.

Research and Market India Ready-To-Eat Food (RTE) Market Study, 2013–2023: Analysis by Segment, Distribution Channel and State, Featuring Profiles of Leading Players 2020.Accessed from www.globenewswire.com/news-release/2020/01/09/1968244/0/en/india-ready-to-eat-food-rte-market-study-2013-2023-analysis-by-segment-distribution-channel-and-state-featuring-profiles-of-leading-players.html

Reshi, M., Shafiq, F., Hussain, S.Z., Naseer, B. (2020). Tawheed amin. *Journal of Food Processing and Preservation*, e14606, 1–13. doi: 10.1111/jfpp.14606.

Rico, D., Barcenilla, B., Meabe, A., González, C., & Martín-Diana, A. B. (2019). Mechanical properties and quality parameters of chitosan-edible algae (Palmaria palmata) on ready-to-eat strawberries. *Journal of the Science of Food and Agriculture*, 99(6), 2910–2921.

Rigo-Adrover, M.D., Mar, K., Knipping, J., Garssen, K., van Limpt, J., Knol, A., Franch, M., Castell, M.J., Rodrıguez-Lagunas, Perez-Cano, F.J. 2019. Prevention of rotavirus diarrhea in suckling rats by a specific fermented milk concentrate with prebiotic mixture. *Nutrients*, 11(1), 189.

Sakata, T., Akagaki, C., Akamatsu, R., Hagiwara, T., Hayashi, N., Li, H., Kato, H., Kurotani, K., Ishikawa-Takata, K., Omoto, S., Tanaka, T., Yokomukai, Y. (2021). Current situation and perspectives of ready-to-eat food/meal Suppliers Takashi Sakata for Working Group 2 of the Healthy Diet Research Committee of International Life Sciences Institute, Japan. *Nutrition Reviews VR*, 78(S3), 27–30.

Sant'Anna, P., Franco, B., Maffei, D. (2020). Microbiological safety of ready-to-eat minimally processed vegetables in Brazil: an overview. *Journal of the Science of Food and Agriculture*, 100: 4664–4670. doi: 10.1002/jsfa.10438.

Sante publique, F. (2023 Update). *Nutri-score*. Retrieved April 25, 2023, from www.santepubliquefrance.fr/import/nutri-score-2023-

Sawant, A.A., Thakor, N.J. Swami, S.B., Divate, A.D. (2013). Physical and sensory characteristics of ready-to-eat food prepared from finger millet based composite mixer by extrusion cooking. *Agricultural Engineering International: CIGR Journal*, 15(1), 100–105.

Sharma, J. (2009). Readymade in India. In Super Market Food Business, Progressive Grocer, April 2009, pp. 56–62. Made-in-india.pdf

Spencer, K.C. (2005). Modified atmosphere packaging of ready-to-eat foods. In *Innovations in Food Packaging*. Elsevier Ltd. ISBN: 0-12-31 1632–5.

Suri, S., Dutta, A., Shah, N.C., Raghuvanshi, R.S., Singh, A., Chopra, C.S. (2020). Numerical optimization of process parameters of ready-to-eat (RTE) iron rich extruded snacks for anemic population. *LWT—Food Science and Technology*, 134, 110164.

Steyn, N.P., Labadarios, D., Nel, J.H. (2011). Factors which influence the consumption of street foods and fast foods in South Africa—a national survey. *Nutrition Journal*, 10, 104.

Technopak. (2009). *Consumer Trends 2009: Changing Food Trends*, January 2009.

Temgire, S., Borah, A., Kumthekar, S., Idate, A. (2021). Recent trends in ready to eat/cook food products: a review. *Pharma Innovation Journal*, 10(5), 211–217.

Thienhirun, S., Chung, S. (2017). Consumer attitudes and preferences toward cross-cultural ready-to-eat (RTE). *Food, Journal of Food Products Marketing*, 24(1): 1–24.

Thienhirun, S., & Chung, S. (2018). Consumer attitudes and preferences toward cross-cultural ready-to-eat (RTE) food. *Journal of Food Products Marketing*, 24(1), 56–79.

Tian, Huaixiang, Yongbo Shen, Haiyan Yu, Yujie He, and Chen Chen. 2017. Effects of 4 probiotic strains in coculture with traditional starters on the flavor profile of yogurt. Journal of Food Science 82 (7):1693–1701. doi: 10.1111/1750-3841.13779.

Toda, Y. (2015). The recent history of the home meal replacement industry in Japan. In *Proceedings of the Conference on the 29th Conference on Japan Society for Distribution Science, Tokyo, Japan*.

Tripathi, M.K., Giri, S.K. (2014). Probiotic functional foods: survival of probiotics during processing and storage. *Journal of Functional Foods*, 9(1), 225–241.

Vijay, D.W. (2018). Ready to use therapeutic food formulation & packaging for malnutrition: an overview. *Nutrition and Food Science International Journal*. 4(5), 555648. doi: 10.19080/NFSIJ.2018.04.555648.

Vinderola, C. G., M. Gueimonde, T. Delgado, J. A. Reinheimer, and CGde los Reyes-Gavilan. 2000. Characteristics of carbonated fer mented milk and survival of probiotic bacteria. International Dairy Journal 10 (3):213–220. doi: 10.1016/S0958-6946(00)00031-5.

Waage, J., Grace, D., Fèvre, E.M., McDermott, J., Lines, J., Wieland, B., Naylor, N.R., Hassell, J.M., Chan, K. (2022). Changing food systems and infectious disease risks in low-income and middle-income countries. *Lancet Planet Health*, 6, e760–e768.

White Paper of Japan Ready-Made Meal Association. (2018). (Digest on-line edition). Japan ready-made meal association. www.nsouzai-kyoukai.or.jp/wp-content/uploads/hpb-media/hakusho2018_digest1.pdf

World Health Organization. (1999). *Management of Severe Malnutrition. Manual for Physicians & Other Senior Health Workers*. Geneva: WHO.

Yaman, M., Sargın, H. S., & Mızrak, Ö. F. (2019). Free sugar content, in vitro starch digestibility and predicted glycemic index of ready-to-eat breakfast cereals commonly consumed in Turkey: An evaluation of nutritional quality. *International Journal of Biological Macromolecules*, 135, 1082–1087.

Yano Research Institute. (2014). Ready-to-eat foods market in Japan: key research findings 2014. Retrieved from the Yano Research Institute website. www.yanoresearch.com/press/pdf/1257.pdf

12 Biogenic Amines in Ready-to-Eat Foods

Dipali Saxena[1], Ira Tripathi[2], Murod Kurbanov[3], and Amritpal Singh[4]*
[1]Assistant Professor, Shri Vaishnav Institute of Home Science, Shri Vaishnav Vidyapeeth Vishwavidyalaya, Indore, Madhya Pradesh, India
[2]Assistant Professor, Department of Home Science, Kashi Naresh Government Post Graduate College Gyanpur, Bhadohi, Uttar Pradesh, India
[3]Head, Professor, Department of Food Technology, Bukhara Engineering Technological Institute, Bukhara, Uzbekistan
[4]Scholar, Department of Food Technology, Mata Gujri College, Fatehgarh Sahib, Punjab, India
*Corresponding Author

12.1 INTRODUCTION

Sustaining life and fostering good health requires having access to an adequate supply of safe and nutritious food. A wide range of illnesses, from cancer to diarrhea, can be spurred on by eating unsuitable food that contains harmful microbes, viruses, parasites, or chemicals. There is an inter-link between food security, nutrition, and food safety. A vicious cycle of illness and starvation can be caused by contaminated food. It is estimated that one in ten individuals worldwide becomes ill after consuming contaminated food (WHO, 2022) despite significant advancements made in recent years in improving food safety. The vulnerability regarding food contamination can emerge in phases of the food chain that is production, distribution, preparation, and consumption. Apart from this, some threats are posed by food from its origin, and it can be in the form of natural toxins, some bioactive compounds, internal physicochemical changes, or deterioration of food nutrients by the actions of specific microbiota. One such hazard processed during nutrient–microbe interaction instigates the production of biogenic amines (BAs).

BAs are nitrogenous organic bases with low molecular weight and non-volatile properties, known as BAs, which are created through the decarboxylation of amino acids (Rêgo et al. 2014). They can be formed as a result of the deamination and transamination of ketones and aldehydes (Figueiredo et al. 2013).

Almost every kind of food, feed, and beverage contains BAs in varying amounts (Bover-Cid et al. 2014). Cheese, seafood, beef products, soy sauces beer, and wine are a few products that are vulnerable sources (Silva and Glória 2002; Savvaidis and Ruiz-Capillas 2009; Gomes et al. 2014; Lee et al. 2015). The scientific literature provides an increasing amount of information about the occurrence of BAs in all dietary matrices, including halal meals, fresh fruit and vegetables, pulses, infant foods, and alcohol.

According to Bjelakovic et al. (2017), BAs play a number of significant biological roles in the body. In addition to being essential regulators of numerous functions like body temperature regulation, brain activity, gastric acid secretion, stomach pH, and immunological reaction, BAs are also required for cellular formation and proliferation (Santosh 1996; Ladero V et al. 2010).

DOI: 10.1201/9781032622408-12

Despite their essential physiological functions, BAs can cause oral burning, headaches, vomiting, abdominal pains and cramps, flushing, diarrhea, gastric complications, hyper- or hypotension, heart palpitations, heart failure, renal intoxication, urticaria, respiratory problems and brain hemorrhage (Kalac and Krausová 2005). The development of BAs in food products is reflected as hazardous to health due to their adverse biochemical, biological, and toxicological effects, caused by BA (Eruck and Omer 2019; Wójcik et al. 2020). Because of this situation, scientists are forced to consider that every member of the population is susceptible to the build-up of BAs due to their dietary choices, meal composition, and of course, susceptibility.

12.2 TYPES OF BAs

Aliphatic, aromatic, and heterocyclic amines are some types of BAs. Generally, amines are classed as monoamines (tyramine and phenylethylamine), diamines (histamine, putrescine, and cadaverine), or polyamines (spermidine and spermine) based on the number of amine groups (Figure 12.1). The common BAs in foods are histamine (HIS), tyramine (TYR), putrescine (PUT), cadaverine (CAD), β-phenylethylamine (PHE), tryptamine (TRP), spermine (SPM), and spermidine (SPD) (Ekici and Omer 2018, 2019). They typically originate when amino acids are decarboxylated by microorganisms (Santos 1996). Tyrosine amino acid is a common precursor of all BAs generated from catechols.

It is important to distinguish between harmful biological amines and biological amines engaged in normal physiological activities, such as adrenaline, histamine, tyramine, and serotonin (Kalac 2014). Furthermore, many BAs typically generate harmful effects rather than just one. Histamine and tyramine are the most hazardous of these BAs and the most significant in terms of food safety. By using gas chromatography and mass spectroscopy or high-performance liquid chromatography, BAs can be precisely identified (Ruiz and Herrero 2019).

12.3 CHARACTERISTICS OF COMMON BAs

12.3.1 Histamine (HIS)

HIS intoxication can occur after consuming food having a high concentration of histamine. Mackerel, tuna, butterfly kingfish, seer fish, saury, and bonito, as well as certain no-scombroid fish

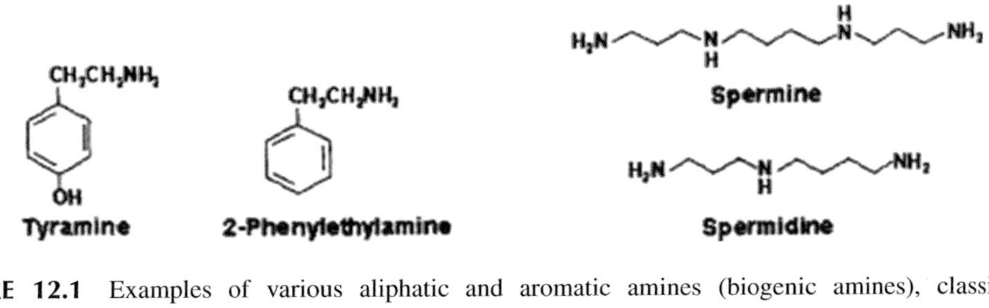

FIGURE 12.1 Examples of various aliphatic and aromatic amines (biogenic amines), classified as monoamines, diamines, and polyamines depending on number of amines attached.

such as anchovies, sardines, mahi-mahi, marlin, and bluefish, are frequently associated with cases of BA intoxication, particularly HIS poisoning (Lehane and Olley 2000). Cheese ranks second in terms of food items linked to flare-ups of HIS intoxication, after fish. Epidemics of HIS poisoning have been linked to the consumption of several types of cheese. Swiss and Cheddar cheeses have been connected to numerous outbreaks of HIS sickness and may contain elevated levels of HIS (Sttratton et al. 1991).

The symptoms of HIS poisoning can appear anywhere from a few minutes to several hours, and they usually remain noticeable for several hours. Localized inflammation, edema (eyelids), rashes, urticaria, and pruritus are symptoms of HIS poisoning that affect the smooth muscles and integumentary system. The symptoms of gastrointestinal illness include nausea, vomiting, stomach upset, and diarrhea. According to Taylor and Eitenmiller (1986), neurological symptoms include flushing, tingling, itching, burning, headache, and palpitations.

12.3.2 Tyramine (TYR), β-Phenylethylamine (PHE), and Tryptamine (TRP)

β-phenylethylamine, tyramine, and tryptamine are vasoactive precursor amines produced through the decarboxylation of amino acid substrates (phenylalanine, tyrosine, and tryptophan, respectively). Hypertension develops by vasoconstriction initiated by TYR, PHE, and TRP, and these BAs can also operate as neurotransmitters or neuromodulators (Tittarelli et al. 2015). Clinical indications develop within 30 minutes to some hours after consuming BAs-dense source foods and generally resolve within hours, with full recovery occurring within 24 hours. The consequences of eating foods containing potentially high levels of TYR may be minimized by MAO inhibitor (MAOI) drugs, which can be utilized to treat individuals with hypertensive emergencies and food-induced headaches (Shalaby 1996).

The most frequent BAs noticed in cheese is TYR. This non-physiological condition is frequently connected with an opposing response between MAOIs and raised dietary TYR concentrations, resulting in hypertensive reactions, shock, and death from a brain hemorrhage or heart failure. Concentrations of TYR in meats and animal products have also been shown to be high.

12.3.3 Putrescine (PUT), Cadaverine (CAD), Spermine (SPM), and Spermidine (SPD)

Polyamine chemical reactions such as CAD, PUT, SPD, and SPM, appear to be less dangerous than those of HIS, TYR, and PHE. These four BAs can also inhibit cell growth and induce apoptosis. Before any symptoms, such as bradycardia, dyspnea, lowered systolic and diastolic blood pressure (hypotension), and paralysis of the extremities, arise, very high doses of these BAs must be taken. (EFSA 2017).

12.4 FACTORS AFFECTING BA FORMATION IN RTE FOODSTUFFS

Food manufacturing processes include sanitary raw material qualities, microbial composition, fermentation conditions, and length if listed among fermented food items, all of which have an impact on the concentration of BAs in foods, particularly ready-to-eat (RTE) food products.

12.4.1 Hygienic Condition of Raw Materials

BAs have been found to be unsurprisingly present in fresh milk, grapes, and raw meats without fermentation. Poorly sanitary meat used in sausage has been shown to drastically limit the capability of the amino acid decarboxylase negative strain Lactobacillus saki CTC494. High levels of cadaverine, putrescine, and spermidine have been found in the berries pericarp during the winemaking practice (Gloria et al. 1998). Fresh Indian anchovies had lower levels of BAs than those held at 35 °C for 8 and 16 hours earlier being used to make fish sauce, with anchovies stored for 16 hours showing the maximum level of BAs (Yongsawatdigul et al. 2019).

12.4.2 Microorganisms with BAs-Producing Ability

Decarboxylase activity has been found in many kinds of microbial organisms. Decarboxylase activity has been demonstrated in various microbes found in meat products, including *Enterobacteriaceae* (Huis et al. 1990), *Enterococci* (Roig-Sagués & Eerola 1997), *Lactobacilli* (Maijala Eerola 1993), and Pseudomonas (Bauer et al. 1996). Most yeasts can produce substantial amounts of cadaverine and putrescine. However, only a small number of yeasts isolated from cheese, such as *Yarrowia lipolytica*, and *Debaryomyces hansenii* may create histamine and tyramine (Gardini et al. 2006).

12.4.3 Temperature and pH

Rising temperatures during long-term storage or processing can have a major impact on the production of BAs in food items. Furthermore, mesophilic bacteria can create considerable amounts of these amines at temperatures ranging from 20 to 37 °C. *Morganella morgani* produces histamine at a temperature of 25 degrees Celsius in fish. Tyramine level in *Carno bacterium* divergens meat-fat combination produced at 25 °C is higher than that formed at 15 °C. Marcobal et al. (2012) revealed that the tyramine concentration of bacteria such as *Enterococcus faecium* and *L. brevis* increases temperature rises. Krizek et al. (2002) reported that increasing the storage temperature increases the concentration of BAs in carp meat.

12.5 BAs AMONG VARIABLY CATEGORIZED RTE STUFF

Although BAs can be identified in lower concentrations in fresh foodstuffs, they can have a greater impact than predicted, particularly when combined with protein-rich animal food (Ten et al. 1990). Low protein foodstuffs such as beer, wine, beer, fruits, and vegetables may also contain substantial amounts of BAs, as may protein-rich foods such as fermented or heat-treated fish and fish products, meat products, beef, milk and milk products (Flick and Granata 2004). Blood levels of BAs can rise to potentially dangerous levels. These BAs can be present in processed fermented foods, created by microbial decarboxylation in foods, and formed by aldehyde and ketone deamination. In accordance with the kind of food product, varying BA levels are present. Additionally, the type and quantity of microbes in meals have an impact on them. Free-form amino acids or proteins in foods can have microbiological or metabolic action. BAs have the potential to be employed in nonfermented goods such as breakdown indicators or marker molecules. Foodstuffs including fermented veggies, cheese, wine, fish and fish items, sausages, and dried meat products contain high concentrations of BAs.

12.5.1 BAs in Fermented Foods

Many foods, including seafood, pork, and cheese products, can contain BAs; however, fermented foods like wines, yogurts, and other similar items are particularly high in these compounds. BAs are produced in substantial amounts by the metabolic pathways found in lactic acid bacteria; they can proliferate and flourish in food and drink. Microbial amine producers can be impacted by the microbiological and physicochemical properties of different meals and beverages as well as other biotic and abiotic variables (Capozzi et al. 2009).

Numerous investigations have documented the presence of histamine-producing bacteria in fermented soybean products, such as miso made from Bacillus amyloliquefaciens, Bacillus subtilis,

Bacillus megaterium, and Staphylococcus pasteuri (Kung et al. 2007), *Bacillus subtilis* in Tofu and *Staphylococcus capitis* in douchi (Tsai et al. 2007). *Streptococcus thermophilus* is responsible for the synthesis of BAs in yogurt, including tyramine and histamine (Gezginc et al. 2013).

12.5.2 Alcoholic Beverages or Fermented Beverages

The beverage industry is keen on knowing where BAs originate and how they might be managed at the lowest levels, given the possible toxicological problems associated with these chemicals in alcoholic beverages. The raw materials utilized in the beverage, the brewing technique, and any potential microbial contamination during the process or storage all affect the types of amines that are created during the brewing process.

The overall concentrations of BAs in wine range from 1 mg to approximately 50 mg, contingent on the quality of the wine. Over 20 amines have been identified in wine. (Kanotor et al. 2015). Besides the BAs identified above, isoamyl amine and agmatine tend to occur in alcoholic beverages. In red wines, the greatest amounts of tyramine, histamine, and putrescine were found to be up to 37, 28, and 122 mg/L, respectively. It has been determined that the two main amines in grape and must are spermidine and putrescine. It has been noted that as malolactic fermentation does not take place during the production of white wine, white wine often has lower levels of BAs than red wine (Smit et al. 2013). Because maceration with grape skins happens exclusively during the manufacturing of red wine and gives rise to large levels of polyphenols and free amino acids, there are occasionally differences in biogenic quantities between white and red wines (Sahu et al. 2015).

Around the world, a popular carbonated alcoholic beverage is beer. The two primary contributors of BAs in beer are the raw material (malt, in particular) and the contaminating microbes (LAB, in most cases). The most often identified BAs generated by decarboxylase-positive microorganisms include histamine, 2-phenylethylamine cadaverine, putrescine, and tyramine; nevertheless, spermidine, 2-phenylethylamine, and spermine are naturally occurring in malt, wort, and hops. Furthermore, prolonged product storage can promote BA production in beer (Lorencová et al. 2020).

In terms of raw material, malt has greater BA levels than hops, particularly, spermidine, spermine, putrescine, and agmatine. The amine concentrations can be affected by barley variety and germination intensity. However, in brewing, the fermentation stage is also accountable for the majority of BA generation. Although yeast appears to have no effect on their levels, various LABs (including *Lactobacillus frigidus, Pediococcus damnosus, L. brevis,* and *Lactobacillus brevissimilis*) have been identified as amine-producing microbes in beer.

12.5.3 Milk and Milk Products

Dairy products, particularly cheese, may contain BAs, with concentrations ranging from negligible to over 1000 mg. There was a consensus that they shouldn't be permitted to build up in food due to their toxicity. The most prevalent type of amino acid in the raw material (milk) is polyamine. However, histamine, putrescine, tyramine, and cadaverine at lesser amounts, phenylethylamine, and tryptamine are all detected in the end product. The research primarily focuses on cheese meals because it is a dairy product that contains the maximum amount of BAs (Table 12.1). Different varieties of cheese have different amounts of BA. However, it can differ across areas of the same kind of cheese as well as within the same variety of cheese (Rodreguez et al., 2003).

Raw milk has detectable levels of histamine, which rise after processing. Histamine levels in pasteurized milk vary from 0.3 to 0.7 ppm and can reach 0.8 ppm in UHT milk, while it is 10 times higher in fermented dairy products.

After fish and fish products, cheese is the second most prevalent edible product known to cause histamine poisoning. The level of BAs in some cheeses rises at the start of the maturation process

TABLE 12.1
Characteristics of the Main Biogenic Amines and Polyamines

S. No.	Amine	IUPAC Name	Classification
Tyramine	4-(2-Aminoethyl)phenol	$C_8H_{11}NO$	Aromatic, monoamine
Phenylethylamine	Phenylethan-2-amine	$C_8H_{11}N$	Aromatic, monoamine
Histamine	2-(1H-Imidazol-4-yl)ethanamine	$C_5H_9N_3$	Heterocyclic, monoamine
Tryptamine	2-(1H-Indol-3-yl)ethanamine	$C_{10}H_{12}N_2$	Heterocyclic, monoamine
Cadaverine	Pentane-1,5-diamine	$C_5H_{14}N_2$	Aliphatic, diamine
Putrescine	Butane-1,4-diamine	$C_4H_{12}N_2$	Aliphatic, diamine
Spermidine	*N*-(3-Aminopropyl)butane-1,4-diamine	$C_7H_{19}N_3$	Aliphatic, polyamine
Spermine	*N*-*N*′-Bis(3-aminopropyl)butane-1, 4-diamine	$C_{10}H_{26}N_4$	Aliphatic, polyamine

and thereafter drops. This shift can be used to gauge the maturity of these items (Joosten and Northolt, 1987). Numerous amines, such as histamine, tyramine, cadaverine, putrescine, tryptamine, and phenylethylamine, have been discovered in different ripening cheeses. Gram-positive bacteria are the primary makers of BA in cheese, while LAB is the primary creator of histamine and tyramine. Some species of Enterococcus, Lactobacillus Leuconostoc, and Streptococcus have been identified as makers of BAs. Before, during, or after the manufacturing of dairy products, they may be introduced via contamination or already present in the milk microbiome. It is possible for BA+-LAB to be included in adjunct or starting cultures. Tyrosine and histamine decarboxylase activity has been reported by a number of authors in strains derived from different starter cultures (Burdychova and Komprda 2007; Komprda et al., 2008). Consequently, strains meant to be used as starters must be required to not produce BA. It goes without saying that utilizing high-quality milk is essential for decreasing the risk of amine development. According to several authors, high amounts of tyramine in cheese are a direct result of high amounts of enterococci in milk (Joosten and Northolt, 1987).

12.5.4 Meat and Meat Products

The good amount of free amino acids and protein dictates that meat and meat products are suitable conditions for the progression of microorganisms creating BAs. The polyamines—spermidine and spermine—are naturally occurring substances found in fresh meat. Spermidine levels seldom go over 10 mg/kg, whereas the spermine content ranges from 20 to 60 mg/kg.

The concentration of spermidine stays consistent as the levels of histamine, putrescine, tyramine, and cadaverine rise in meat after storage. Food rotting can be tracked by looking at the correlation between the amount of BAs in the foodstuffs and how long they have been stored. A rise in the concentration of BAs brought on by microbes is thought to be a sign of low-quality items that contain them, particularly nonfermented food.

For meat products, the biogenic amines index (BAI) can be used as a freshness indicator. It is calculated by adding the components histamine, cadaverine, putrescine, and tyramine. There is no BAI criterion for assessing the superiority of raw-ripening meat products; the BAI is mostly useful for fresh meat and heat-treated products. Fresh meat has a minimum BAI value of 5 mg/kg; a value within 20 and 50 mg/kg indicates poor quality meat; and a value above 50 mg/kg indicates bacteria utilized in sausage fermentation, such as coagulase-negative cocci, yeast, and mold, which are starter cultures that can produce BAs. The two most prevalent BAs in dry sausages are putrescine and tyramine. Throughout sausage fermentation and drying, the non-protein nitrogen fraction—which is made up of free amino acid precursors of BAs—rises. The proteolytic enzyme activity of the meat-based bacteria is often linked to the synthesis of beef amino acids (BAs) in dry sausages. It has been discovered that coagulase-negative staphylococci and lactic acid bacteria create tyramine (Talon R, Leroy et al. 2011).

According to Latorre-Moratalla et al. (2010), amino acid decarboxylation is possible in 13% of staphylococci and 48% of lactic acid bacteria that were recovered from sausages that spontaneously fermented. The amount of BAs in sausages differs according to the manufacturer and kind of food. The biological quality of the raw materials, additives, components, banger diameter, and ripening method are the main elements linked to this variation (Komprda et al. 2004).

12.5.5 Fish and Fish Products

The most frequent BAs produced in seafood are cadaverine, histamine, putrescine, and tyramine (Prester 2011). Temperature and length of storage have a major influence on the development of BAs in fish products. According to some investigators, samples of mackerel and herring kept at 10 °C for the same amount of time showed levels of histamine, cadaverine, putrescine, and spermidine that were 2–20 times greater than those of samples kept at 2 °C (Klausen and Lund 1986). Fish products like dried fish, fish sauce, fish paste, fermented anchovies, prawn paste, and mildly cured horse mackerel have been found to have a significant level of histamine. The FDA imposed a 50-ppm limit for the extreme amount of histamine in fish products.

12.6 BAs' IMPACT ON HUMAN HEALTH

BAs have been connected to multiple outbreaks of foodborne infections that have a variety of deleterious consequences on human health. BAs have the potential to produce carcinogenic nitrosamines when they react with nitrites. High dietary concentrations of BAs are frequently associated with toxicological risks and health issues, as well as psychotropic, vasoactive, and hypertensive effects. High levels of BAs in food can have a variety of physiological, pharmacological, and toxic consequences that can result in foodborne illnesses such as TYR toxicity and HIS intoxication.

Excessive levels of BAs can cause a number of physiological symptoms, including headaches, sweating, heart palpitations, hyper- or hypotension, nausea, and respiratory discomfort. They can also reduce the quality of food. When combined with alcohol and acetaldehyde, the primary symptoms of several BAs, like HIS and TYM, include nausea, headaches, and respiratory issues (Lee et al. 2015).

Polyamines are involved in several physiological processes that can be both beneficial and detrimental to human health (Kalac 2014). According to Hui and Sherkat (2005), BAs may be neuro- or vasoactive, even though polyamines are crucial for growth. For example, HIS causes flushing, hypotension, urticaria, and headaches by increasing vaso-permeability and vasodilatation. Additionally, HIS causes the smooth muscle in the intestines to contract, which results in vomiting, diarrhea, and cramping in the abdomen (Lehane and Olley 2000). Polyamines appear to speed up the growth of tumors rather than cause cancer (Kalac 2014).

According to Guo et al. (2015), certain people have sensitivity to BAs, which causes symptoms that resemble an allergic reaction. Skin (rash, inflammation, etc.), gastrointestinal (vomiting, nausea, diarrhea), neurological (headache, burning, itching), and circulatory (hypotension) symptoms can all result from HIS poisoning. Hui and Sherkat (2005) claims that once generated in seafood, beta-amylases (BAs) are heat-resistant and won't be harmed by baking, frying, or even canning. Certain migraines have been linked to tyramine and PHE. However, migraine is a complicated problem that is impacted by numerous environmental, psychological, and physiological factors in addition to one particular food (Bover-Cid et al. 2014). A "cheese reaction" is seen in cheese with high TYM levels, which is what causes headaches, migraines, and indications of elevated blood pressure (Ruiz-Capillas and Jiménez-Colmenero 2004).

12.7 SIGNIFICANCE OF BAs IN RTE FOODS

As long as these amines exist in low amounts and their metabolism isn't hindered or genetically altered, they are not thought to pose a significant risk to humans (Vidal-Carou et al. 1990). Under normal circumstances, some enzymes called monoamine (MAO) and diamine (DAO) oxidases rapidly metabolize dietary BAs in healthy persons (Prester 2011). As a result, BAs consumed with food are quickly detoxified by these intestinal mucosa enzymes. But when this system is unable to get eliminate the large amounts of bile acids (BAs) that are consumed with some fermented or spoiled foods (ten Brink et al. 1990), or if these enzymes are dysfunctional because of a genetic defect or because they are consuming inhibitors like alcohol or some antidepressants, the bile acids enter the bloodstream and cause major health issues for humans by acting toxically on various organs (Alvarez and Moreno-Arribas 2014).

12.8 CONCLUSION

BAs are found in a wide variety of meals and beverages, with concentrations varying greatly between and within food types. Over the last few decades, the quest for safer foods has spurred, resulting in further research on BAs.

Evolving regulatory strategies for suspending BA development, such as high hydrostatic pressure, irradiation packaging, microbiological modeling, and the usage of preservatives or food additives must be explored. However, refrigeration is a possible approach for reducing BAs in foodstuffs. Although certain bacteria that produce BAs can upsurge at temperatures below 5 °C, hence refrigeration alone is not permanently adequate. Even though facts on BAs are not presently included in food composition inventories, evidence of their presence, dispersal, and concentration in fermented foods is critical and might be valuable to the food business, health experts, and consumers. Insights on the enhanced knowledge of the mechanisms by which microbes produce BAs in foodstuff is required to avoid their development. Good sanitary practices from manufacturer to consumer can control the occurrence of BAs in food products both at the receiving stage and during processing. Awareness regarding the most common products that contain BAs manufacturers and consumers and the associated factors that facilitate their formation can also help in avoiding food poisoning resulting from bioaccumulation. Understanding the level of BAs is an essential and promising feature of future foods as they can be regarded as freshness markers to assess how much food has degraded.

REFERENCES

Alvarez, MA, Moreno-Arribas, MV. "The problem of biogenic amines in fermented foods and the use of potential biogenic amine-degrading microorganisms as a solution," Trends Food Sci. Technol. 39(2014): 146–155.

Bardócz, S. "Polyamines in food and their consequences for food quality and human health," Trends Food Sci. Technol. 6(10)(1995), 341–346.

Bauer, F, Potzelberge, D, Hellwing, E, Paulsen, P. "Formation of biogenic amines in fresh meat packed in oxygen permeable foil," Proceedings of the 42nd International Congress of Meat Science and Technology (1996), 552–553.

Bjelakovic, G, Stojanovic, I, Stoimenov, TJ, Pavlovic, D, Kocic, G, Bjelakovic, GB, et al. "Polyamines, folic acid supplementation and cancerogenesis," Pteridines. 28 (2017): 115–131.

Bover-Cid, S, Latorre-Moratalla, ML, Veciana-Nogués, MT, Vidal-Carou, MC. "Biogenic amines," Encyclopedia of Food Safety. Academic Press, Switzerland (2014): 381–391.

Burdychova, R, Komprda, T. "Biogenic amine-forming microbial communities in cheese," FEMS Microbiol. Lett. 276(2007): 149–155. 10.1111/j.1574-6968.2007.00922.x

Capozzi, V., Fiocco, D, Amodio, ML, Gallone, A, Spano, G. "Bacterial stressors in minimally processed food," Int. J. Mol. Sci. 10(2009), 3076–3105.

Ekici, K, Omer, AK. "The determination of some biogenic amines in Turkish fermented sausages consumed in Van," Toxicol. Rep. 5(2018): 639–643.

Ekici, K, Omer, AK. Biogenic Amines in Plant Food. CRC Press, Boca Raton, FL (2019): 305–330.

European Food Safety Authority. "Assessment of the incidents of histamine intoxication in some EU countries," EFSA J. 14(2017): 1301E.

Figueiredo, TC, Viegas, RP, Lara, LJC, Baião, NC, Souza, MR, Heneine, LGD, Cancado, SV. "Bioactive amines and internal quality of commercial¸ eggs," Poult. Sci. 92(2013): 1376–1384.

Flick GJ, Granata LA. Biogenic Amines in Foods. CRC Press (2004): 121–154.

Gardini, F, Tofalo, R, Belletti, N, Iucci, L, Suzzi, G, Torriani, S, Guerzoni, ME, Lanciotti, R. "Characterization of yeasts involved in the ripening of Pecorino Crotonese cheese," Food Microbiol. 23(2006): 641–648.

Gezginc, Y, Akyol, I, Kuley, E, Özogul, F. "Biogenic amines formation in *Streptococcus thermophilus* isolated from home-made natural yogurt," Food Chem. 138(2013): 655–662.

Glória, MBA, Watson, BT, Simon-Sarkardi, L, Daeschel, MA. "A survey of biogenic amines in Oregon Pinot noir and Cabernet sauvignon wines," Am. J. Enol. Vitic. 49(1998): 279–282.

Gomes, MB, Pires, BAD, Fracalanzza, SAP, Marin, VA. "The risk of biogenic amines in food," Ciencia Saude Coletiva 19(2014): 1123–1134.

Guo, X, Guan, X, Wang, Y, Li, L, Wu, D, Chen, Y, Pei, H, Xiao, D. "Reduction of biogenic amines production by eliminating the PEP4 gene in *Saccharomyces cerevisiae* during fermentation of Chinese rice wine," Food Chem 178(2015): 208–211.

Hui, YH, Sherkat, F. Handbook of Food Science, Technology, and Engineering - 4 Volume Set. In *CRC Press eBooks*. (2005). https://doi.org/10.1201/b15995

Huis YH et al. "Occurrence and formation of biologically amines in food," Int. J. Food Microbiol. 11(1990): 73–84. doi: 10.1016/0168-1605(90)90040-C

Joosten, HMLJ, Northolt, MD. "Conditions allowing the formation of biogenic amines in cheese. 2. Decarboxylative properties of some non-starter bacteria," Neth. Milk Dairy J. 41(1987): 259–280.

Kalač, P. "Health effects and occurrence of dietary polyamines: a review for the period 2005–mid 2013," Food Chem.161(2014): 27–39.

Kalac, P, Krausová, P. "A review of dietary polyamines: formation, implications for growth and health and occurrence in foods," Food Chem. 90(2005): 219–230.

Kántor, A, Kačániová, M, Pachlová, V. "Biogenic amines content in different wine samples," *Journal of Microbiology, Biotechnology and Food Sciences*, *4*(special issue 1 (Microbiology) (2015): 37–40. https://doi.org/10.15414/jmbfs.2015.4.special1.37-40

Kiss, J, Korbasz, M, Sass-Kiss, A. "Study of amine composition of botrytized grape berries," J. Agr. Food Chem. 53(2006): 8909–8918.

Klausen, NK, Lund, E. "Formation of biogenic amines in herring and mackerel," Z. Lebensm. Unters. F. A. (2016): 459–463.

Komprda, T, Smela, D, Pechova, P, Kalhotka, L, Stencl, J, Klejdus, B. "Effect of starter culture, spice mix and storage time and temperature on biogenic amine content of dry fermented sausages," Meat Sci. (2004): 607–616.

Krizek M, Pavlicek T, Vacha F. Formation of selected biogenic amines in carp meat. J. Sci. Food Agr. 82(2002): 1088–1093. doi: 10.1002/jsfa.1154

Kung, HF, Lee, YH, Chang, SC, Cheng-I Wei, CI, Tsai, YH. "Histamine contents and histamine-forming bacteria in sufu products in Taiwan," Food Control. 18(2007): 381–386.

Kung, HF, Tsai, YH, Wei, CI. "Histamine and other biogenic amines and histamine-forming bacteria in miso products," Food Chem. 101(2007): 351–356.

Ladero V, Calles M, Fernandez M, Alvarez MA. "Toxicological effects of dietary biogenic amines," Curr. Nutr. Food Sci. 6(2010): 145–156.

Lee, S, Yoo, M, Shin, D. The identification and quantification of biogenic amines in Korean turbid rice wine, Makgeolli by HPLC with mass spectrometry detection," Lebensmittel–Wissenschaft Technologie. 62(2015):, 350–356.

Lehane, L, Olley, J, "Histamine fish poisoning revisited," Int. J. Food Microbiol. 58(2000): 1–37 doi:10.1016/S0168-1605(00)00296-8.

Lorencová E, Salek RN, Černíková M, Buňková L, Hýlková A, Buňka F. "Biogenic amines occurrence in beers produced in Czech microbreweries" Food Control. 117(2020)117:107335. doi: 10.1016/j.foodcont.2020.107335

Maijala, R, Eerola, S. "Contaminant lactic acid bacteria of dry sausages produce histamine and tyramine," Meat Sci. 35(1993): 387–395.

Marcobal, A, de Las Rivas, B, Landete, JM, Tabera, L, Muñoz, R, Tyramine and phenylethylamine biosynthesis by food bacteria. Cri. Rev. Food. Sci. Nutr. 52(2012): 448–467.

Pessione, E, Mazzoli, R, Giuffrida, MG, Lamberti, C, Garcia-Moruno, E, Barello, C, Conti, A, Giunta, C. "A proteomic approach to studying biogenic amine producing lactic acid bacteria," Proteomics. 5(2005): 687–698.

Prester, L. "Biogenic amines in fish, fish products and shellfish: a review," Food Add. Contamin. 28(2011): 1547–1560.

Rêgo, IOP, Menezes, LDM, Figueiredo, TC, Oliveira, DD, Rocha, JSR, Lara, LJC, Lima, AL, Souza, MR, Cançado, SV. "Bioactive amines and microbiological quality in pasteurized and refrigerated liquid whole egg," Poult. Sci. 93(2014): 1018–1022.

Rodriguez, A, Gonzalez, C, Diaz, V, Vecchionacce, H, Hurtado, E. "Effect of lipids and zeolite incorporation on total apparent digestibility of diets with sweet potato (Ipomoea batatas l.) foliage in pigs," Cuban J. Agric. Sci. 37(): 421–424.

Roig-Sagués, AX, Eerola, S. "Biogenic amines in meat inoculated with *Lactobacillus*'s sake starter strains and an amine-positive lactic acid bacterium," Z. Lebensm. Unters. F. A. 205(1997): 227–231.

Ruiz-Capillas C, Herrero AM. "Impact of biogenic amines on food quality and safety," Foods. 8(2)(2019): 62.

Ruiz-Capillas C, Jiménez-Colmenero F. "Biogenic amines in meat and meat products. Crit. Rev. Food Sci. Nutr. 44(2004): 89–499. doi:10.1080/10408690490489341

Sahu L, Panda SK, Paramithiotis S, Zdolec N, Ray RC, Biogenic amines in fermented foods: Overview." In: Montet D., Ramesh C.R., editors. *Fermented Foods—Part I: Biochemistry and Biotechnology*. CRC Press; Boca Raton, FL: 2015.

Santos, MHS. "Biogenic amines: their importance in foods," Int. J. Food Microbiol. 29(1996): 213–231.

Savvaidis, I, Ruiz-Capillas, C. "Biogenic amines in fresh and processed meat and meat products," Ed. G Dandrifosse (2009): 231–266.

Shalaby, AR. "Significance of biogenic amines to food safety and human health," Food Res. Int. 29(1996): 675–690.

Silla Santos, MH. "Biogenic amines: their importance in foods," Int. J. Food Microbiol. 29(1996): 213–231.

Silva, CMG, Glória, MBA. "Bioactive amines in chicken breast and thigh after slaughter and during storage at 41degree C and in chicken-based meat products," Food Chem. 78(2002): 241–248.

Smit, AY, du Toit, WJ, Stander, M, du Toit, M. "Evaluating the influence of maceration practices on biogenic amine formation in wine," LWT Food Sci. Technol. 53(2013): 297–307.

Sttratton, JE, Hutkins, RW, Taylor, SL. Biogenic amines in cheese and other fermented foods: a review," J. Food Prot, 54(1991):460–470.

Talon, R, Leroy, S. "Diversity and safety hazards of bacteria involved in meat fermentation," Meat Sci. 89(2011): 303–309.

Taylor, SL, Eitenmiller, RR. "Histamine food poisoning: toxicology and clinical aspects," Crit. Rev. Toxicol. 17(1986): 91–128.

Ten Brink, B, Damink, C, Joosten, HM, Huis in 't Veld, JH. "Occurrence and formation of biologically active amines in foods," Int. J. Food Microbiol. 11(1)(1990): 73–84.

Tittarelli, RG, Mannocchi, F, Pantano, FS. "Recreational use, analysis and toxicity of tryptamines," Curr. Neuropharmacol. 13(2015): 26–46.

Tsai, YH, Chang, SC, Kung, HF. "Histamine contents and histamine- forming bacteria in natto products in Taiwan," Food Control. 18(2007): 1026–1030.

Tsai, YH, Kung, HF, Chang, SC, Lee, TM, Wei, CI. "Histamine formation by histamine-forming bacteria in douchi, a Chinese traditional fermented soybean product," Food Chem. 103(2007): 1305–1311," J. Micro. Biotech. *Food Sci.* 4(2015): 37–40.

Vidal-Carou MC, Izquierdo-Pulido ML, Martín-Morro MC, Mariné-Font A. "Histamine and tyramine in meat products: relationship with meat spoilage," Food Chemistry 37(1990): 239–249. doi:10.1016/0308-8146(90)90104-C

WHO. Food safety. 2022. www.who.int/news-room/fact-sheets/detail/food-safety

Wójcik, W., Łukasiewicz, M., Puppel, K. "Biogenic amines: formation, action and toxicity—a review," J. Sci. Food Agric. 101(2020): 2634–2640.

Yongsawatdigul, J, Choi, YJ, Udomporn, S. "Biogenic amines formation in fish sauce prepared from fresh and temperature-abused Indian Anchovy (*Stolephorus indicus*)," Food Chem. Toxicol. 101(4) (2004): 312–319.

13 Foodborne Pathogens cum Contamination, Hygiene Practices, and Other Associated Issues in Ready-to-Eat Food

Swati Ray[1], Joyati Das[2], Ranjana Pande[3], and A. Nithya[2]*
[1]Department of Food Technology, Haldia Institute of Technology, Haldia, West Bengal, India
[2]Agricultural and Food Engineering Department, Indian Institute of Technology Kharagpur, Kharagpur, West Bengal, India
[3]Life Sciences Department, School of Basic Science and Research, Sharda University, Greater Noida, India
[*]Corresponding Author

13.1 INTRODUCTION

Foodborne pathogens are microorganisms that can cause illness when consumed through contaminated food. Common pathogens include bacteria, viruses, parasites, and fungi. In ready-to-eat (RTE) foods, the risk of contamination is significant, as these products are often consumed without further cooking. Contamination can occur at various stages of the food production process. It may result from inadequate hygiene practices, contaminated raw materials, cross-contamination during processing, or poor storage conditions. It is essential to implement stringent control measures. Hygiene is crucial in preventing contamination and ensuring food safety. Proper handwashing, use of protective clothing, and sanitization of equipment and surfaces are fundamental practices. Additionally, training food handlers on safe food handling procedures is essential to minimize the risk of introducing pathogens into RTE foods during preparation. Cross-contamination occurs when harmful microorganisms are transferred from one surface or food item to another. In RTE food preparation, it is essential to separate raw and cooked ingredients, use dedicated utensils, and sanitize cutting boards and equipment thoroughly to prevent cross-contamination.

Proper temperature control is critical in preventing the growth of pathogenic microorganisms. RTE foods should be stored and served at safe temperatures to inhibit bacterial growth. Refrigeration and hot-holding equipment must be maintained and monitored regularly to ensure food safety. The packaging of RTE foods plays a crucial role in preventing contamination post-processing. Airtight and tamper-evident packaging helps maintain the quality and safety of the product. Understanding the shelf life of RTE foods is vital to prevent the consumption of expired products that may pose health risks.

DOI: 10.1201/9781032622408-13

Adherence to food safety regulations and guidelines is essential for the production and sale of RTE foods. Compliance ensures that food businesses meet the necessary standards, undergo regular inspections, and implement corrective measures to address potential risks.

Ongoing research and advancements in technology contribute to the continuous improvement of food safety. Detection methods for pathogens, traceability systems, and the development of novel preservation techniques are areas where innovation is making significant strides.

Understanding and addressing the challenges associated with foodborne pathogens, contamination, and hygienic practices in the production of RTE foods are paramount to ensuring the safety and well-being of consumers. Proactive measures, continuous education, and compliance with regulations are essential components of a robust food safety program in the RTE food industry.

13.2 INDUSTRY-BASED PATHOGENIC CONTAMINATION AND SPOILAGE OF READY-TO-EAT FOOD PRODUCTS

13.2.1 Ready-to-Eat Fruit- and Vegetable-Based Products

The first point of concern in the production of RTE fruit- and vegetable-based products is the potential contamination of raw materials. Fruits and vegetables can be contaminated with pathogens during cultivation, harvesting, and transportation. Agricultural practices, water quality, and soil conditions all play a role in determining the microbial safety of the raw produce. The processing stage introduces additional risks. Cross-contamination can occur if processing equipment is not adequately cleaned and sanitized. Shared equipment or processing lines for raw and RTE products may lead to the transfer of pathogens. Strict hygiene practices, regular equipment cleaning, and proper sanitation protocols are crucial to mitigate these risks. Water is a critical component in the production of fruit- and vegetable-based products. Contaminated water used for washing, rinsing, or processing can introduce pathogens to the products. Regular testing of water quality and adherence to water safety standards are essential to prevent the spread of contaminants. Inadequate packaging or storage conditions can contribute to both spoilage and pathogenic contamination. Packaging materials should provide a protective barrier against external contaminants and maintain the freshness of the product. Improper storage temperatures can lead to spoilage, and if the products are stored in conditions favorable for pathogen growth, it can pose a serious health risk. Cross-contamination within the processing facility is a significant concern. Separate processing lines for raw and RTE products, dedicated utensils, and effective cleaning protocols help minimize the risk. Attention should be given to high-risk areas such as cutting and chopping stations, where pathogens can be easily transferred between products. A variety of microbiological agents, such as bacteria, parasites, viruses, fungi, and mycotoxins, can lead to outbreaks of foodborne illnesses. Of the 600 million instances of foodborne pathogen-related illnesses reported worldwide in 2010, the WHO linked 120 million of those cases to norovirus (Havelaar et al. 2015). The main contaminant was norovirus accountable for fruit and vegetable-related foodborne illness outbreaks in the European Union (EU) (53%) and the USA (59%) between 2004 and 2012 (Callejón et al. 2015). However, bacterial pathogens are the second major contributor to outbreaks, constituting 36% and 42% of the outbreaks related to fruit and vegetables in the USA and EU, respectively, between 2004 and 2012 (Callejón et al. 2015). In the USA, between 2009 and 2015, foodborne illness outbreaks resulted in 82% of hospitalizations and deaths due to three commonly associated bacterial species which were *Salmonella enterica*, *Escherichia coli*, and *Listeria monocytogenes* (Dewey-Mattia et al. 2018). Other bacteria, such as *Bacillus cereus, Vibrio cholerae, Campylobacter* spp., *Shigella* spp., and *Clostridium* spp., have also been reported. Furthermore, food surveillance sampling conducted in the EU revealed that *Salmonella* was detected in up to 0.84% of RTE fruits and vegetables (EFSA 2018). Maintaining the cold chain is crucial for RTE fruit- and vegetable-based products. Temperature fluctuations during transportation and storage can accelerate spoilage and create conditions for the growth of pathogens. Implementing robust temperature monitoring systems and ensuring compliance with recommended storage temperatures are essential preventive measures. Regular testing for pathogens and spoilage organisms is vital for quality assurance. Batch testing of final products, as well as environmental monitoring within the processing facility, helps identify potential issues before they reach the market. Rapid detection methods can aid in early intervention and prevent widespread contamination.

The fruit and vegetable processing industry must adhere to food safety regulations and standards. Regular inspections, audits, and certifications ensure that processing facilities are following best practices. Compliance with regulations helps build consumer trust and ensures the safety of RTE products. Educating consumers about the proper handling and storage of RTE fruit- and vegetable-based products is crucial. Clear labeling and instructions can help consumers understand the importance of refrigeration and the shelf life of the product, thereby reducing the risk of spoilage and potential pathogenic contamination.

Mitigating pathogenic contamination and spoilage in the production of RTE fruit- and vegetable-based products requires a comprehensive approach encompassing the entire supply chain. From the cultivation of raw materials to processing, packaging, storage, and distribution, strict adherence to hygiene practices, quality control measures, and regulatory compliance are essential to ensure the safety and quality of these products for consumers.

13.2.2 Ready-to-Eat Meat-Based Products

The quality and safety of RTE meat products start with the sourcing of raw materials. Contamination can occur at the farm level using contaminated feed, exposure to diseased animals, or unhygienic processing practices. Quality assurance measures in the selection and inspection of meat suppliers are critical to ensuring a safe starting point. Pathogenic contamination is a significant concern during the processing of meat-based products. Cross-contamination can occur if processing equipment is not adequately cleaned and sanitized between batches. Strict hygiene protocols, including employee training, equipment sanitation, and separation of raw and cooked processing areas, are essential to prevent the introduction of pathogens. *L. monocytogenes* can be found in meat products, RTE meat products with a lengthy refrigerated shelf life, and are mostly linked to human listeriosis (Sofos 2008). Previous research has explicitly shown the presence of *L. monocytogenes* in food processing facilities. Approximately 83% of human cases of listeriosis in the USA have been linked to deli meats, especially those that are sliced and packed (Buchanan et al. 2017). *L. monocytogenes* is the most relevant pathogen for processed, fully cooked, RTE meat products. Processing factories continue to harbor *L. monocytogenes*, which can lead to postprocessing contamination. To mitigate this, strict plant cleanliness measures and appropriate post-packaging treatments are necessary (Capita et al. 2001; Gómez et al. 2015). RTE pork products, such as hot dogs in the USA and rillettes and jellied pork tongue in European countries, have been linked to multiple outbreaks of *L. monocytogenes* contaminations, which have been linked to listeriosis, a serious infection with a high death rate (Baer et al. 2013; De Valk et al. 1999). *L. monocytogenes* may grow and survive in refrigerated environments, reaching high cell counts in retail and storage settings. About 35.0% of retail meat samples, including RTE meat, had *Staphylococcus aureus* contamination in them when they were collected from 39 cities across China between 2011 and 2016 (Wu et al. 2018). Notably, a sizable fraction (20%) of bacterial foodborne outbreaks worldwide is caused by the coliform bacterium strain *E. coli* (Getaneh et al. 2021). Reports are there addressing *E. coli* contamination in RTE red meat products (Ronald et al. 2023). Maintaining proper temperature control throughout the processing, storage, and transportation of RTE meat products is crucial. Temperature abuse can lead to the growth of pathogenic bacteria, such as *Salmonella* spp. and *Listeria* spp. Continuous monitoring and adherence to cold chain requirements are essential to prevent spoilage and ensure the safety of the products. Cross-contamination is a potential risk during various stages of processing, especially in facilities that handle both raw and cooked products. Proper segregation of processing lines, equipment, and utensils is necessary to prevent the transfer of pathogens from raw to cooked products. Strict hygiene practices and the use of color-coded tools can help minimize this risk. To extend shelf life and enhance safety, RTE meat products often contain additives and preservatives. Improper use or inadequate control of these substances can lead to spoilage or compromise the quality of the product. Careful formulation and adherence to regulatory guidelines are essential to ensure the effectiveness and safety of these additives. Packaging plays a crucial role in preventing both spoilage and contamination. Inadequate packaging or storage conditions can compromise the quality and safety of RTE meat products. Proper packaging materials, including vacuum-sealing or modified atmosphere packaging, can help extend shelf life and reduce the risk of contamination during storage. Regular testing for pathogens, spoilage organisms, and product quality is a fundamental aspect of the meat processing industry. Batch testing of final products and environmental monitoring within the processing facility help identify potential issues before products reach the market. Rapid detection methods are valuable for early intervention. Adherence to food safety regulations and standards is non-negotiable in the meat processing industry. Compliance with guidelines such as Hazard Analysis and Critical Control Points (HACCP) is essential for ensuring a systematic approach to identifying and controlling potential hazards. Educating consumers about the proper handling, storage, and reheating of RTE meat products is crucial. Clear labeling with instructions on safe handling and storage conditions can help consumers minimize the risk of foodborne illnesses and spoilage.

Ensuring the safety and quality of RTE meat-based products involves a comprehensive approach, including careful sourcing of raw materials, rigorous processing hygiene, effective temperature control, and adherence to regulatory standards. Regular testing and ongoing quality assurance efforts are vital to identifying and addressing potential issues throughout the production process.

13.2.3 Ready-to-Eat Dairy-Based Products

The quality of RTE dairy products starts with the raw milk used in their production. Contamination can occur at the farm level due to poor animal hygiene, improper milking practices, or environmental contamination. Stringent quality control measures and regular testing for pathogens are essential to ensure the safety of raw milk. Pathogenic contamination is a significant concern during the processing of dairy-based products. Cross-contamination can occur if processing equipment is not adequately cleaned and sanitized. Strict hygiene protocols, including employee training, equipment sanitation, and segregation of raw and finished product areas, are essential to prevent the introduction of pathogens. Dairy products may contain bacteria as pathogenic agents or undesired food deterioration. *L. monocytogenes*, psychrotrophic bacteria like *Pseudomonas fluorescens* and *Pseudomonas putrefaciens*, *Salmonella* spp., *E. coli*, *S. aureus*, *Clostridium botulinum*, *Clostridium perfringens*, *Clostridium tyrobutyricum*, *V. cholera* are examples of these unwanted bacteria (Giraffa 2003; Oliver et al. 2005; Gálvez et al. 2008; Machado et al. 2013). Bitterness and off-flavors in dairy products can be caused by proteases and lipases secreted by psychrotrophic bacteria in milk, such as *P. fluorescens* and *P. putrefaciens*. Due to the heat stability of these enzymes, off-flavors can emerge in fermented dairy products even at conventional pasteurization temperatures of 72–74°C for 15–30 seconds (Diarmuid and Sheehan 2013). Proper pasteurization or heat treatment is crucial for eliminating harmful bacteria in dairy products. Inadequate heat treatment can lead to the survival of pathogens, compromising the safety of the product. Regular monitoring and validation of pasteurization processes are necessary to ensure effective pathogen reduction. In the dairy industry, pathogenic bacteria that can endure the production process of cheeses, yogurts, and soured creams are the primary cause of worry. Gram-positive *L. monocytogenes* is the causative agent of gastroenteritis and listeriosis, both of which can lead to sepsis or meningitis, two potentially fatal illnesses. Extreme listeriosis has a 20–30% death rate (WHO 2004; Carpentier and Cerf 2011). Gram-negative

E. coli is a normally occurring bacteria found in the lower intestine, although certain strains can lead to urinary tract infections and gastroenteritis. *E. coli* is a byproduct of post-pasteurization contamination in dairy products like yogurt. Studies have demonstrated that *E. coli* O157 can endure the acidic conditions involved in yogurt production, posing a significant health risk to consumers (Cirone et al. 2013). Several gram-negative species of *Salmonella* spp. are known to cause foodborne illnesses. *Salmonella typhimurium* and *Salmonella enteritidis* are two of the most prevalent types discovered in dairy products (Leyer and Johnson 1992). Gram-positive bacteria called *S. aureus* is one of the main global sources of foodborne illnesses, including gastroenteritis, which is brought on by consuming enterotoxins, the bacteria generate. Thus, the heat-resistant enterotoxin persists even after the bacteria are destroyed by heat or pressure treatment (Gálvez et al. 2008; Fleurot et al. 2014). The fermentation of dairy products creates an acidic atmosphere, therefore manufacturing facilities must maintain high standards of cleanliness and hygiene practices. Correct storage of dairy products results in preventing the development and contamination of goods with these undesired pathogenic and food spoilage microbes (Hickey et al. 2015). Cross-contamination is a potential risk during various stages of processing, especially in facilities that handle both raw and pasteurized products. Proper segregation of processing lines, equipment, and utensils is necessary to prevent the transfer of pathogens from raw to finished products. Strict hygiene practices and the use of color-coded tools can help minimize this risk. Maintaining proper temperature control during storage and transportation is crucial for RTE dairy products. Temperature fluctuations can lead to spoilage and create conditions for the growth of pathogens. Continuous monitoring and adherence to cold chain requirements are essential to prevent quality deterioration and ensure product safety. Packaging plays a critical role in preventing spoilage and contamination in dairy-based products. Inadequate packaging materials or storage conditions can compromise the quality and safety of RTE dairy products. Packaging should provide a protective barrier against external contaminants and maintain the freshness of the product. Regular testing for pathogens, spoilage organisms, and product quality is fundamental in the dairy processing industry. Batch testing of final products and environmental monitoring within the processing facility help identify potential issues before products reach the market. Rapid detection methods are valuable for early intervention. Adherence to food safety regulations and standards is crucial in the dairy processing industry. Compliance with guidelines such as HACCP is essential for ensuring a systematic approach to identifying and controlling potential hazards.

Educating consumers about the proper handling, storage, and consumption of RTE dairy products is crucial. Clear labeling with instructions on safe handling and storage conditions can help consumers minimize the risk of foodborne illnesses and spoilage.

Ensuring the safety and quality of RTE dairy-based products involves careful attention to raw milk quality, processing hygiene, pasteurization, temperature control, packaging, and adherence to regulatory standards. Regular testing and ongoing quality assurance efforts are vital to identifying and addressing potential issues throughout the production process.

13.2.4 Ready-to-Eat Fat and Oil-Based Products

Commercial fat and oil-based products are often used as cooking aids or as part of the preparation of certain specialty foods in which some products have a certain degree of readiness to be considered RTE food. These include salad dressings, mayonnaise, spreads, butter, ghee, margarine, confectionery fillings, dark chocolate, cocoa butter substitutes, etc. Due to their high-fat, low water activity, and high acid content, mostly they do not encourage microbial growth (Tareq et al. 2023). Sometimes these products may appear visibly safe and edible but may harbor excessive microbes including pathogens.

The basic microorganisms that are found in contaminated fat and oil-based products mentioned above include yeasts, molds, and coliforms especially *E. coli*, *S. aureus*, etc. (Anna et al. 2022;

Hassan et al. 2017; Sevmiş et al. 2020; Yang and Lee 2019; Windyga et al. 1978; Sharifzadeh et al. 2016). Salad dressing, mayonnaise, and spreads also reported incidences of the presence of *Enterobacter* spp., *Lactobacillus* spp. along with pathogens such as *L. monocytogenes*, *Salmonella typhi* (Łepecka et al. 2022), and *Bacillus amyloliquefaciens* (Pereira et al. 2018). Butter usually gets contaminated by unsterile packaging-hall air (Memisi et al. 2019), and in some samples, psychrotrophic *Pseudomonas* spp. (El-Hajjaji et al. 2021) were isolated. An indigenous popular recipe called Gimbichu butter was found contaminated with the pathogenic organism *L. monocytogenes* (Asrat et al. 2023). Dark chocolate and coca substitutes come mainly under high-fat confectionary products and are mostly safe except for a few incidences of *Salmonella* contamination due to a lack of personal hygiene (Iacumin et al. 2022).

13.2.5 Ready-to-Drink Beverage Products

A wide range of ready-to-drink beverages both in liquid and instant powders are available in the market. Mainly these drinks can be dairy-based (badam milk, flavored or non-flavored lassi, flavored or non-flavored milkshake, flavored milk, etc.), non-dairy or vegan (almond, soy, peanut, oat milk, etc.), chocolate-based (chocolate drink, hot chocolate, etc.), fruit and vegetable-based (fruit juice, squash, sherbet, vegetable juice, etc.), leaf or plant part extract based (aloe juice, peepal juice, neem juice, tulsi juice, etc.), seed-based (flavored seed drink), carbonated (carbonated cola and fruit beverage, tonic water, soda water, etc.), non-carbonated (instant tea or coffee, iced tea, infusion tea, flavored tea, etc.), and spice-based (jeera water, shikanji, jaljeera, etc.). Besides instant drinks like flavored or infused water, cocktail drinks, energy drinks, sports drinks, vitamin water, ayurvedic or herbal drinks, electrolyte drinks, or isotonic solutions, and Nata de coco drinks have gained immense popularity. Alcoholic beverages are considered ready to serve in some countries but mostly they are considered otherwise. The high-water activity and the nutritional availability of most ready-to-serve beverages can make them very much prone to microbial spoilage.

The dairy-based product contamination has already been discussed previously in this chapter. Instant non-dairy or vegan milk samples were mostly found to be satisfactory in terms of microbial quality except in a few isolated cases of *Geobacillus stearothermophilus* (Misiou et al. 2023), yeast, mold, and *B. cereus* (Giugliano et al. 2023) contamination. Improper handling of some soy milk samples in *Nigeria* resulted in contamination by *E. coli*, *Bacillus subtilis*, *Aspergillus niger*, *S. aureus*, *Klebsiella pneumonie*, *Streptococcus faecalis*, *Aspergillus flavus*, *Fusarium oxysporum*, *S. typhi*, and *Penicillium caseicolum* (Adegbehingbe et al. 2022). Chocolate-based drinks suffer microbial spoilage from contaminated cocoa powders which may harbor a higher population of mesophilic and thermophilic spore formers (Iacumin et al. 2022). A study on vended instant hot chocolate samples reported the presence of pathogenic *B. cereus* (Hall et al. 2012). The high acidity and sugar content of fruit juices can protect them from getting spoiled quickly while low-acid vegetable juices are more prone to microbial invasion. Any contamination in raw material, additives, water, or improper storage can lead to the lavish growth of fungal species of *Aspergillus*, *Fusarium*, yeasts, and *Penicillium* along with some fecal Streptococci, *S. aureus*, and *B. cereus* (Divyashree et al. 2013; Lee et al. 2021). The contaminated fruit juices, squash, and sherbet may allow the growth of many microbes dominated by heat and preservative resistant, mycotoxin producers along with bacterial and viral pathogens such as *Salmonella*, *Shigella*, *C. botulinum*, *E. coli*, *L. monocytogenes* and hepatitis A virus, Norwalk virus respectively. Off-flavor generation can be caused by the growth of some lactic acid, acetic acid, and *Alicyclobacillus* bacterial populations (Ahmed et al. 2019). Shelf-stable sherbet and squashes often get mixed with contaminated sugar syrups during processing which makes them spoiled (Jazmati et al. 2022). Contaminated vegetable juices were also reported with the presence of *B. cereus*, *E. coli* O157:H7, *S. typhi*, *S. aureus*, and *L. monocytogenes* (Lee et al. 2021) along with deadly pathogens like *Campylobacter jejuni*, *C.botulinum*, *Shigella* spp., and *V. cholera* that get entry from unhygienic raw materials (Balali et al. 2020). The juices extracted from various plant parts such as leaves, stems, etc., mostly have in-built antimicrobial activities and are self-preserving as proven by various researchers. Among them, some ready-to-drink untreated aloe-vera juice may contain gram-negative *Rahnella aquatilis*, yeast *Rhodotorula mucilaginosa* and gram-positive *Arthrobacter* spp. and *Micrococcus* spp. causing spoilage (Reyes et al. 2012). Carbonated beverages get spoiled by organisms like *Streptococcus* spp., *Pseudomonas aeruginosa*, *Bacillus stearothermophilus*, *Lactobacillus* spp., *Salmonella* spp. (Muhammad et al. 2009), *Penicillium glabrum*, etc. (Peter and Davenport 2007). Very high incidences of coliform, especially fecal coliform were detected in many samples in various regions of the world (Muhammad et al. 2009). Soda water samples also reported high total microbial count and coliform count that is used for cocktail preparation (Kregiel 2015). Most Green and black tea samples have higher microbial load predominant with *Pseudomonas psychrotolerans*, *Staphylococcus warneri*, *Pantoea gaviniae*, *A. niger*, *Aspergillus tubingensis* and *C. perfringens* that may transfer themselves in the ready to drink instant teas, tea infusions, iced tea, etc. (Carratur et al. 2018). While proper processing technologies may eliminate most of them, strains of *Lactobacillus* spp. especially *Lactobacillus plantarum* were isolated in large numbers (Vanos et al. 1987). Fecal coliforms like *Klebsiella* and *Enterobacter*; and some species of *Salmonella* were detected in instant iced tea, green tea infusion, and tea infusions of Chamomile, Peppermint, etc. (Carraturo et al. 2018). Roasting and brewing coffee beans significantly reduce the microbial load in coffee. Instant coffee spoilage occurs by using contaminated coffee powders containing strains of *Enterococcus* spp. and *Staphylococcus* spp. (Shivakumar and Muralidharan 2017). Infused and flavored water contamination is mostly dependent on the source of water that has been used for making the product which harbors fecal coliforms such as *Staphylococcus* spp., *Salmonella* spp., *Shigella* spp., *Klebshiella* spp., etc. (Rodríguez et al. 2017). Improper storage conditions of sports and energy drinks may allow *Bacillus* spp. to grow and cause defects. Instant herbal preparations also show the presence of high coliform load along with *Bacillus* spp. and *P. aeruginosa* (Darkwah et al. 2022). Incidences of contamination are not yet reported for flavored seed drinks, tonic water, instant herbal tea, spiced drinks like jeera water, shikanji, jaljeera, cocktail drinks, vitamin water, electrolyte or isotonic drinks, and Nata de coco drinks.

13.2.6 Ready-to-Eat Cereal Products

Classification of RTE cereals is composed of 12 categories which is shown in Figure 13.1.

RTE wheat and oat flakes get contaminated during processing and packaging with fungal microflora such as *Eurotium* spp., *Aspergillus* spp., *Penicillum* spp. (Weidenbrner and Kunz 1995), while the raw materials are reported with *Salmonella altona* contamination (Eran et al. 2017). Corn flakes and popcorn may contain species of *Penicillium*, *A. niger*, *Fusarium*, *Cladosporium herbarum*, *Aspergillus tiresias*, etc. that mostly enter during handling or lack of proper hygiene (Shaimaa et al. 2023). *B. cereus* forms the main contaminating agent in breakfast cereals, rice cakes, pasta, rice, and infant formulas as studies suggest (Anna et al. 2019; Jun et al. 2016). Additionally, traces of opportunistic pathogens like *Cronobacter sakazakii*, *Acinetobacter* spp., *Pantoea* spp., and enterotoxigenic *Bacillus wiedmannii* aerobic mesophilic bacteria, molds, Enterobacteriaceae, *Cronobacter*

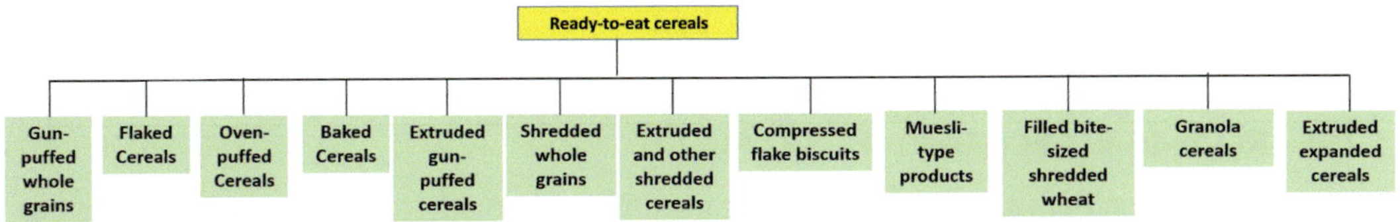

FIGURE 13.1 Categories of ready-to-eat processed cereal products. (Robert 2000.)

spp., etc. were found often to spoil cereal-based infant formulas (Christina et al. 2022). Rice flakes–jaggery instant mix in religious festivals caused outbreaks of *S. aureus*, and *C. perfringens* in Tamil Nadu (Stanislaus et al. 2022) as reported by the public health department. Foodborne outbreaks upon consumption of spicy puffed rice contaminated with multidrug-resistant foodborne bacteria were also reported in a study in Bangladesh (Khandoker et al. 2015). Street-vended pani-puri harbor species such as *Mucor*, *Rhizopus*, *Salmonella* spp., *E. coli*, *Pseudomonas* spp., *Enterobacter* spp., and *Bacillus* spp. that get transmitted through the hands of the servers (Rakshitha et al. 2022). Ready-to-eat rice and millet-based products grow coliforms like *E. coli, S. aureus*, and *B. cereus* during handling and packaging (Rizwanullah et al. 2020; Christopher et al. 2020). Ready-to-cook as well as instant noodle samples were detected with bacteria like *Bacillus*, *E. coli*, *Pseudomonas*, *Staphylococcus*, *Aeromonas*, *Streptococcus,* and *Staphylococcus* spp. along with fugal populations such as *Aspergillus*, Mucor, *Penicillium*, *Rhodotorula,* and *Candida* which may cause intestinal disorder (Akhigbemidu et al. 2015, Farah et al. 2021). Fast and instant packaged cereal-based foods like chips, biscuits, muesli, salted peanuts, papad, and ready pasta may allow growth of *Salmonella* spp., *E. coli*, *Pseudomonas* spp., *Staphylococcus* spp. and *Shigella* spp. if hygienic quality standards fail to satisfy (Ahmed et al. 2020; Gori et al. 2019).

13.2.7 Ready-to-Eat Bakery Products

Most of the bakery products are ready to eat and they can vary from pure freshly baked products to bakery confectionery products. The categories of products can be various kinds of breads, biscuits, brownies, cookies, cakes, muffins, pastries, puff rolls, burgers, pies, bagels, croissants, tortillas, pizza, donuts, sandwiches, etc. Both packed as well as unpacked slices of bread may get contaminated during cooling, slicing, handling, and packaging which results in a higher total plate count and yeast and mold count along with the presence of *S. aureus* (Manal et al. 2023). A 10-year survey of bakery food products from a canteen that includes sandwiches, frozen and baked pastries, and cakes with dairy creams showed the prevalence of foodborne pathogens such as *B. cereus*, *Salmonella* spp., presumptive *E. coli*, *S. aureus* and *L monocytogenes* (Parthena et al. 2013). Bakery products like burgers, puff rolls, croissant, pizza, and donuts were reported with higher than usual microbial counts than normal on various samples collected on different occasions (Al-Nasiry 2020).

13.2.8 Ready-to-Eat Sugar and Confectionary Products

Sugar confectionary products can be boiled sweet or fondant. Most confectionaries have high sugar and low water activity which ensures low microbial count. Though rare, but ready to eat chocolate bars may show incidences of *Salmonella* which causes severe health hazards (Geertrui et al. 2016). Chewing and bubble gums may contain trapped microbes detected by scanning electron microscopy (Stefan et al. 2015). *S. enteritidis* phage type 4 caused infection in children consuming marshmallows which is suspected to be made from raw egg white (Lewis et al. 1996). Sweet-based products like mithai may contain species of *Streptococcus*, *Klebsiella*, *E. coli*, *Proteus*, *Bacillus,* and *Corynebacterium* which may enter due to poor handling and lack of hygiene (Nayem and Ahmed 2013). The most preferred caramel coating in confectionaries may harbor *L. monocytogenes* which is a pathogen causing diseases (Kathleen et al. 2015). Confectionaries like colored lollipops or cotton candies may cause cancer-causing substances and chemicals that may cause foodborne illness and outbreaks in children.

13.2.9 Ready-to-Eat Egg-Based Products

The bacteria that contaminate the egg surface are mostly gram-positive, but it is gram-negative bacteria that are largely involved in the internal infection of eggs. This effect may be partly due to the high motility of gram-negative bacteria and their ability to resist egg white antimicrobial activities (Techer et al. 2014; Legros et al. 2021; Liu et al. 2021). *S. enteritidis* (gram-negative bacteria), have been described as more resistant than other microorganisms to the egg's natural defenses (Gantois et al. 2009; Lock et al. 1992; Clavijo et al. 2006). When eggs are processed into egg products, the risk of contamination of the egg contents is increased, and so egg products are systematically pasteurized by heat treatment. Unfortunately, the heat sensitivity of egg white proteins restricts the pasteurization process to a modest heat regime that is insufficient for the effective eradication of all bacteria. Fortunately, egg white possesses various intrinsic antimicrobial activities and only a few gram-negative bacteria (in particular, *S. enteritidis*), can survive in egg white. The resistance of *S. enteritidis* may explain the relatively high incidence of *Salmonella* in food poisoning outbreaks (Gantois et al. 2009).

Salmonellosis is characterized by cramping in the abdomen, nausea, vomiting, and diarrhea. Eggs are still one of the foods consumed on a weekly basis (e.g., 2–3 times/week) (Hessel et al. 2019) or even daily (one egg/day) (Islam, Sayeed, Akhtar, Hossain and Liza 2018). These consumers include the elderly (>65 years old), children, pregnant women, and immune-compromised persons (CDC (Centers for Disease Prevention and Control) 2021; Mihalache et al. 2022). And those involved in food safety scandals involving the egg industry from previous years, such as the *Salmonella* outbreak in the USA in 2015 (Whiley and Ross 2015), the EU in 2017, and the EU multi-country salmonellosis outbreak (838 confirmed cases) that occurred during 2016–2018 and originated from Poland (Pijnacker er al. 2019). The most common products containing raw eggs that are linked to salmonellosis are spread and sauces (like mayonnaise and egg butter), desserts (like tiramisu and chocolate mousse) that are made without a cooking step, and beverages (like eggnog and raw egg Hogg protein smoothies) that contain raw eggs (NSW). As consumers are the last line of defense in the food chain, their food safety knowledge and behaviors during meal preparation have a significant role in minimizing foodborne infections that occur at home (Redmond and Griffith 2003). Inadequate food handling techniques such as not heating meals properly cooked or occasionally not heating milk sufficiently have been linked in the past to the occurrence of foodborne incidents (Chen et al. 2018).

Egg products that are "prepared" and ready to eat, including scrambled eggs, omelets, and patties, are usually cooked to a temperature of up to 85°C (185°F). This heat treatment offers an acceptable reduction of *L. monocytogenes* and is far beyond the pasteurization treatment advised by the USDA FSIS (2020) for a five-log reduction of *Salmonella* in liquid egg products. To reduce risk, appropriate postprocess actions are necessary as *L. monocytogenes* postprocess contamination is a possibility. Currently, these prepared egg products are sold frozen; however, due to consumer demand for "fresh," unfrozen items, these products must now be kept in a refrigerator, raising the risk of food safety. A US egg processor's refrigeration-distributed hard-boiled eggs were connected to a multistate epidemic of *L. monocytogenes* infections between 2017 and 2019 (Centers for Disease Control and Prevention 2020). Therefore, it would be wise to incorporate antimicrobials into RTE egg products intended for refrigerated distribution and storage, just as with cooked meat and poultry products. The literature lacks information regarding the antimicrobials' effectiveness in RTE egg products. According to Jung et al. (1992), nisin is one of the most researched and effective antibacterial agents against gram-positive spoilage and pathogenic bacteria, including *L. monocytogenes* (Shrestha et al. 2023) have demonstrated nisin's ability to inhibit *L. monocytogenes* in RTE meat, poultry, liquid eggs, and dairy products. In the USA, it is often called and labeled as nisin preparation, and it is a commercial extract that is sold under different brand names. According to USDA FSIS Directive 7120.1, Safe and Suitable Ingredients not to exceed 250 ppm (equal to 6.25 ppm of nisin) in egg products, nisin preparation is permitted as an antibacterial (U.S. Department of Agriculture, Food Safety and Inspection Service 2021). In the USA, it is often called and labeled as nisin preparation, and it is a commercial extract that is sold under different brand names. According to USDA FSIS Directive 7120.1, Food products have been preserved using organic acids including propionic, lactic, and acetic, as well as corresponding salts (Ray and Sandine 1992). Shrestha et al. (2023) studies have indicated that in the fight against *L. monocytogenes*, organic acids and nisin may have either an additive or synergistic effect.

13.2.10 Ready-to-Eat Fish and Seafood-Based Products

Nethathe et al. (2023) studied throughout the production and supply chain, that fish and fish products are susceptible to the introduction of pathogenic and spoilage bacteria. Most regions of the world are seeing an increase in RTE consumption, and this has been linked to major health issues (Food and Drug Administration Act (FDA) 1997). According to Monday et al. (2014), high bacterial count levels in RTE foods are a sign of improper food handling and storage practices as well as a possible means of disease transmission. The most important variables in determining the safety and caliber of food are the amounts of bacteria in water and cooked foods. Bacterial counts also show the degree of hygienic techniques used by food handlers during food preparation. *E. coli* and other fecal coliforms are employed as markers of sanitary conditions; however, their identification may also signal the presence of other dangerous microbes in humans (Osakue et al. 2016). Aches, vomiting, fever, diarrhea, and stomach pains are common signs of foodborne illnesses (Darwish 2018; Zohreh et al. 2015). *Salmonella* spp., a frequent cause of the foodborne illness salmonellosis, has been linked to contaminated water used in the production of RTE meals (Sudad 2018). Particularly prevalent in undercooked poultry and eggs, *Salmonella* spp. is the source of several outbreaks worldwide. The presence of pathogenic microorganisms on raw vegetables, such as lettuce used in sandwiches, can be caused by contaminated irrigation water, which raises the risk of contracting enteric pathogens. According to Osakue et al. (2016), the high frequency of *S. aureus* in raw foods can result from food handlers or environmental sources and can cause a variety of diseases. Additionally, reports have indicated that various food items such as meat, vegetable ingredients, raw and cooked foods, chicken shawarmas, raw chicken, sandwiches, beef burgers, commercial mayonnaise, RTE salad vegetables, frozen chicken, seafood, poultry products, and lastly the hands of food handlers have high prevalences of *Salmonella* spp., *Campylobacter* spp., *E. coli*, *Staphylococcus* spp., *Listeria*, and *Yersinia* spp. Most people who handle fast food are unaware of the causes of foodborne illnesses (Adebukola et al. 2015). To prevent cross-contamination, the FDA advises that food be prepared with the least amount of human touch possible, using clean, appropriate utensils, and sanitizing surfaces. Sometimes infections from rodents and insects—not human or portable water—contaminate utensils and equipment. Food handlers at restaurants or processing facilities can control most contamination issues. According to Rane (2011), to eradicate harmful bacteria from food, it is necessary to cook it to the proper temperature. Food cooked for at least two minutes at 70°C is safe to eat, according to studies. Before serving, chilled RTE food must be kept below 5°C, while heated food must be kept above 60°C (Sudad 2018). Microbiological quality assessment of RTE foods is necessary because it shows the hygienic conditions at the time of production and distribution. Distinct nations have distinct lists of antibiotics that are permitted for use in aquaculture (Preena et al. 2020; Sheng et al 2020). For instance, 13 antibiotics, including doxycycline, enrofloxacin, florfenicol, flumequine, neomycin, norfloxacin, oxalinic acid, sulfadiazine, sulfamethazine, sulfamethoxazole, sulfamonomethoxine, thiamphenicol, and trimethoprim, have been approved for use in aquaculture in China, the world's top producer of aquaculture (Liu et al. 2017; Sheng and Wang 2021).

13.2.11 Ready-to-Eat Frozen Foods

Kotzekidou (2013) studied along with chilled RTE items, such as sandwiches and desserts, food service organizations also offer their patrons hot goods, primarily oven-baked pastries that are made from frozen pastries that are heat-processed at the point of sale. A growing share of the food market, including supermarkets, restaurants, catering companies, and institutional businesses, is made up of ready-to-bake frozen pastries (Kennedy 2000). Before baking, they are kept under freezing conditions for an extended period of time. Due to the fact that RTE foods, particularly salads and sandwiches, are frequently made by hand and this close contact may raise the risk of contamination with possible foodborne pathogens, these foods have been linked to outbreaks of foodborne disease (Colombari et al. 2007). The opportunistic pathogen *S. aureus* prefers to grow on foods that have a higher percentage of water-phase salt and less water activity, such as cheese, cooked pork products, and dry-fermented sausage. According to Perez-Rodríguez et al. (2007), the presence of *S. aureus* is linked to cross-contamination or raw material contamination that results from improper handling during processing and storage. A common pathogenic bacterium that forms spores, *B. cereus*, could be dangerous when present in baked goods that have been roasted in the oven. Owing to their resistance, endospores can endure in starchy products that have undergone heat treatment. In these cases, the organism may develop because of inappropriate product cooling (Rosenquist et al. 2005). According to Chan and Wiedmann (2008), *L. monocytogenes* can cause significant sickness in sensitive individuals, such as expectant mothers, those with cancer and AIDS, recipients of organ transplants, and persons over 65. The risk that *L. monocytogenes*, if present, will multiply to the point where it could cause human disease is greatly increased when refrigerated RTE foods stored at slightly abusive temperatures (e.g., 7–10°C) allow *L. monocytogenes* to grow (i.e., products with pH > 4.4 or aw > 0.92, products with pH > 5.0 and aw > 0.94, and goods that, in accordance with European Commission Regulations 2073/2005 & 1441/2007, have a shelf life longer than five days. The permissible lots for RTE foods in the European Union have a microbiological limit of 100 CFU/g, meaning that *L. monocytogenes* growth will not occur during product storage. Rather, the microbiological limit for RTE foods that can support the growth of *L. monocytogenes* is "absence in 25 g." Regarding Listeria in RTE meals, the United States has implemented a zero-tolerance policy. However, the U.S. Food and Drug Administration has released a proposal that would allow *L. monocytogenes* to grow up to 100 CFU/g in RTE foods that are unable to support this pathogen's growth (FDA 2012). As most RTE products have a shelf life of less than five days, the European Union generally permits a tolerance of up to 100 *L. monocytogenes* CFU/g being accepted in items supplied by catering venues.

Numerous frozen snack products can become contaminated by microorganisms during the harvesting, processing, post processing, transportation, retailing, storing, or handling stages (Sultana et al. 2014; Akter et al. 2019). Furthermore, because they contain sufficient nutrients to support the growth of microorganisms, items derived from chicken are regarded as among the most perishable meals (Sultana et al. 2014). The main contaminants of these food products are bacteria that cause spoilage and disease, such as *C. botulinum*, *C. perfringens*, *Salmonella* species, *Jovibacter campylobacter*, *Parahaemolyticus Vibrio*, *Enterocolitica Yersinia*, *Pseudomonas* species, *Campylobacter* species, and *Mucor* species, *Moraxella* spp., lactic acid bacteria, and different genera of the Enterobacteriaceae family (Sultana et al. 2014; Enayat 2012). Foodborne pathogenic bacteria produce a variety of foodborne illnesses and intoxications, including signs of enteric illness, fever, hemorrhagic colitis, meningitis, joint infection, renal disease, enteric symptoms, and infertility (Sultana et al. 2014). Unwanted tastes, textures, and scents are produced when microorganisms that cause spoilage break down frozen food products (Chakraborty et al. 2015).

13.3 OTHER ASSOCIATED ISSUES IN READY-TO-EAT FOOD PRODUCTS

In simple terms, food safety can be defined as measures taken to ensure that the food produced does not present any risk to its consumers. Assessing the risk in food safety starts by identifying potential food hazards. Food safety hazards can be defined as any element found in food with the capacity to risk the well-being of the consumer, leading to illness or injury. The hazards can be acutely toxic, cause harm, promote diseases, and contribute to poor nutrition. Hazards can be introduced into food at various stages, including harvesting, formulation, processing, packaging, labeling, transportation,

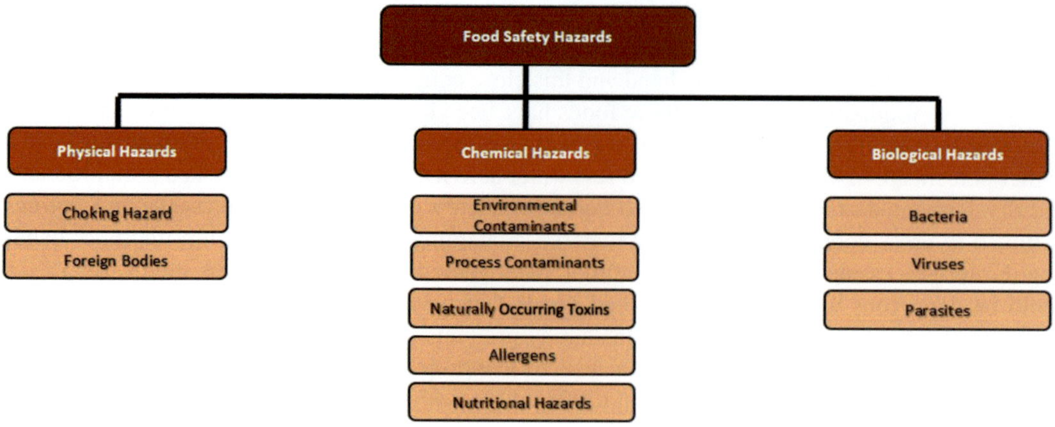

FIGURE 13.2 Food safety hazards classification.

storage, preparation, and serving. Based on the type of agents causing the hazards, hazards can be classified as physical, chemical, and biological hazards. Apart from the microbiological agents which were discussed extensively in the previous sections, chemical and physical agents also affect the wholesomeness of RTE foods, thus making them unsafe for consumption. Like microbial contamination, physical and chemical hazards are the results of cross-contamination, failure in prerequisite programs, good manufacturing practices, and poor handling of the food. Figure 13.2 shows the different types of food safety hazards that pose potential risks to consumers.

13.3.1 Physical Hazards

Physical hazards are either unintentionally introduced foreign bodies such as metal fragments or shattered glass, or naturally occurring objects, such as stones and pits in fruits or bones in meat-based products, that pose a health risk to consumers. Any material from outside sources that can cause injury or illness should be deemed hazardous. According to the 2022 report from the United States Department of Agriculture (USDA), the prevalence of foreign materials continues to be the second most frequent cause of food recalls, and in 2019, it held the top position as the primary reason for food recalls. As per Kleter et al. (2009), the primary reported physical hazards include the presence of foreign objects such as glass in products, choking hazards, and the identification of unspecified materials and metal fragments. Choking hazards are associated with product dimensions, shape, and texture. Confectionery accounts for the majority about 92% of the reports concerning choking, making it the largest product category with physical hazards overall.

The risks associated with physical agents are cuts and injuries, broken teeth, and choking which could cause serious illness to death. Kenner (2001) listed that the most reported illnesses or injuries linked to physical contaminants are problems concerning dental issues, oral injuries or lacerations, and trauma to the esophagus, abdomen, or other interconnected organs of the alimentary canal. Features that determine the potential risks of the physical agent to consumers in food products include the size of the physical agent, the nature of the product such as beverage or infant formulas, the target consumers such as infants, and elderly, and the physical characteristics of the agents such as hardness, sharpness, etc., affect the risk level of the hazards. The key sources of the origin of physical hazards can be traced to the raw materials, ingredients, manufacturing environment, plant equipment, and shop floor employees. Different approaches implemented to address the physical hazards are diverse, spanning online visual inspection, in-line metal detection, the utilization of magnets, online automated vision systems, and X-ray technology, as well as screens, filters, and sieves. Employee training programs and adherence to Good Manufacturing Practices (GMPs) are integral components of a comprehensive strategy for controlling physical hazards.

13.3.2 Chemical Hazards

Chemical agents may get introduced into the food products at any stage of food production, and when they exceed the safety limits become chemical hazards. In contrast to physical hazards, chemicals pose a subtler challenge as they are more elusive and challenging to identify and eliminate from the manufacturing process. Chemical contaminants are generally the contaminants from environment or processing, adulterants and food additives, and monomers migrating from the packaging material of the food. It is noteworthy that many ingredients, additives, process aids, flavor, and color compounds, as well as commonly used sanitizers, lubricants, and adhesives in the food industry, harbor the potential for danger, capable of causing human illness. Certain hazards associated with ingredients, like food allergens or naturally occurring mycotoxins, are intrinsic to the food itself or generated in the natural environment. On the other hand, hazards linked to ingredients, such as pesticides, drug residues, heavy metals, and environmental contaminants, stem from the raw materials used. Process-related chemical hazards can either be intentionally included in product formulation, as seen with sulfites affecting sensitive consumers or inadvertently introduced during food production. Industrial chemicals used for non-food purposes in a facility represent an example of unintentional introduction. Additionally, process contaminants, like acrylamide, may form during the processing. Approved levels for the use of permissible food additives can be found in government food regulations, such as FSSAI Regulations or in Codex Alimentarius. Based on the type and source of the chemical contamination, they can be categorized as naturally occurring components, allergens, mycotoxins, heavy metals, pesticides, industrial chemicals, and miscellaneous contaminants. The list of the chemical components is provided in Table 13.1.

13.3.3 Other Biological Hazards

Biological hazards include the hazards involving biological agents such as bacteria, viruses, molds, protozoa, and parasites. A larger part of the food poisoning incidents reported was due to the contamination of the food with pathogenic bacteria such as *Salmonella* spp., *E. coli* O157:H7, *L. monocytogenes*, *C. perfringens*, *C. botulinum*, *S. aureus*, and *Campylobacter jejune*. The hazards associated with these pathogenic bacteria for every category of RTE foods are discussed in detail in the previous parts of this chapter.

13.3.3.1 Virus

Apart from the pathogenic bacteria, food can be contaminated by viruses such as hepatitis A and E viruses, the Norwalk group of viruses, and rotavirus, which can lead to food safety outbreaks. Pathogenic viruses can be grouped into three categories namely, gastroenteritis-causing viruses, enterically transmitted viruses, and viruses that cause illness by migrating to other organs of the body from the intestine. In contrast to foodborne pathogens, foodborne viruses are not dependent on the food environment for their survival, as they do not propagate within food systems. They are stable outside their host and resistant to acidic conditions. It is also highly challenging to detect and recover these viruses from contaminated food.

13.3.3.2 Parasites

Another type of biological hazard that could compromise consumer safety are parasites. Some of the common foodborne pathogenic parasites are tapeworms such as *Diphyllobothrium* spp., and *Taenia* spp., protozoa such as *Toxoplasma gondii*, *Giardia intestinalis*, *Cryptosporidium* spp., and *Cyclospora cayetanensis*, and nematodes such as *Trichinella* spp., and *Anisakids* spp. Parasites are

TABLE 13.1
Types and Sources of Chemical Hazards

Type of the Chemical Hazard	Contaminants Examples	Source of the Contamination
Pesticides	Insecticides, fungicides, rodenticides, insect repellents, herbicides or weed killers, and some antimicrobials. Common groups are organophosphate, carbamate, pyrethroid, organochlorine pesticides, etc.	Improper agricultural practices. May enter the produce before, or during harvesting operations.
Heavy metals	Arsenic, cadmium, chromium, copper, lead, mercury, zinc, nickel, etc.	Enters the food system through industrially polluted soil and water
Migration compounds	Plasticizers (like epoxidized soy bean oil (ESBO) or phthalates), unreacted monomers and oligomers (like styrene, bisphenol, etc.), additives, adhesives, and printing inks	From the packaging material to the food products
Persistent organic pollutants	Dioxins, organochlorine pesticides such as dichloro diphenyltrichloroethane, industrial chemicals like polychlorinated biphenyls, etc.	They are persistent contaminants that enter the soil and water through industrial pollutions and as pesticides, through which it enters the food system
Veterinary drug residues	Growth hormones such as estradiol-17β, zeranol, progesterone, testosterone, antibiotics such as fluoroquinolones, penicillin, tetracyclines and sulfonamides, anti-parasitic drugs such as avermectins, benzimidazoles.	Using veterinary drugs inappropriately on animals raised for food may lead to the accumulation of drug residues in their tissues and organs thus resulting in the food products prepared from them.
Prohibited food additives	Artificial flavorings such as benzophenone, coumarin, artificial colorings such as sudan dye, red 1, artificial sweeteners such as p-ethoxy-phenylurea, flour improver such as potassium bromate, and preservatives such as formalin, etc.	Intentional addition by corrupted food manufacturers to improve the appearance of food and to obtain financial gain.
Mycotoxins	Aflatoxins, deoxynivalenol, fumonisins, patulin, and ochratoxin A	Contamination by the fungi spores before harvesting or during storage.
Naturally occurring toxins	Marine toxins such as ciguatoxins, tetrodotoxin, phytotoxins like cyanogenic glycosides, furocoumarins, lectins, solanines, etc.	Either accumulated to the contamination of the surroundings by toxin carrying algae and protozoan, or integral component of the produces.
Process contaminants	Acrylamide, ethyl carbamate, furan, nitrosamines, polycyclic aromatic hydrocarbons, etc.	Undesirable chemical by-products formed during food processing, particularly during heating, drying, or fermenting of foods.
Allergens	Common allergens are milk, eggs, fish, Crustacean shellfish, tree nuts, peanuts, wheat, and soybeans	Integral component of the food that affects only a particular set of population that are sensitive to it. The allergens trigger the vulnerable groups immune system thus causing symptoms that can vary from mild discomfort to severe reactions
Adulterants	Water in milk, prohibited colors in food, papaya seeds in pepper, spent spices, etc.	Intentional addition by corrupted food manufacturers for financial gain
Nutritional hazards	Minerals (iron, calcium, sodium, etc.) and vitamins (vitamin A, D, E, and K)	Excess addition due to failure in GMP or addition below the required limit resulting in malnutrition.

more commonly found in raw meat, seafood, fruits, and vegetables. They might be present in RTE products when the processing conditions required for food safety are not met.

13.3.3.3 Prions

Prions are abnormally folded proteins that are recently identified as a food safety hazard, linked to the consumption of bovine-derived products. Prions are protein molecules that are glycosylated and can replicate without DNA or RNA related to them. These eccentric infective agents are associated with several brain diseases such as Creutzfeldt–Jakob disease (CJD), Gerstmann–Sträussler–Scheinker syndrome (GSS), fatal familial insomnia (FFI), and Kuru, all being a type of fatal neurological disorder. The extent of their infective virulence remains poorly understood; however, they have consistently been associated with the fatal brain disease, Creutzfeldt–Jakob disease (vCJD) identified in 1996. It is believed to be a human version of bovine spongiform encephalopathy (BSE or 'mad cow disease'), a transmissible spongiform encephalopathy (TSE) affecting cattle. It is hypothesized that human vCJD cases may arise from consuming prion-infected meat products from BSE-affected cattle in the food supply, raising significant concerns about prions' potential to cross species barriers. The prion-causing foodborne vCJD in humans is found in specific tissues of BSE-infected cattle. These tissues, known as specified risk material (SRM), are banned from the food supply in many countries due to their high prion levels. Previously, these materials were present in some meat products such as pies and burgers, but stricter controls have been implemented.

13.3.4 Other Issues

Other issues identified with ready-to-food products include food adulteration, food additives misappropriation, genetically modified foods, mislabeling, and foods sold past their expiry dates, etc.

13.3.4.1 Food Adulteration

Food adulteration refers to the addition of harmful or unnecessary substances to food, thus diminishing its quality. It is an act of debasing a product, aiming to counterfeit purity or substitute an inferior item for financial gain. Woldemariam et al. (2015), in their paper listed four ways of adulteration, the addition of unwanted substances such as the addition of water to milk to increase the quantity, mixing of sub-standard material with a superior one like the addition of spent tea leaf residues to fresh tea leaves, addition of prohibited dye and preservatives to superficially improve the quality of substandard products, like dyes in chilli powder to make it appear bright red, and removal of valuable substance from the standard product like extraction of oleoresins from the spices and selling the spent spices. The adulterants resulting from the malicious intentions, and intentional or unintentional addition may result in any one or combination of physical, chemical, or biological hazards (FSPCA 2019).

13.3.4.2 Food Additives Misappropriation

The colorants are the most misused food additives commonly found in RTE food products, which include the use of prohibited food colorants, the addition of permitted food colorants above the prescribed limit, and the addition of food colorants in products in which food colorants are not permitted by the government regulations. In a 2004 study by Jonnalagadda et al. on RTE foods in Hyderabad, India, among 545 analyzed items, 90% contained permitted colors, 2% mixed permitted and non-permitted colors, and 8% had only non-permitted colors (Jonnalagadda et al. 2004). Of the RTE foods with permitted colors, 73% exceeded the 100-ppm limit under the Prevention of Food Adulteration Act, while 27% complied. Sweet meats and non-alcoholic beverages showed the highest color concentrations, reaching 18,767 ppm. Some foods like savories and miscellaneous items, though not supposed to contain colors as per the Prevention of Food Adulteration Act, were found to contain them. Commonly reported non-permitted colorants and sweeteners in literature are amaranth, auramine, carmoisine, cyclamate, malachite green, metanil yellow, orange II, quinoline yellow, rhodamine B, and Sudan dyes.

13.3.4.3 Genetically Modified Foods

Genetically modified (GM) foods result from altering an organism's genetic material using techniques like genetic engineering. This involves introducing genes from other species or modifying existing ones to confer traits like pest resistance, improved nutrition, or longer shelf life. GM foods are an emerging food safety concern as per the reports published by the European Community's Rapid Alert System for Food and Feed (RASFF). Although some scientists argue there is inadequate proof regarding the risks of GM foods to public health and the environment, the precautionary principle advises against adopting such technology in the face of potential but uncertain impacts. There have also been certain concerns of an increase in incidences of food allergies after the introduction of these foods and there is another concern of genetic modification introducing an allergenic gene to a non-allergenic product. However, the National Academy of Sciences reported no correlation between GE food intake and an increase in the occurrence of food allergies. The GM products could be incorporated to produce food ingredients, microorganisms, feeds, etc. Thus, unless communicated, GM foods must be treated as an adulterant and a food safety hazard.

13.3.4.4 Mislabeling and Selling of Expired Foods

Zemichael (2019) reported in his review of public health risks related to the food market, many food samples gathered from supermarkets, stores, and restaurants were genetically found to be different from the labeled species, indicating widespread mislabeling. Seafood emerged as the most mislabeled food category in these studies. Apart from that, the stores, shops, and restaurants were found to sell expired food products adding to significant public health and environmental issues.

13.4 HYGIENIC PRACTICES FOR THE PRESERVATION OF READY-TO-EAT FOOD PRODUCTS

Food safety incidences such as foodborne illness outbreaks and cross-contamination are reported to be the outcome of the unhygienic practices of the food establishment or the food industries. They are a direct result of in-adequate strategizing, developing, and effectively executing a robust Food Safety Management System (FSMS) and its fundamental components by food industries. This global issue is prevalent across all food processors, particularly prominent in medium- and small-scale food processing facilities. Lee et al. (2021) reported in their review that the absence of adequate prerequisite programs (PRPs) was the number one obstacle in implementing FSMS, and lack of knowledge of HACCP was ranked as the second reason. This section discusses the different hygienic components of the FSMS required to ensure food safety and mitigate risks.

13.4.1 Food Safety and Hygiene

The World Health Organization defines food hygiene as encompassing all essential conditions and practices across food production, processing, storage, distribution, and preparation. These measures are crucial in ensuring that food remains safe, wholesome, and suitable for human consumption. Food hygiene is mainly focused on the control and mitigation of the risks posed by food pathogens, while food safety encompasses all three types of hazards. HACCP is a risk-based systematic approach to prevent food safety hazards during the production process and stands as a critical standard, both obligatory and voluntary, in global food trade. Proper implementation of prerequisite programs such as Good Hygiene Practices (GHPs), Good Agricultural Practices (GAPs), and GMPs, coupled with training and traceability, lays the groundwork for an efficient HACCP system.

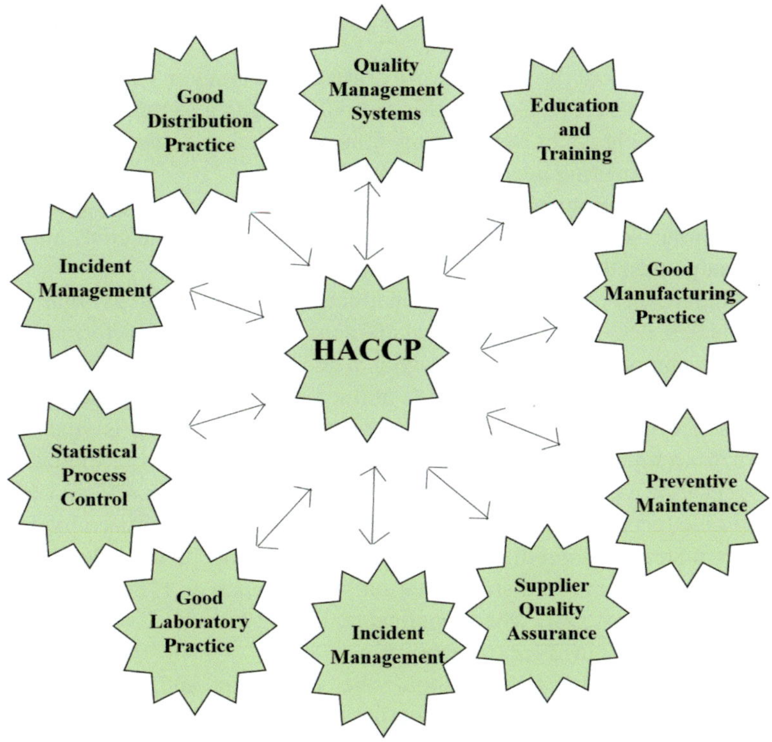

FIGURE 13.3 HACCP support systems. (Lee et al. 2021.)

13.4.1.1 Pre-Request Programs (PRPs)

PRPs serve as the essential cornerstone supporting the HACCP system. The World Health Organization (WHO 1998) defines PRP as "practices and conditions needed prior to and during the implementation of HACCP and which are essential for food safety." They apply across all phases of the worldwide food supply chain and encompass crucial practices for cultivation, harvesting, manufacturing, storage, distribution, retail, food service, and small-scale industries. These comprehensive programs incorporate various fundamental conditions and actions crucial for upholding cleanliness and managing hazards in food production.

Mortimore et al. (2013) describe PRPs as the support network of HACCP in their book. The support network includes the GMPs, process control, supplier quality assurance, preventive maintenance, training, quality management system, etc., as shown in Figure 13.3. The number of systems required will depend on the size and type of the establishment that produces the RTE. This encompasses practices such as GMP, GHPs, GAPs, sanitation protocols, pest control, personnel hygiene, environmental monitoring, and more.

The Codex general principles of Food Hygiene (2003) serves as the reference document for the guidelines and standards for food manufacturers. In India, the Food Safety and Standards Authority of India (FSSAI) has laid down the regulations and requirements for different aspects and categories of food, including the general hygienic and sanitary practices to be followed by food business operators (FBO) in the Schedule 4 of the Food Safety and Standards (Licensing and Registration of Food Businesses) Regulation, 2011. Every FBO in the country is required to have a proper registration or valid license before commencing their operations. The possession of a valid license is a prerequisite for initiating any food-related enterprise. Adherence to safety, sanitation, and hygiene standards is imperative for all operators, an unyielding requirement across all operational periods.

13.4.1.1.1 Guidelines for Food Manufacturers

a. Facility Design: The layout should minimize cross-contamination, ensuring segregation of raw materials and finish products. Consideration should be given to the availability of necessary services such as potable water for cleaning. It is also vital to assess the staff and equipment movement patterns, ensuring adequate hygiene facilities.
b. Equipment Design: Equipment should be designed as per hygienic engineering principles such as smooth surface, contact surfaces made of food grade material, absence of hollow spaces and sharp ends, easy access for cleaning, etc.
c. Preventive Maintenance: The integrity of buildings and equipment is vital for food safety. Preventive maintenance ensures operational efficiency and machine performance. The program should include a comprehensive list of equipment and facility areas needing maintenance, along with documented schedules and procedures, utilities, spare parts, lubricants, calibration, etc.
d. Temperature Control: Consistent and monitored temperature control is vital across the entire supply chain, from sourcing to consumer delivery, covering chilling, transportation, and storage. For RTE street food vendors, perishables must be stored between 4°C and 6°C. Other food businesses should receive potentially high-risk food below 5°C and frozen food below −18°C.
e. Storage: Raw materials, ingredients, work-in-progress, and processed/cooked or packaged food products must adhere to a stock rotation system, such as FIFO (First in, First Out) or FEFO (First Expire First Out) where applicable, during storage. Proper stock inventory and records should be maintained for traceability.
f. Sanitation: The sanitation program should be systematic and should be based on risk assessment, documentation, validation of effectiveness, and regular verification. The sanitation method should be standardized and validated for the type of hazards specific to the food industry such as allergen for cereals or *Salmonella* for salads, etc.
g. Pest Control: Pest acts as vectors of microbiological hazards and fragments of their body parts could become potential foreign bodies. A proper pest management program should detail target pests, control methods, plant schematics, chemical use, inspection frequency, and training needs.
h. Environmental Monitoring: The environmental monitoring program (EMP) entails verifying cleaning and sanitation protocols and other pathogen control measures through diverse sampling analyses. Its objective is to avert environmental contamination of the final product. *Listeria* and/or *Salmonella* EMP are mandatory requirements for an RTE food production unit depending on the type of sanitation program available.
i. Personal Hygiene: Handwashing is the very basic yet effective means of ensuring food safety in an RTE establishment. Proper handwashing and entry routines should be in place with required facilities, and the personnel involved in handling food should be free of any contagious diseases or cuts. Jeweler policy and proper uniforms should be enforced to prevent cross-contamination.
j. Training: Education and training stand as pivotal elements in food safety programs and promoting a food safety culture. All food industry staff should undergo general hygiene training, comprehending their crucial role. This training must be recurrent and duly validated and verified.

All the programs mentioned should be standardized and formulated based on risk analysis. The programs must be validated before implementation and verified periodically using appropriate methods to ensure efficient functioning. Standard operating procedures (SOP) and sanitation standard operating procedures (SSOP) for the implementation and execution of the programs need to be prepared and used to train the employees. All the processes and procedures need to be documented properly for future reference and auditing purposes.

13.4.1.2 Hazard Analysis and Critical Control Point (HACCP)

Pillsbury food scientists collaborated with researchers from Natick Research Laboratories and the National Aerospace Agency (NASA), leading to the creation of a quality assurance system that focused on preventive measures, controlling processes, and maintaining records from raw material sourcing to end-product distribution. Originating from the space program's food safety project, this non-destructive quality assurance system laid the groundwork for the present-day HACCP plan system.

HACCP is a proactive food safety system that scientifically evaluates and regulates every phase of food manufacturing, storage, and distribution, to control the occurrence of biological, chemical, and physical hazards. The well-known seven principles of the HACCP system as taken from the FDA are given below:

Principle 1: Conduct a hazard analysis.
Principle 2: Determine the critical control points (CCPs).
Principle 3: Establish critical limits.
Principle 4: Establish monitoring procedures.
Principle 5: Establish corrective actions.
Principle 6: Establish verification procedures.
Principle 7: Establish record-keeping and documentation procedures.

HACCP-based food safety systems have found success in food processing facilities, retail stores, and food service establishments. The seven HACCP principles enjoy global acceptance among government bodies, trade associations, and the worldwide food industry. Studies have shown that the food hygiene in food establishments depends significantly on their HACCP status. As mentioned earlier, for the success of the HACCP proper PRPs should be in place followed by education and training of the management as well as the employees. During the development and execution of each HACCP plan, it is crucial to evaluate the presence and efficiency of prerequisite programs. The steps involved in making a HACCP plan are provided in the Figure 13.4. The implemented HACCP system in a food facility should undergo regularly scheduled reviews and, if necessary, be amended appropriately.

Commitment from top management to food safety is necessary for the successful implementation of the HACCP plan. Resolute allegiance from the top management to a food safety culture, reinforces the importance of producing safe food among all employees, guiding its success throughout the organization.

13.5 CONCLUSION

The presence of foodborne pathogens and contamination in RTE foods poses a significant risk to public health, emphasizing the crucial need for stringent hygienic practices throughout the food production and distribution chain. The potential for rapid and widespread outbreaks underscores the importance of proactive measures, such as thorough food safety protocols, regular testing, and stringent quality control. Hygienic practices, including proper handwashing, sanitation, and adherence to food safety guidelines, are fundamental in preventing the introduction and proliferation of harmful microorganisms in RTE foods. The onus falls not only on food producers but also on consumers to handle and store these products with care to minimize the risk of contamination. Ongoing education and awareness campaigns are essential to inform both industry professionals and the general public about the risks associated with foodborne pathogens and the critical role they play in preventing outbreaks. Regulatory bodies must continue to enforce and update standards to reflect the latest scientific knowledge and technological advancements in food safety. In navigating the complex landscape of RTE food safety, collaboration among stakeholders, including government agencies, food producers, and consumers, is paramount. By collectively prioritizing and implementing stringent measures, we can mitigate the risks of contamination, ensuring that ready-to-eat foods remain a safe and convenient option for consumers worldwide.

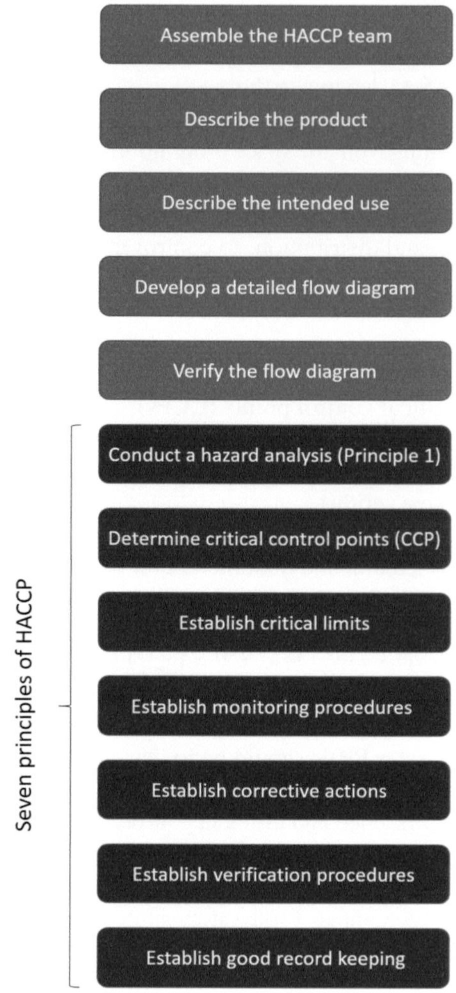

FIGURE 13.4 12-Step guidelines for HACCP plan.

REFERENCES

Abu, Fahmi, Abu Fahmi, Fahmi Abu Al-Rub, Penelope Shibhab, Paulo Baptista. 2020. Food Safety Management Systems. Gavin Publications. https://bit.ly/3qjI0En

Adegbehingbe, K. T., B. S. Adeleke, and D. Adejoro. 2022. "Microbiological assessment of soya milk samples from Akoko areas of Ondo State, Nigeria." *Biotechnology*. 21(3): 120–126. http://dx.doi.org/10.3923/biotech.2022.120.126.

Ahmed, T., A. Sharmin, A. Paul, M. R. Dipu, and T. Islam. 2020. "Microbiological quality analysis of different types of popular dried food items." *Food Research*. 4(4): 1297–1302. https://doi.org/10.26656/fr.2017.4(4).080.

Ahmed, Tasnia, Kamal Kanta Das, and Md Aftab Uddin. 2019. "The microbiological quality of commercial fruit juices—current perspectives." *Bangladesh Journal of Microbiology*. 35(2): 128–133. http://dx.doi.org/10.3329/bjm.v35i2.42643.

Akhigbemidu, W., A. Musa, and O. Kuforiji. 2015. "Assessment of the microbial qualities of noodles and the accompanying seasonings." *Nigerian Food Journal*. 33(1): 48–53. https://doi.org/10.1016/j.nifoj.2015.04.005.

Akond, Muhammad Ali, Saidul Alam, S. M. R. Hasan, Sanzida Mubassara, Sarder Nasir Uddin, and Momena Shirin. 2009. "Bacterial contaminants in carbonated soft drinks sold in Bangladesh markets." *International Journal of Food Microbiology*. 130(2): 156–158. http://dx.doi.org/10.1016/j.ijfoodmicro.2009.01.014.

Akter, Moushumi, Selina Sultana, and Saurab Kishore Munshi. 2019. "Microbiological quality assessment of ready-to-eat fried chicken and chicken soup samples sold in Dhaka Metropolis, Bangladesh." *Sumerianz Journal of Biotechnology*. 2(7): 48–54. www.sumerianz.com.

Ali, Manal A., Mona H. Hashish, and Marwa M. Fekry. 2023. "Microbiological quality of some packed and unpacked bread products in Alexandria, Egypt." *Journal of the Egyptian Public Health Association*. 98(1): 16. https://doi.org/10.1186/s42506-023-00141-9.

Ali, Shaimaa AM, Shahad R. Mutlaq, and Amin S. Badwy. 2023. "Detection of deoxynivaenol (DON) toxin in some samples of corn chips and corn flex collected from local markets." *IOP Conference Series: Earth and Environmental Science*. 1214: 012034. https://doi.org/10.1088/1755-1315/1214/1/012034.

Aljaloud, Sulaiman O. 2016. "Microbiological quality and safety of energy drink available in the local markets in Saudi Arabia." *International Journal of Food Science, Nutrition and Dietetics*. 5(4): 287–289. http://dx.doi.org/10.19070/2326-3350-1600051.

Alli, I. 2003. Food Quality Assurance: Principles and Practices. CRC Press. https://doi.org/10.1201/9780203484883.

Al-Nasiry, B. S. A. N. 2020. "Detection of bacterial contamination in filled and dried biscuit products of young children." *Annals of Tropical Medicine and Public Health*. 23(16): 9–13. http://dx.doi.org/10.36295/ASRO.2020.231602.

Baer, Arica A., Michael J. Miller, and Anna C. Dilger. 2013. "Pathogens of interest to the Pork Industry: A review of research on interventions to assure food safety." *Comprehensive Reviews in Food Science and Food Safety*. 12(2): 183–217. https://doi.org/10.1111/1541-4337.12001.

Balali, Gadafi Iddrisu, Denis Dekugmen Yar, Vera Gobe Afua Dela, and Priscilla Adjei-Kusi. 2020. "Microbial contamination, an increasing threat to the consumption of fresh fruits and vegetables in today's world." *International Journal of Microbiology*. 2020: 3029295. https://doi.org/10.1155/2020/3029295.

Bansal, Sangita, Apoorva Singh, Manisha Mangal, Anupam K. Mangal, and Sanjiv Kumar, (2017). Food adulteration: Sources, health risks, and detection methods. *Critical Reviews in Food Science and Nutrition*. 57(6): 1174–1189. doi: 10.1080/10408398.2014.967834.

Berthold-Pluta, Anna, Antoni Pluta, Monika Garbowska, and Ilona Stefańska. 2019. "Prevalence and toxicity characterization of *Bacillus cereus* in food products from Poland." *Foods*. 8(7): 269. https://doi.org/10.3390%2Ffoods8070269.

Buchanan, Robert L., Leon G.M. Gorris, Melinda M. Hayman, Timothy C. Jackson, and Richard C. Whiting. 2017. "A review of *Listeria monocytogenes*: An update on outbreaks, virulence, dose–response, ecology, and risk assessments." In *Food Control*. Elsevier Ltd. https://doi.org/10.1016/j.foodcont.2016.12.016.

Callejón, Raquel M., M. Isabel Rodríguez-Naranjo, Cristina Ubeda, Ruth Hornedo-Ortega, M. Carmen Garcia-Parrilla, and Ana M. Troncoso. 2015. "Reported foodborne outbreaks due to fresh produce in the United States and European Union: Trends and causes." *Foodborne Pathogens and Disease*. 12(1): 32–38. https://doi.org/10.1089/fpd.2014.1821.

Capita, Rosa, Carlos Alonso-Calleja, Benito Moreno, and Marıa Camino Garcıa-Fernandez. 2001. "Occurrence of *Listeria* species in retail poultry meat and comparison of a cultural immunoassay for their detection." *International Journal of Food Microbiology*. 65. www.elsevier.nlrlocaterijfoodmicro.

Carpentier, Brigitte, and Olivier Cerf. 2011. "Review—persistence of *Listeria monocytogenes* in food industry equipment and premises." *International Journal of Food Microbiology*. https://doi.org/10.1016/j.ijfoodmicro.2011.01.005.

Carraturo, Federica, Olga De Castro, Jacopo Troisi, Adriana De Luca, Armando Masucci, Paola Cennamo, Marco Trifuoggi, Francesco Aliberti, and Marco Guida. 2018. "Comparative assessment of the quality of commercial black and green tea using microbiology analyses." *BMC Microbiology*. 18(1): 1–12. https://doi.org/10.1186/s12866-017-1142-z.

Centres for Disease Control and Prevention. Outbreak of *Listeria* infections linked to hard-boiled eggs. Final Update. Retrieved from www.cdc.gov/listeria/outbreaks/eggs-12-19/index.html#:~:text=CDC%2C%20several%20states%2C%20and%20the,Almark%20Foods%20of%20Gainesville%2C%20Georgia.&text=Do%20not%20eat%2C%20sell%2C%20or,recalled%20hard%2Dboiled%20egg%20products accessed December 09, 2023.

Chan, Yvonne C., and Martin Wiedmann. 2008. "Physiology and genetics of *Listeria monocytogenes* survival and growth at cold temperatures." *Critical Reviews in Food Science and Nutrition.* 49: 237–253. doi: 10.1080/10408390701856272.

Chen, Yujuan, Yufeng Wen, Jiangen Song, Baifeng Chen, Shushu Ding, Lei Ding, and Jiajia Dai. 2018. "The correlation between family food handling behaviours and foodborne acute gastroenteritis: A community-oriented, population-based survey in Anhui, China." *BMC Public Health.* 18(1): 1–10.

Christopher, M. A., V. F. Nyoyoko, U. E. Antia, and I. E. Eyo. 2020. "Evaluation of bacteria associated with ready to eat rice in the Niger Delta south–south Nigeria." *Annals of Microbiology and Immunology.* 3(1): 1021.

Cirone, K., Y. Huberman, C. Morsella, L. Méndez, M. Jorge, and F. Paolicchi. 2013. "Growth of *Mycobacterium avium* subsp. paratuberculosis, *Escherichia coli*, and *Salmonella enteritidis* during preparation and storage of yogurt." *ISRN Microbiology.* (December): 1–7. https://doi.org/10.1155/2013/247018.

Clavijo, Raul I., Cindy Loui, Gary L. Andersen, Lee W. Riley, and Sangwei Lu. 2006. "Identification of genes associated with survival of *Salmonella enterica* serovar *enteritidis* in chicken egg albumen." *Applied Environmental Microbiology.* 72(2): 1055–1064. doi: 10.1128/AEM.72.2.1055-1064.2006.

Codex Alimentarius: Food hygiene, basic texts. (2003). Italy: Food and Agriculture Organization of the United Nations.

Colombari, Viviane, Mariana D.B. Mayer, Zaira M. Laicini, Elza Mamizuka, Bernadette D.G.M. Franco, Maria Teresa Destro, and Mariza Landgraf. 2007. "Foodborne outbreak caused by *Staphylococcus aureus*: phenotypic and genotypic characterization of strains of food and human sources." *Journal of Food Protection.* 70(2): 489–493. doi: 10.4315/0362-028x-70.2.489.

Darkwah, Samuel, Doris Agbettor, Francis Codjoe, and Eric S. Donkor. 2022. "Microbial contamination of herbal preparations on the Ghanaian market, Accra." *Microbiological Insights.* 15: 11786361221139602. https://doi.org/10.1177%2F11786361221139602.

Darwish, Al-Zahraa M. 2018. "Foodborne pathogens of fast food and ready-to-eat foods in Tabuk city and evaluating hazard for food quality." *International Journal of Healthcare and Biomedical Research.* 6(02): 149–158.

Dewey-Mattia, Daniel, Karunya Manikonda, Aron J. Hall, Matthew E. Wise, and Samuel J. Crowe. 2018. "Surveillance for foodborne disease outbreaks—United States, 2009–2015." *MMWR Surveillance Summaries.* 67(10): 1.

Diarmuid, J J (, and) Sheehan. 2013. "Milk Quality and Cheese Diversification." *Irish Journal of Agricultural and Food Research.* 52. www.jstor.org/stable/23631035.

Diriba, Asrat, Mitiku Eshetu, and Yonas Hailu. 2023. "Production practice, microbial quality and consumer acceptability test of traditionally produced butter in North Shoa Zone, Oromia Regional State, Ethiopia." *Heliyon.* 9(6): e17510. https://doi.org/10.1016/j.heliyon.2023.e17510.

Divyashree, S., Jamuna Prakash and Prabahavathis. N. 2013. "Microbial quality of selected commercial fresh fruit juices sold in Mysore City." *Journal of Food Science and Technology Nepal.* 8: 30–34. http://dx.doi.org/10.3126/jfstn.v8i0.11746.

Djekic, Ilija, Nada Smigic, Eleni P. Kalogianni, Ada Rocha, Lamprini Zamioudi, and Rita Pacheco. (2014). Food hygiene practices in different food establishments. *Food Control.* 39: 34–40. https://doi.org/10.1016/j.foodcont.2013.10.035.

El-Hajjaji, Soundous, Amaury Gérard, Juliette De Laubier, Aurélie Lainé, Viviane Patz, and Marianne Sindic. 2021. "Study of the bacterial profile of raw milk butter, made during a challenge test with *Listeria monocytogenes*, depending on cream maturation temperature." *Food Microbiology.* 98: 103778. https://doi.org/10.1016/j.fm.2021.103778.

Enayat, Kalantar, Amin Mansour, Bahmani Nasrin, Tamimi Mohammad, Hashemzadeh Mohammad, and Narenji Hanar. 2012. "Antibiotic resistance pattern in bacterial isolates obtained from frozen food samples of animal origin in Sanandaj and Ahvaz." *African Journal of Bacteriology Research.* 4(3): 38–41. https://doi.org/10.5897/JBR12.012.

European Food Safety Authority and European Centre for Disease Prevention and Control (EFSA and ECDC). 2018 "The European Union summary report on trends and sources of zoonoses, zoonotic agents and food-borne outbreaks in 2017." *EFSA Journal.* 16(12): e05500

Fast, Robert B. 2000. "Manufacturing technology of ready-to-eat cereals." In Breakfast Cereals and How They Are Made, (second edition, pp. 17–54).

FDA, 2012. CPG Sec. 555.320 *Listeria monocytogenes*. www.fda.gov/ICECI/ComplianceManuals/CompliancePolicyGuidanceManual/ucm136694.htm.

Fleurot, Isabelle, Marina Aigle, Renaud Fleurot, Claire Darrigo, Jacques Antoine Hennekinne, Alexandra Gruss, Elise Borezée-Durant, and Agnès Delacroix-Buchet. 2014. "Following pathogen development and gene expression in a food ecosystem: The case of a *Staphylococcus aureus* isolate in Cheese." *Applied and Environmental Microbiology*. 80(16): 5106–5115. https://doi.org/10.1128/AEM.01042-14.

Food Safety Preventive Controls Alliance. 2019. FSPCA Intentional Adulteration Conducting Vulnerability Assessments Participant Manual. Version 1.0. Food Safety Preventive Controls Alliance. IIT IFSH (Lesson 1).

Gálvez, Antonio, Rosario Lucas López, Hikmate Abriouel, Eva Valdivia, and Nabil Ben Omar. 2008. "Application of bacteriocins in the control of foodborne pathogenic and spoilage bacteria." *Critical Reviews in Biotechnology*. https://doi.org/10.1080/07388550802107202.

Ganjyal, Girish M. 2020. Food Safety Hazards: An Overview. Washington State University Extension.

Gantois, Inne, Richard Ducatelle, Frank Pasmans, Freddy Haesebrouck, Richard Gast, Tom J. Humphrey, and Filip Van Immerseel. 2009. "Mechanisms of egg contamination by *Salmonella enteritidis*." *FEMS Microbiology Review*. 33(4): 718–738. doi: 10.1111/j.1574-6976.2008.00161.x.

Getaneh, Dawit Kassaye, Lemessa Oljira Hordofa, Desalegn Admassu Ayana, Tesfaye Sisay Tessema, and Lemma Demissie Regassa. 2021. "Prevalence of *Escherichia coli* O157:H7 and associated factors in under-five children in eastern Ethiopia." *PLoS One*. 16(1 January). https://doi.org/10.1371/journal.pone.0246024.

Giraffa, Giorgio. 2003. "Functionality of enterococci in dairy products." *International Journal of Food Microbiology*. 88(2–3): 215–222. Elsevier. https://doi.org/10.1016/S0168-1605(03)00183-1.

Giugliano, Roberta, Noemi Musolino, Valentina Ciccotelli, Carla Ferraris, Valentina Savio, Barbara Vivaldi, Carlo Ercolini, Daniela Manila Bianchi, and Lucia Decastelli. 2023. Soy, rice and oat drinks: Investigating chemical and biological safety in plant-based milk alternatives." *Nutrients*. 15(10): 2258. https://doi.org/10.3390/nu15102258.

Gizaw, Zemichael. 2019. "Public health risks related to food safety issues in the food market: a systematic literature review." *Environmental Health and Preventive Medicine*. 24: 1–21. https://doi.org/10.1186/s12199-019-0825-5.

Glass, Kathleen A., Max C. Golden, Brandon J. Wanless, Wendy Bedale, and Charles Czuprynski. 2015. "Growth of *Listeria monocytogenes* within a caramel-coated apple microenvironment." *Journal of Clinical Microbiology*. 6(5): e01232-15. https://doi.org/10.1128%2FmBio.01232-15.

Gómez, Diego, Laura Pilar Iguácel, Ma Carmen Rota, Juan Josécarramiñana, Agustín Ariño, and Javier Yangüela. 2015. "Occurrence of *Listeria monocytogenes* in ready-to-eat meat products and meat processing plants in Spain." *Foods*. 4(3): 271–282. https://doi.org/10.3390/foods4030271.

Gori, Francesco, C. Cuccaro, M. Santamaria, R. Cardaci, S. Burgassi, M. A. Battaglia, and G. Messina. 2019. "Could chips and nuts be vector of food-borne outbreaks during happy hour?" *European Journal of Public Health*. 29(4): ckz185.312. https://doi.org/10.1093/eurpub/ckz185.312.

Hall, Andrew, Hugh Griffiths, Mike Saltmarsh, Adrian Peters, and Louise Fielding. 2012. "Profiling *Bacillus cereus* populations in a traditional style, hot-drinks vending machine and vended hot chocolate drink using polymerase chain reaction (PCR) and random amplified polymorphic DNA (RAPD) techniques." *Food Control*. 27(1): 127–131. https://doi.org/10.1016/j.foodcont.2012.02.028.

Hassanzadazar, Hassan, Mahrokh Forghani, Ahdieh Salim, and Majid Aminzare. 2017. "An investigation of microbial contamination of animal butter at the market level in Zanjan." *Journal of Human, Environment and Health Promotion*. 3(1): 43. http://dx.doi.org/10.29252/jhehp.3.1.8.

Havelaar, Arie H., Martyn D. Kirk, Paul R. Torgerson, Herman J. Gibb, Tine Hald, Robin J. Lake, Nicolas Praet, et al. 2015. "World Health Organization global estimates and regional comparisons of the burden of foodborne disease in 2010." *PLoS Medicine*. https://doi.org/10.1371/journal.pmed.1001923.

Hessel, Claudia Titze, Susana de Oliveira Elias, João Pedro Pessoa, Laís Mariano Zanin, Elke Stedefeldt, and Eduardo Cesar Tondo. 2019. "Food safety behavior and handling practices during purchase, preparation, storage and consumption of chicken meat and eggs." *Food Research International*. 125: 108631. https://doi.org/10.1016/j.foodres.2019.108631.

Hickey, Cian D., Jeremiah J. Sheehan, Martin G. Wilkinson, and Mark A. E. Auty. 2015. "Growth and location of bacterial colonies within dairy foods using microscopy techniques: A review." *Frontiers in Microbiology. Frontiers Research Foundation*. https://doi.org/10.3389/fmicb.2015.00099.

Iacumin, Lucilla, Michela Pellegrini, Andrea Colautti, Elisabetta Orecchia, and Giuseppe Comi. 2022. "Microbial characterization of retail cocoa powders and chocolate bars of five brands sold in Italian supermarkets." *Foods*. 11(18): 2753. https://doi.org/10.3390/foods11182753.

Islam, Md Jakiul, Md Abu Sayeed, Shakil Akhtar, Md Sakhawat Hossain, and Afroza Akter Liza. 2018. "Consumers profile analysis towards chicken, beef, mutton, fish and egg consumption in Bangladesh." *British Food Journal*. 120(12): 2818–2831. https://doi.org/10.1108/BFJ-03-2018-0191.

Jazmati, Fatema Nour, Saleh Trefi, Ali Ibrahim, and Yaser Bitar. 2022. "Microbial evaluation of some syrups in Syrian pharmacies." *Heliyon*. 8(5): e09366. https://doi.org/10.1016/j.heliyon.2022.e09366.

Jonnalagadda, Padmaja R., Pratima Rao, Ramesh V. Bhat, and A. Nadamuni Naidu. 2004. Type, extent and use of colours in ready-to-eat (RTE) foods prepared in the non-industrial sector—a case study from Hyderabad, India. *International Journal of Food science & Technology*. 39(2): 125–131. https://doi.org/10.1046/j.0950-5423.2003.00749.x.

Jung, Dong-Sun, Floyd W. Bodyfelt, and Mark A. Daeschel. 1992. "Influence of fat and emulsifiers on the efficacy of nisin in inhibiting *Listeria monocytogenes* in fluid milk." *Journal of Dairy Science*. 75(2): 387–393. doi: 10.3168/jds.S0022-0302(92)77773-X.

Keener, Larry. 2001. Chemical and physical hazards: The "other" food safety risk. *Food Testing and Analysis Magazine*. doi: 10.1016/B978-0-08-100596-5.03437-5.

Kennedy, Christopher J., 2000. Managing Frozen Foods (pp. 143–144). Woodhead Publishing Limited and CRC Press LLC, Cambridge, UK.

Khandoker, A., M. A. Islam, M. M. Rahman, A. A. Husna, S. Das, and M. M. Khatun. 2015. "Bacterial contamination of street-vended spicy puffed-rice sold at Bangladesh Agricultural University campus." Bangladesh Veterinarian. 31(1): 20–26. http://dx.doi.org/10.3329/bvet.v31i1.22839.

Kleter, G. A., A. L. D. O. Prandini, L. A. U. R. A. Filippi, and H. J. P. Marvin. 2009. "Identification of potentially emerging food safety issues by analysis of reports published by the European Community's Rapid Alert System for Food and Feed (RASFF) during a four-year period." *Food and Chemical Toxicology*. 47(5): 932–950. https://doi.org/10.1016/j.fct.2007.12.022.

Kopel, Eran, Nadav Davidovitch, and Hagai Levine. 2017. "Lessons from a 2016 large-scale contamination of cereals with *Salmonella altona* in Israel." *American Journal of Public Health*. 107(2): 243–245. https://doi.org/10.2105/AJPH.2016.303567.

Kotzekidou, Parthena. 2013. "Microbiological examination of ready-to-eat foods and ready-to-bake frozen pastries from university canteens." *Food Microbiology*. 34(2): 337–343. https://doi.org/10.1016/j.fm.2013.01.005.

Kregiel, Dorota. 2015. "Health safety of soft drinks: Contents, containers, and microorganisms." *BioMed Research International*. 128697. https://doi.org/10.1155%2F2015%2F128697.

Lawley, Richard, Laurie Curtis, and Judy Davis. 2012. *The Food Safety Hazard Guidebook*. Royal Society of Chemistry. https://doi.org/10.1039/9781849734813.

Lee, Jocelyn C., Aura Daraba, Chrysa Voidarou, Georgios Rozos, Hesham A. El Enshasy, and Theodoros Varzakas. 2021. "Implementation of food safety management systems along with other management tools (HAZOP, FMEA, Ishikawa, Pareto). The case study of *Listeria monocytogenes* and correlation with microbiological criteria." *Foods*. 10(9): 2169. https://doi.org/10.3390/foods10092169.

Lee, Soyul, Areum Han, Suyoung Jo, Hyewon Cheon, Hana Song, A-Ra Jang, Danbi Kim, and Sun-Young Lee. 2021. "Microbiological quality and safety of commercial fresh fruit and vegetable juices in Korea." *LWT*. 152(2021): 112432. https://doi.org/10.1016/j.lwt.2021.112432.

Legros, Julie, Sophie Jan, Sylvie Bonnassie, Michel Gautier, Thomas Croguennec, Stéphane Pezennec, Marie-Françoise Cochet, Françoise Nau, Simon C. Andrews, and Florence Baron. 2021. "The role of ovotransferrin in egg-white antimicrobial activity: A review." *Foods*. 10(4): 823. https://doi.org/10.3390/foods10040823.

Łepecka, Anna, Dorota Zielińska, Piotr Szymański, Izabela Buras, and Danuta Kołożyn-Krajewska. 2022. "Assessment of the microbiological quality of ready-to-eat salads—are there any reasons for concern about public health?" *International Journal of Environmental Research and Public Health*. 19(3):1582. https://doi.org/10.3390/ijerph19031582.

Lewis, D. A., R. Paramathasan, D. G. White, L. S. Neil, A. C. Tanner, S. D. Hill, J. C. Bruce, J. M. Stuart, A. M. Ridley, and E. J. Threlfall. 1996. "Marshmallows cause an outbreak of infection with *Salmonella enteritidis* phage type 4." Communicable Disease Report. *CDR Review.* 6(13): R183-6.

Leyer, Gregory J., and Eric A. Johnson'. 1992. "Acid adaptation promotes survival of *Salmonella* spp. in Cheese." *Applied and Environmental Microbiology.* 58.

Liu, Wukang, Jing Zhang, Ailing Guo, Qian Chen, Lihong Gu, Yao Ruan, and Xinshuai Zhang. 2021. "The specific biological characteristics of spoilage microorganisms in eggs." *LWT—Food Science and Technology.* 135: 110069. https://doi.org/10.1016/j.lwt.2020.110069.

Liu, Xiao, Joshua Caleb Steele, and Xiang-Zhou Meng. 2017. "Usage, residue, and human health risk of antibiotics in Chinese aquaculture: A review." *Environmental Pollution.* 223: 161–169. https://doi.org/10.1016/j.envpol.2017.01.003.

Lock, J. L., and R. G. Board. 1992. "Persistence of contamination of hens' egg albumen in vitro with *Salmonella* serotypes." *Epidemiology and Infection.* 108(3): 389–396. doi: 10.1017/s095026880004989x.

Machado, Solimar Gonçalves, Denise Mara Soares Bazzolli, and Maria Cristina Dantas Vanetti. 2013. "Development of a PCR method for detecting proteolytic psychrotrophic bacteria in raw milk." *International Dairy Journal.* 29(1): 8–14. https://doi.org/10.1016/j.idairyj.2012.09.007.

Memisi, N., S. Veskovic Moracanin, and M. Milijasevic. 2019. "The influence of air hygiene on microbiological safety of butter." *IOP Conference Series: Earth and Environmental Science.* 333(1), 012076. https://doi.org/10.1088/1755-1315/333/1/012076.

Mihalache, Octavian Augustin, Paula Teixeira, and Anca Ioana Nicolau. 2022. "Raw-egg based-foods consumption and food handling practices: A recipe for foodborne diseases among Romanian and Portuguese consumers." *Food Control.* 139: 109046. https://doi.org/10.1016/j.foodcont.2022.109046.

Misiou, Ourania, Konstantinos Koutsoumanis, and Jeanne-Marie Membré. 2023. "Quantitative microbial spoilage risk assessment of plant-based milk alternatives by *Geobacillus stearothermophilus* in Europe." *Food Research International.* 166: 112638. https://doi.org/10.1016/j.foodres.2023.112638.

Mohammed, Sudad Jasim. 2018. "Quality and quantity microbial assessment of the mobile restaurants (caravans) in Baghdad." *Journal of Pharmaceutical Sciences and Research.* 10(9): 2354–2355.

Monday, I. E., J. I. Francis, and S. U. Mohammad. 2014. "Microbiological quality of ready-to-eat foods (rice and Moimoi) sold by food vendors in Federal Polytechnic Bali, Taraba State Nigeria." *IOSR Journal of Environmental Science, Toxicology and Food Technology.* 8(2): 145–149. doi: 10.9790/2402-0824145149.

Mortimore, Sara, Carol Wallace, Sara Mortimore, and Carol Wallace. 2013. Prerequisites for food safety: PRPs and operational PRPs. HACCP: A Practical Approach (pp. 113–154).

Mota, Juliana De Oliveira, Geraldine Boue, Herve Prevost, Aurelien Maillet, Emmanuel Jaffres, Thomas Maignien, Nathalie Arnich, Moez Sanaa, and Michel Federighi. 2021. "Environmental monitoring program to support food microbiological safety and quality in food industries: A scoping review of the research and guidelines." *Food Control.* 130: 108283. https://doi.org/10.1016/j.foodcont.2021.108283.

Mumu, Farah-Tuz, Mrityunjoy Acharjee, Ifra Tun Nur, Tamanna Islam, and Mohammad Maksudul Hassan. 2021. "Evaluation of food borne pathogens in ready-to-eat noodles collected from the different shops in the Dhaka City, Bangladesh." *SVOA Microbiology.* 2(2): 32–40.

Rakshitha, N. R., Bhavya, D. L., Swetha, S. A., Anusha, M., and Sheema Kauser. 2022. "Microbiological quality of street-vended pani puri sold in different cities of India." *International Journal for Research in Applied Science & Engineering Technology.* 10(3): 633–638. https://doi.org/10.22214/ijraset.2022.47068.

Nayem, Nesbahuddin Md Rezwanun, and Zakaria Ahmed. 2013. "Microbiological status of different sweet and sweet products in Dhaka city, Bangladesh." *International Journal of Bio-Technology and Research.* 3(1): 1–10.

Nethathe, Bono, Phato Avheani Matsheketsheke, Mpho Edward Mashau, and Shonisani Eugenia Ramashia. 2023. "Microbial safety of ready-to-eat food sold by retailers in Thohoyandou, Limpopo province, South Africa." *Cogent Food & Agriculture.* 9(1): 2185965. https://doi.org/10.1080/23311932.2023.2185965.

Oladoyinbo, Catherine Adebukola, Oluwafunke Opeyemi Akinbule, and Israel Ayodeji Awosika. 2015. "Knowledge of food borne infection and food safety practices among local food handlers in Ijebu-Ode Local Government Area of Ogun State." *Journal of Public Health and Epidemiology.* 7(9): 268–273. doi: 10.5897/JPHE2015.0758.

Oliver, S. P., B. M. Jayarao, and R. A. Almeida. 2005. "Review foodborne pathogens in milk and the dairy farm environment: Food safety and public health implications." *Foodborne Pathogens and Disease*. 2. www.liebertpub.com.

Osaili, Tareq M., Fayeza Hasan, Anas A. Al-Nabulsi, Amin N. Olaimat, Mutamed Ayyash, Reyad S. Obaid, and Richard Holley. 2023. "A worldwide review of illness outbreaks involving mixed salads/dressings and factors influencing product safety and shelf life." *Food Microbiology*. 112: 104238. https://doi.org/10.1016/j.fm.2023.104238

Osakue, O. P., J. O. Igene, P. A. Ebabhamiegbebho, and S. E. Evivie. 2016. "Proximate analysis and microbial quality of ready-to-eat fried Chicken part." *Journal of Food and Industrial Microbiology*. 2(107): 2. doi: 10.4172/2572-4134.1000107.

Pereira, Karen Signori, Caroline Alves Cayres, Jeane Quintanilha Chaves, Josiane Teixeira de Brito, Leon Rabinovitch, and Adriana Marcos Vivoni. 2018. "Salad dressing spoilage by *Bacillus amyloliquefaciens* with gas formation." *Brazilian Journal of Food Technology*. 21: e2017025. https://doi.org/10.1590/1981-6723.2517.

Pérez-Rodríguez, F., A. Valero, E. C. D. Todd, E. Carrasco, R. M. García-Gimeno, and G. Zurera. 2007. "Modeling transfer of *Escherichia coli* O157:H7 and *Staphylococcus aureus* during slicing of cooked meat product." *Meat Science*. 76(4): 692–699. doi: 10.1016/j.meatsci.2007.02.011.

Pijnacker, Roan, Timothy J. Dallman, Aloys SL Tijsma, Gillian Hawkins, Lesley Larkin, Saara M. Kotila, Giusi Amore et al. 2019. "An international outbreak of *Salmonella enterica* serotype *enteritidis* linked to eggs from Poland: A microbiological and epidemiological study." *The Lancet Infectious Diseases*. 19(7): 778–786. doi: 10.1016/S1473-3099(19)30047-7.

Preena, Prasannan Geetha, Thangaraj Raja Swaminathan, Vattiringal Jayadradhan Rejish Kumar, and Isaac Sarojini Bright Singh. 2020. "Antimicrobial resistance in aquaculture: A crisis for concern." *Biologia*. 75:1497–1517. https://doi.org/10.2478/s11756-020-00456-4.

Rizwanullah, Rafed and Muhammad Hamed Etemadi. 2020. "Microbial analysis of commercially processed millet products." *International Journal of Science and Research*. 9(5): 1737–1740. www.ijsr.net/getabstract.php?paperid=SR20302170830.

Rane, Sharmila. 2011. "Street vended food in developing world: hazard analyses." *Indian Journal of Microbiology*. 51(1):100–106. doi: 10.1007/s12088-011-0154-x.

Rasschaert, Geertrui, Koen De Reu, Marc Heyndrickx, and Lieve Herman. 2016. "Case report of *Salmonella* cross-contamination in a food laboratory." *BMC Research Notes*. 9: 156. https://doi.org/10.1186/s13104-016-1969-7.

Rather, Irfan A., Wee Yin Koh, Woon K. Paek, and Jeongheui Lim. 2017. The sources of chemical contaminants in food and their health implications. *Frontiers in Pharmacology*. 8: 830. doi: 10.3389/fphar.2017.00830.

Ray, Bibek, and William E. Sandine. 1992. Acetic, propionic, and lactic acids of starter culture bacteria as biopreservatives. In B. Ray & M. Daeschel (Eds.), *Food Biopreservatives of Microbial Origin* (pp. 103–136). CRC Press.

Redmond, Elizabeth C., and Christopher J. Griffith. 2003. "Consumer food handling in the home: a review of food safety studies." *Journal of Food Protection*. 66(1): 130–161. doi: 10.4315/0362-028x-66.1.130.

Rehagel, Christina, Ömer Akineden, and Ewald Usleber. 2022. "Microbiological and mycotoxicological analyses of processed cereal-based complementary foods for infants and young children from the German market." *Journal of Food Science*. 87(4): 1810–1822. https://doi.org/10.1111/1750-3841.16106.

Reyes, Juan E., Manuel I. Guanoquiza, Gipsy Tabilo-Munizaga, Antonio Vega-Galvez, Margarita Miranda, and Mario Pérez-Won. 2012. "Microbiological stabilization of aloe vera (*Aloe barbadensis Miller*) gel by high hydrostatic pressure treatment." *International Journal of Food Microbiology*. 158(3): 218–224. http://dx.doi.org/10.1016/j.ijfoodmicro.2012.07.019.

Rodríguez, Valadez Francisca, María Cristina Hernández Jiménez, Esperanza Milagros García Oropesa, Laura Yanneth Ramírez Quintanilla, José Francisco Flores Gómez, Guadalupe Concepción Rodríguez Castillejos, and Rosa Issel Acosta González. 2017. "Microbiological quality of ready-to-eat street-vended flavored waters and fruits salads in Reynosa, Tamaulipas, Mexico." *Acta Universitaria*. 27(6): 3–9. https://doi.org/10.15174/au.2017.1304.

Ronald, Chepkwony, Joseph Wafula Matofari, and John Masani Nduko. 2023. "Antimicrobial resistance of *E. coli* strains in ready-to-eat red meat products in Nakuru County, Kenya." *The Microbe*, December, 100022. https://doi.org/10.1016/j.microb.2023.100022.

Rosenquist, Hanne, Lasse Smidt, Sigrid R. Andersen, Gert B. Jensen, and Andrea Wilcks. "Occurrence and significance of *Bacillus cereus* and *Bacillus thuringiensis* in ready-to-eat food." *FEMS Microbiology Letters*. 250(1):129–136. https://doi.org/10.1016/j.femsle.2005.06.054.

Sevmiş, Evin, Seval Andiç, and O. Ğ. U. Z. Şehriban. 2020. "Comparison of the compositions of butter and butter oil produced in the province of Hakkari." *YYU Journal of Agricultural Science*. 30: 928–937. https://doi.org/10.29133/yyutbd.717665.

Sharifzadeh, Ali, Mohammad Hajsharifi-Shahreza, and Payam Ghasemi-Dehkordi. 2016. "Evaluation of microbial contamination and chemical qualities of cream-filled pastries in confectioneries of Chaharmahal Va Bakhtiari Province (Southwestern Iran)." *Osong Public Health and Research Perspectives*. 7(6): 346–350. https://doi.org/10.1016%2Fj.phrp.2016.09.004.

Sheng, Lina, and Luxin Wang. 2021. "The microbial safety of fish and fish products: Recent advances in understanding its significance, contamination sources, and control strategies." *Comprehensive Review of Food Science and Food Safety*. 20(1): 738–786. doi: 10.1111/1541-4337.12671.

Sheng, Lina, Zheng Zhang, Gang Sun, and Luxin Wang. 2020. "Light-driven antimicrobial activities of vitamin K3 against *Listeria monocytogenes*, *Escherichia coli* O157:H7 and *Salmonella enteritidis*." *Food Control*. 114:107235. https://doi.org/10.1016/j.foodcont.2020.107235.

Shrestha, Subash, Jerry J. Erdmann, Michelle Riemann, Kevin Kroeger, Vijay K. Juneja, and Ted Brown. 2023. "Ready-to-eat egg products formulated with nisin and organic acids to control *Listeria monocytogenes*." *Journal of Food Protection*. 86(5):100081. doi: 10.1016/j.jfp.2023.100081.

Shukla, Seema, Ravi Shankar, and Surya Prakash Singh. 2014. Food safety regulatory model in India. *Food Control*. 37: 401–413. https://doi.org/10.1016/j.foodcont.2013.08.015.

Shuvho, Chakra Borty, Chowdhury Mohammad Nizam Uddin, Khan Md Hiron, Monir Md Shirajum, and Islam Md Alimul. 2015. "Isolation and identification of bacteria from four different frozen snacks of Gazipur district of Bangladesh." *International Journal of Research in Biosciences*. 4(2).

Singh, Pradeep Kumar, Rajat Pratap Singh, Pankaj Singh, and Ram Lakhan Singh. 2019. Food hazards: physical, chemical, and biological. In Food Safety and Human Health (pp. 15–65). Academic Press. https://doi.org/10.1016/B978-0-12-816333-7.00002-3.

Sivakumar, Prema and Muralidharan. 2017. "Demonstration of microbial load in coffee powders." *International Journal of Current Advanced Research*. 6(4): 3048–3050. http://dx.doi.org/10.24327/ijcar.2017.3050.0177.

Sofos, John N. 2008. "Challenges to meat safety in the 21st century." *Meat Science*. 78(1–2): 3–13. https://doi.org/10.1016/j.meatsci.2007.07.027.

Stanislaus, Abishek, Polani Rubeshkumar, Manickam Ponnaiah, and T. S. Selvavinayagam. 2022. "An outbreak of acute gastroenteritis attributed to consumption of contaminated rice-flakes-Jaggery in a religious festival, Coimbatore, Tamil Nadu." *Indian Journal of Public Health*. 66(4): 529–530. https://doi.org/10.4103/ijph.ijph_762_22.

Sultana, Fouzia, Hafsa Afroz, Afroz Jahan, Md Fakruddin, and Suvamoy Datta. 2014. "Multi-antibiotic resistant bacteria in frozen food (ready to cook food) of animal origin sold in Dhaka, Bangladesh." *Asian Pacific Journal of Tropical Biomedicine*. 4: S268–S271. doi: 10.12980/APJTB.4.2014B85.

Techer, C., Florence Baron, L. Delbrassinne, R. Belaïd, N. Brunet, Angélique Gillard, Fabienne Gonnet, et al. 2014. "Global overview of the risk linked to the *Bacillus cereus* group in the egg product industry: Identification of food safety and food spoilage markers." *Journal of Applied Microbiology*. 116(5): 1344–1358. doi: 10.1111/jam.12462.

U.S. Department of Agriculture, Food Safety and Inspection Service. (2021). Safe and suitable ingredients used in the production of meat, poultry, and egg production. Retrieved from www.fsis.usda.gov/policy/fsis-directives/7120.1 accessed December 09, 2023.

U.S. Department of Agriculture, Food Safety and Inspection Service. (2020). FSIS food safety guideline for egg products. Retrieved from www.fsis.usda.gov/guidelines/2020-0005 accessed October 14, 2021.

De Valk, H., V. Vaillant, C. Jacquet, J. Rocourt, F. Le Querrec, F. Stainer, N. Quelquejeu, et al. 1999. "Two consecutive nationwide outbreaks of listeriosis in France." Vol. 154. http://aje.oxfordjournals.org/.

Vanos, V., S. Hofstaetter, and L. Cox. 1987. "The microbiology of instant tea." *Food Microbiology*. 4(1): 19–33.

Wang, Jun, Joong-Hyun Park, Na-Jung Choi, Sang-Do Ha, and Deog-Hwan Oh. 2016. "Microbiological analysis of rice cake processing in Korea." *Journal of Food Protection*. 79(1): 157–162. http://dx.doi.org/10.4315/0362-028X.JFP-15-237.

Wareing, Peter William, and Davenport RR. 2007. "Microbiology of soft drinks and fruit juices." In Chemistry and Technology of Soft Drinks and Fruit Juices (second edition, pp. 279–297). http://dx.doi.org/10.1002/9780470995822.ch11.

Weidenbörner, M., and B. Kunz. 1995. "Mycoflora of cereal flakes." *Journal of Food Protection*. 58(7): 809–812.

Wessel, Stefan W., Henny C. van der Mei, David Morando, Anje M. Slomp, Betsy van de Belt-Gritter, Amarnath Maitra, and Henk J. Busscher. 2015. "Quantification and qualification of bacteria trapped in chewed gum." *PLoS One*. 10(1): e0117191. https://doi.org/10.1371/journal.pone.0117191.

Whiley, Harriet, and Kirstin Ross. 2015. "*Salmonella* and eggs: From production to plate." *International Journal of Environmental Research and Public Health*. 12(3): 2543–2556. doi: 10.3390/ijerph120302543.

Windyga, B., M. Borowiak, B. Ciesielska, J. Chybowska, K. Debeska, Z. Dziurowicz, B. Frasunkiewicz, M. Głowacki, H. Gomułkiewicz, J. Kłos, T. Kozłowska, S. Krzyzanowski, W. Kubiak, M. Lipińska, A. Maciaszek, S. Meinhardt, S. Nowotko, I. Ostrowska, B. Smykał, K. Szyszło, B. Tyśnicka, I. Wanacka-Twardela, K. Wolańczyk, and S. Zerger. 1978. "Microbiological contamination of margarine." *Roczniki Państwowego Zakładu Higieny* 29(1): 23–31.

Woldemariam, Henock Woldemichael, and Biresaw Demelash Abera. 2015. The extent of adulteration of selected foods at Bahir Dar. Ethiopia. 1(6): 1–6. doi: 10.13140/RG.2.1.1874.5769.

World Health Organization (WHO). 1998. Guidance on Regulatory Assessment of HACCP. Report of a joint FAO/WHO consultation on the role of government agencies in accessing HACCP, WHO/FSF/FOS/98.5, Geneva.

World Health Organization. 2004. Risk assessment of *Listeria monocytogenes* in ready-to-eat foods: Technical report no. 5. Food & Agriculture Organization.

Wu, Shi, Jiahui Huang, Qingping Wu, Jumei Zhang, Feng Zhang, Xiaojuan Yang, Haoming Wu, et al. 2018. "*Staphylococcus aureus* isolated from retail meat and meat products in China: Incidence, antibiotic resistance and genetic diversity." *Frontiers in Microbiology*. 9. https://doi.org/10.3389/fmicb.2018.02767.

www.campdenbri.co.uk/white-papers/chemical-hazards-food.php#

www.fda.gov/animal-veterinary/biological-chemical-and-physical-contaminants-animal-food/chemical-hazards

www.fda.gov/food/hazard-analysis-critical-control-point-haccp/haccp-principles-application-guidelines#:~:text=Seven%20basic%20principles%20are%20employed,and%20record%2Dkeeping%20and%20documentation.

www.fsis.usda.gov/food-safety/recalls-public-health-alerts/annual-recall-summaries/summary-recall-cases-calendar-8

www.gov.mb.ca/agriculture/food-safety/at-the-food-processor/food-safety-program/pubs/physical-hazards.pdf

Yang, Jiyun, and Jeehyun Lee. 2019. "Application of sensory descriptive analysis and consumer studies to investigate traditional and authentic foods: A review." *Foods*. 8(2): 54. https://doi.org/10.3390%2Ffoods8020054.

Zohreh, Mashak, Mojaddar Langroodi Ali, Ehsani Ali, Ilkhanipoor Ashkan, and Ebadi Fathabad Ayub. 2015. "Microbiological quality of ready-to-eat foods of Tehran province." *African Journal of Food Science*. 9(5): 257–261. doi: 10.5897/AJFS2015.1260.

14 Sustainable cum Innovative Packaging and Labeling Approach in Ready-to-Eat Food

A Global Perspective

Rakesh Mohan Shukla[1*], *Chandrakant Genu Dalbhagat*[2], *Siddharth Vishwakarma*[3], *Shubham Mandliya*[3], *and Hari Niwas Mishra*[3]
[1]Processing and Food Engineering Division, College of Agricultural Engineering and Technology, SKUAST-K, Shalimar Campus, Srinagar (UT of J&K), India
[2]Department of Food Process Engineering, National Institute of Technology Rourkela, Odisha, India
[3]Agricultural and Food Engineering Department, Indian Institute of Technology Kharagpur, Kharagpur, West Bengal, India
*Corresponding Author

14.1 INTRODUCTION

14.1.1 Overview of Ready-to-Eat Packaging and Labeling

Ready-to-eat (RTE) products have emerged as a cornerstone of modern consumer convenience in the ever-changing landscape of the food industry. RTE foods are characterized by their ready-to-consume nature, eliminating the need for additional preparation, and thereby aligning seamlessly with the fast-paced lifestyles of contemporary consumers. The intricately designed packaging and labeling strategies that ensure not only the preservation of freshness and quality but also the dissemination of vital information to consumers are critical to the success of these products.

RTE packaging is designed with a holistic approach, considering factors such as barrier properties, shelf stability, and visual appeal. Modified atmosphere packaging (MAP) and vacuum packaging are prevalent techniques aimed at preserving the freshness and flavor of RTE foods by controlling the atmospheric conditions within the packaging. Additionally, the choice of packaging materials, such as barrier films and laminates, plays a pivotal role in protecting against external elements like moisture, light, and oxygen.

This chapter delves into the complex world of RTE packaging and labeling, dissecting the various elements that contribute to the overall consumer experience. In RTE foods, accurate and transparent

DOI: 10.1201/9781032622408-14

FIGURE 14.1 RTE packaging material.

labeling is paramount. Labeling serves not only as a means of communication between the manufacturer and the consumer but also as a tool for ensuring food safety and regulatory compliance. Key information on RTE labels includes nutritional facts, ingredient lists, allergen information, and storage instructions. Navigating through the intricate web of international and regional labeling regulations demands meticulous attention to detail to meet the diverse requirements of different markets.

RTE packaging and labeling encompass the delicate balance between preserving food quality and meeting sustainability goals. Minimizing packaging waste, addressing concerns related to recyclability, and incorporating eco-friendly materials pose ongoing challenges. Innovations in active and intelligent packaging systems offer potential solutions by providing real-time monitoring of food conditions, extending shelf life, and enhancing traceability.

The future of RTE packaging and labeling converges toward smart technologies, sustainable practices, and heightened consumer engagement. Emerging trends include the integration of QR codes for access to product information, the utilization of edible packaging materials, and advancements in antimicrobial packaging to ensure food safety [1]. Figure 14.1 highlights variable plastic packaging material employed in food packaging

14.1.2 Current Global and Domestic (India) Market Status of Packaging

The packaging industry, both globally and in India, is undergoing transformative changes driven by technological advancements, sustainability concerns, and shifting consumer preferences. In the global context, the packaging market continues to expand, driven by the growth in e-commerce, the food and beverage sector, and increasing awareness about eco-friendly packaging solutions.

14.1.2.1 Global Market Status

Globally, the packaging market is witnessing substantial growth. According to a report by Smithers, the global packaging market was valued at $917.3 billion in 2020 and is projected to reach $1.05 trillion by 2024 [2]. This growth is attributed to factors such as increased consumer demand for convenience, the rise of online shopping, and innovations in sustainable packaging materials.

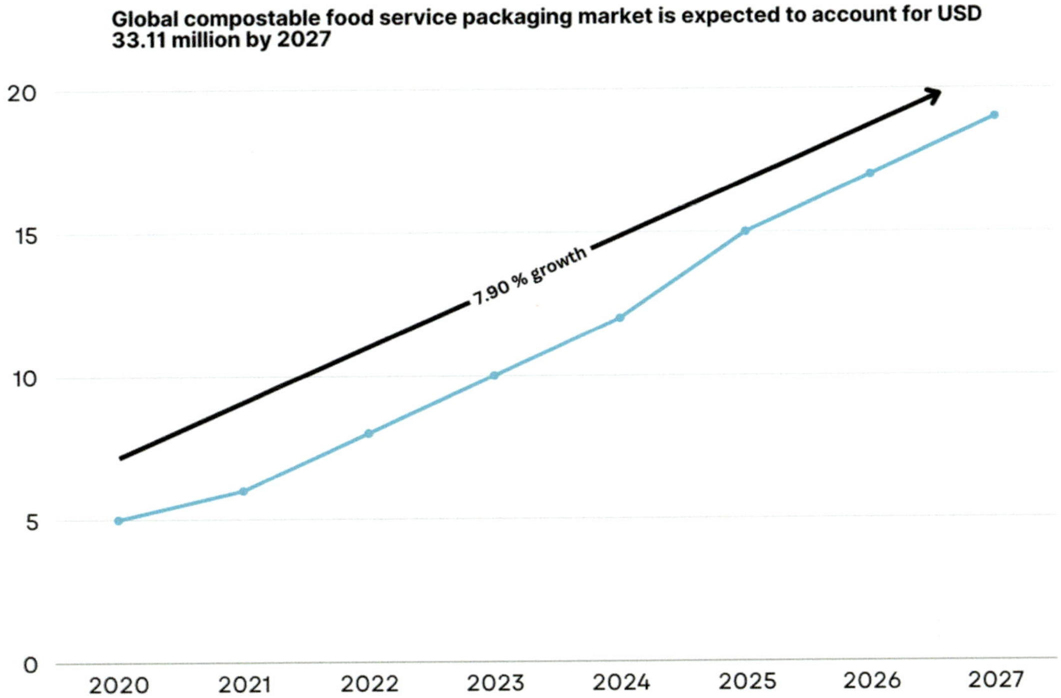

FIGURE 14.2 Global compostable packaging market. (Data bridge market research market study analysis study 2020 [4].)

14.1.2.2 Indian Market Status

In India, the packaging industry is experiencing significant developments due to economic growth, urbanization, and a burgeoning middle class. According to a report by the Federation of Indian Chambers of Commerce and Industry (FICCI), the Indian packaging industry is expected to reach $72.6 billion by 2020, growing at a CAGR of 18.1% [3]. This growth is fueled by the food and beverage sector, pharmaceuticals, and the increasing adoption of flexible and sustainable packaging solutions. Figure 14.2 highlights the global estimate of compostable packaging material market trend. Figure 14.3 is an estimate of trend for plastic production as packaging constituent.

14.1.3 Packaging Requirements of RTE Food

Packaging requirements for RTE food play a crucial role in ensuring the safety, quality, and shelf life of the product. The packaging must provide a protective barrier against external factors such as moisture, light, oxygen, and contaminants while also meeting regulatory standards. Several considerations come into play in the context of processing and food engineering when determining the packaging requirements for RTE food.

Barrier Properties: Packaging materials should possess appropriate barrier properties to prevent the ingress of moisture, oxygen, and other gases, as well as the egress of flavors and aromas from the food [6]. (See reference [6] for insights into barrier properties.)

Microbial Safety: Packaging must prevent microbial contamination and maintain the microbiological safety of the RTE food. Materials with antimicrobial properties or MAP can be considered [7].

Sealing Integrity: A secure and tamper-evident sealing mechanism is essential to ensure the integrity of the package and prevent unauthorized access or contamination.

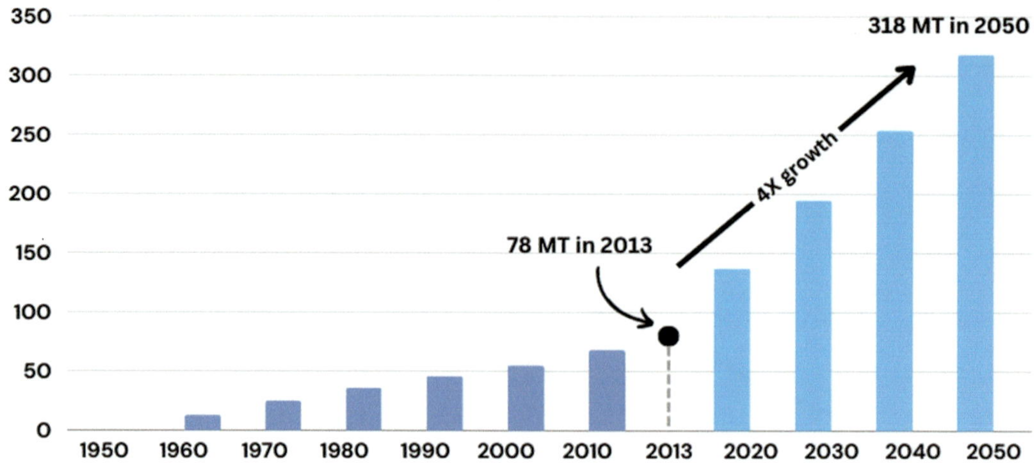

FIGURE 14.3 Plastic packaging production is predicted to quadruple by 2050 and the plastics peak only to be hit by 2100. (www.greenpeace.org © Zero Waste Living Lab by enviu [5].)

Physical Protection: Packaging should provide physical protection to prevent damage during handling, transportation, and storage. Considerations for the fragility, shape, and weight of the RTE product are essential [8]. (Reference [8] is a comprehensive guide to understanding physical protection in packaging.)

Temperature Control: For RTE foods that require temperature control, both insulation and temperature-resistant packaging materials may be necessary. Insulated packaging, phase change materials, and other technologies can be explored [9].

Regulatory Compliance: Ensure compliance with regulatory standards and guidelines for food packaging, including those related to food contact materials. Packaging must comply with local and international regulations regarding food contact materials, labeling, and other relevant standards.

Convenience and Portability: Packaging design should consider consumer convenience and portability, allowing for easy handling and consumption without compromising food safety.

Labeling and Information: Accurate and comprehensive labeling is crucial, including information on ingredients, allergens, nutritional content, expiration date, storage instructions, and any other required details.

Sustainability: Consider sustainable packaging options to align with environmental concerns and consumer preferences. Explore biodegradable, compostable, or recyclable materials [10]. (Reference [10] is good reference for sustainable packaging.)

14.1.4 Labeling and Coding Units of Packaging during Production

Labeling and coding units of packaging during food production is a critical aspect of ensuring food safety, traceability, and compliance with regulations. Here is a detailed explanation of how labeling and coding are typically managed in the context of food production:

Labeling:

1. ***Product Information:*** Labels on food packaging contain essential information such as product name, ingredients, nutritional facts, allergen information, net weight, usage instructions, and manufacturer's details.
2. ***Batch Number and Expiry Date:*** Each batch of food products is assigned a unique batch number, allowing for traceability. The packaging also includes an expiry date or use-by date to inform consumers about the product's shelf life.

3. ***Barcoding:*** Barcodes are used for automated data capture. They encode information like batch numbers, expiry dates, and product details. Barcoding aids in inventory management and traceability throughout the supply chain.
4. ***Country of Origin:*** For international trade, packaging labels often specify the country of origin, providing consumers with information about where the food product was produced.
5. ***Allergen Warnings:*** Allergenic ingredients such as nuts, dairy, soy, and gluten are highlighted on labels to alert consumers with allergies.
6. ***Dietary and Religious Certifications:*** Symbols or text indicating vegetarian/vegan status, non-vegetarian products, halal/kosher certifications, organic status, and GMO-free claims. These help consumers quickly identify products aligning with their dietary preferences or religious requirements.

Coding:

1. ***Inkjet Printing:*** Inkjet printers are commonly used for coding on food packaging. They can print batch numbers, expiry dates, barcodes, and other relevant information directly onto various packaging materials.
2. ***Laser Marking:*** Laser technology offers precise and permanent coding on packaging materials, ensuring better readability and durability, even in challenging environments.
3. ***Thermal Transfer Printing:*** Thermal transfer printers are used for high-resolution printing of barcodes, text, and graphics. They are suitable for a wide range of packaging materials.
4. ***Embossing:*** In some cases, embossing is used to create raised characters or codes on packaging. This method provides a tactile and visually distinct marking solution.

Quality Control:

1. ***Automated Inspection Systems:*** Automated systems equipped with cameras and sensors inspect labels and codes for accuracy, alignment, and legibility. Any discrepancies trigger alerts for manual inspection and correction.
2. ***Sampling and Manual Checks:*** Quality control personnel conduct random sampling and manual checks to ensure the accuracy and quality of labels and codes.

Regulatory Compliance:

1. ***Food Safety Standards:*** Labels and codes must comply with food safety standards set by regulatory authorities to ensure that consumers receive safe and accurately labeled products. For example, vegetarian/non-vegetarian symbol etc.
2. ***Documentation:*** Detailed records of labeling and coding practices are maintained. These records are crucial for demonstrating compliance during inspections and audits conducted by regulatory bodies.

By implementing robust labeling and coding practices, food manufacturers can uphold food safety standards, enable efficient supply chain management, and provide consumers with clear and accurate information about the products they consume. Compliance with regulations and consistent quality control measures are paramount in the food production industry [11, 12].

14.2 FACTORS AFFECTING THE CHOICE OF A PACKAGING MATERIAL

14.2.1 Mechanical Damage

Packaging serves as the first line of defense against external forces that can compromise the integrity of products during their lifecycle, from manufacturing to consumer use. A variety of physical stresses, including compression, vibration, impact, and flexural forces, are included in mechanical damage. To minimize these mechanical stresses and guarantee the safety of the packed goods, the packing material selection is essential. Research in this area often explores material science aspects to optimize packaging for specific mechanical stressors.

14.2.1.1 Mechanical Properties of Packaging Materials

1. **Tensile Strength:** Tensile strength is a fundamental mechanical property that quantifies a material's resistance to a force attempting to pull it apart. Packaging materials with high tensile strength, such as oriented polymers and reinforced composites, are preferred to withstand the stresses encountered during transportation and handling. This critical aspect of packaging performance can be further optimized through the incorporation of nanomaterials. By leveraging the unique properties of nanomaterials, packaging films can be fortified to better resist forces that may compromise their structural integrity, ultimately contributing to the development of more robust and durable packaging solutions [13].
2. **Impact Resistance:** Impact resistance measures a material's ability to withstand sudden forces or shocks. Packaging materials like corrugated cardboard and certain polymer blends are designed to absorb and distribute impact energy, minimizing the risk of damage to the packaged product. This packaging performance can be further enhanced by exploring the impact resistance of biodegradable polymers. Understanding how these polymers respond to sudden forces not only contributes to the development of eco-friendly packaging but also ensures that the materials maintain their protective qualities, providing a balance between environmental considerations and the strong protection of packaged goods [14].
3. **Puncture Resistance:** Puncture resistance denotes the relative ability of a material or object to inhibit the intrusion of a foreign object. Puncture resistance is crucial, especially for packaging materials used in industries such as food and pharmaceuticals. Laminated films and coatings containing materials like ethylene vinyl alcohol (EVOH) enhance puncture resistance, safeguarding against potential damage from sharp objects.
4. **Flexural Strength:** Flexural strength determines a material's ability to bend without breaking. It is defined as the maximum bending stress that can be applied to a material before it yields. For packaging materials to survive bending pressures during handling and storage, they must have optimal flexural strength, such as mixes of polymers and particular thermoplastics. To protect goods along the supply chain, high mechanical property packing materials are essential. Packaging solutions with enhanced tensile strength, impact resistance, puncture resistance, and flexural strength effectively lower the chances of mechanical damage.

14.2.2 Permeability Characteristics

Permeability is a decisive consideration in packaging material selection as it directly influences the shelf life and quality of packaged goods. Permeability refers to the ability of gases, liquids, or other substances to pass through the packaging material. It is typically measured as the rate of transmission of gases or vapors through a unit area of the material over a specific time period. Common units include Barrer for gases and gram/m^2/day for water vapor. Controlling permeability is essential for preserving the freshness of food products, preventing spoilage, and maintaining the efficacy of a product. Permeability characteristics are essential when packaging perishable or moist products. The preservation of the product's quality, freshness, and safety is ensured by the choice of packaging material.

The permeability of packaging materials is influenced by various factors, including the material's composition, thickness, temperature, and the specific gases or vapors involved. Understanding these factors is crucial for selecting materials tailored to the unique requirements of different products. These factors include:

- **Packaging Material Properties:** The physical properties of the packaging material, such as its porosity and barrier properties, play a significant role in determining its suitability for a particular product.
- **Product Properties:** The permeability characteristics of the packaging material should be compatible with the product's requirements. For example, if the product contains high levels of moisture, a material with high water vapor permeability would be more suitable than one with low permeability.
- **Processing Conditions:** The processing conditions during the manufacturing of the packaging material, such as heat sealing or stretch wrapping, can affect the permeability characteristics of the material. Therefore, it is important to consider these factors when selecting a packaging material.
- **Packaging Requirements:** The packaging requirements of the product, such as the need for barrier protection, oxygen barrier, or gas tightness, can also influence the selection of the packaging material.

The choice of a packaging material depends on a complex interaction of these factors. When choosing the best packaging material for a given application, engineers and technologists need to consider the characteristics of the product, the processing environment, and the particular packaging specifications. For example, when packaging perishable food products, it is important to choose a material that offers a combination of good barrier protection and high-water vapor permeability. A material like polyethylene, which has good water barrier properties, is often used for packaging products with high moisture content, such as cheese or sausages.

Prasad et al. published a review on the permeability of polymers in the *International Journal of Polymer Analysis and Characterization* [15]. The authors discussed the importance of controlling the extent of permeant transport through a polymer in packaging applications. They provided a critical analysis of different methods of controlling polymer permeability reported in the literature and recommend fresh approaches for modification of polymer permeability when considering large-scale manufacturing of packages with controlled permeability properties [15].

It is also important to remember that developments in packaging and materials science are constantly creating new materials with better permeability properties. Nanotechnology has emerged as a promising avenue for enhancing the barrier properties of packaging materials [16]. Multilayer films are commonly used in packaging to achieve optimal gas barrier properties. Also, mathematical modeling is employed to predict and understand permeation characteristics. These materials offer better barrier protection, increased oxygen transmission rates, and enhanced flexibility and strength, providing a wider range of options for packaging different types of products.

14.2.3 Grease Proofness

Grease-proofness is a vital property in packaging, especially for products containing fats, oils, or greasy substances. The term refers to the resistance of packaging materials to the penetration or absorption of greases and oils, preventing undesirable effects such as structural degradation, discoloration, or compromise of the package integrity. The selection of packaging materials with optimal grease-proofness is paramount in ensuring the visual appeal, functionality, and overall quality of packaged items. Material selection, incorporating barrier coatings, and adherence to testing standards are essential components in achieving optimal grease resistance. As the industry progresses, the integration of sustainable and biodegradable options further underscores the multi-faceted nature of greaseproof packaging.

Packaging materials with inherent hydrophobic characteristics and low porosity are often preferred for their natural resistance to grease. Common materials chosen for their grease-proofness include certain types of paper, films, and laminates that incorporate barrier coatings [17].

The integration of barrier coatings and additives plays a significant role in enhancing the grease-proofness of packaging materials. Materials like fluorochemicals and certain polymers are incorporated to create a barrier against oil penetration [18]. In fact, the global functional additive and barrier coating sales for plastic packaging are forecast to grow during 2022–2027 at an annual average rate of 5.1% to just over $5.1 billion. The demand for packaging materials that give greater protection to their contents is increasing, especially in the food, beverage, and pharmaceutical industries [19].

Traditional greaseproof packaging often involves the use of wax-coated paper. The wax creates a hydrophobic barrier, preventing the migration of grease and oils into the paper structure.

"There is a functional element to grease resistance – it essentially protects the package from its contents – but there is an aesthetic reason too," says Foster, The Industry Marketing Manager for Printing and Packaging BASF. "If something on the package looks discolored, less vibrant, or less 'premium' than the package beside it, consumers may make a different choice. It raises questions about the quality of the product and the quality of the company" [20].

14.2.4 Temperature

Temperature is a significant factor influencing the choice of packaging materials, as it directly affects the physical and chemical properties of both the packaged product and the packaging material itself. From manufacturing processes to storage, transportation, and end-use, understanding the impact of temperature is crucial for selecting materials that ensure product safety, integrity, and overall quality.

RTE products often have specific temperature requirements for optimal quality and safety. The chosen packaging material must accommodate the anticipated temperature range throughout the product's lifecycle, from production and storage to transportation and retail. Ready-to-use retort pouches are flexible packages made from multilayer plastic films with or without aluminum foil as one of the layers. Unlike the usual flexible packages, they are made of heat-resistant plastics, thus making them suitable for processing in the retort at a temperature of around 121 °C. For legumes or pulses, LDPE and PP pouches have been found to offer 1½ to 3 months of storage life under high temperature/RH conditions, which would correspond to 5 to 6 months under normal conditions.

The thermal stability of packaging materials is vital. Materials must maintain their structural integrity, barrier properties, and overall functionality under the temperature conditions encountered during various stages, including processing, hot-filling, refrigeration, and even potential exposure to microwave or oven heating by the end consumer [22]. Temperature-regulating material (TRM)-based food packaging is recently trending in the food science and technology sectors. Although this technology is still not fully commercially viable, it has good potential to control the temperature of perishable and high-value agricultural products during transport. Packaging materials for RTE products may incorporate insulation layers or technologies that regulate temperature. This is particularly relevant for products requiring maintenance of a specific temperature range, such as chilled or frozen meals.

Different polymer types demonstrate diverse behaviors at varying temperatures. For example, polyethylene terephthalate (PET) remains stable at low temperatures and is often used for refrigerated RTE products, while polypropylene (PP) may be chosen for its high-temperature resistance in microwaveable packaging [22]. Packaging materials for RTE products that may undergo cooking or reheating require high-temperature-resistant properties. Heat-resistant plastics or laminates with specific coatings are chosen to ensure the packaging maintains its structural integrity and safety during exposure to elevated temperatures.

14.2.5 Light and Other Packaging Materials

The interplay between light and packaging materials is a nuanced consideration in RTE packaging. The selection of packaging material plays a crucial role in both enhancing the visual appeal and preserving the quality of the product. Striking a delicate balance, it must showcase the product while also protecting it from any potential harm caused by exposure to light.

Light is a maestro orchestrating the visual symphony of RTE products. Transparent packaging enhances the sensory experience by allowing customers to focus their attention on the vivid colors and textures. However, this visual indulgence must be carefully balanced with the need to shield the contents from specific wavelengths of light that can induce degradation. The exposure of certain RTE products to light, especially ultraviolet (UV) rays, can trigger photodegradation. This process may lead to changes in color, flavor, and nutritional content, compromising the overall integrity of the product. Packaging materials must act as guardians, mitigating the impact of photodegradation [23].

Materials equipped with UV-blocking additives play a crucial role in intercepting the potentially harmful effects of UV light. These additives, integrated into packaging materials like films or coatings, selectively absorb or reflect UV radiation, shielding the product from degradation [24]. UV-blocking films are an emerging technology currently being utilized in food packaging applications. Incorporating various UV-blocking agents is the current trend in active food packaging applications. These agents absorb, reflect, or scatter the incident UV light, reducing the transmittance of UV radiation through packaging film [25]. The most commonly used UV absorbers are benzotriazoles, benzophenones, and organic nickel compounds.

The assessment of light transmittance and barrier properties involves a meticulous examination of haze and luminous transmittance. Standardized tests, such as ASTM D1003, provide a structured approach to evaluating the optical characteristics of transparent plastics in response to light exposure [26].

14.3 MATERIALS AND CONTAINERS USED FOR PACKAGING FOODS

14.3.1 Papers, Paperboards, and Fiberboards

Papers, paperboards, and fiberboards form a diverse category of packaging materials, each offering distinct characteristics. Papers are commonly lightweight and flexible, providing an excellent medium for packaging products frequently used for packaging organic and natural foods like dry goods and baked items. Paper is generally termed board when its grammage exceeds 250 GSM. Paperboards, being thicker and sturdier, are often employed for items requiring additional structural support, such as cereal boxes and frozen food packaging where branding and shelf appeal are paramount. Fiberboards, known for their strength and durability, find utility in heavier applications like bulk packaging of fruits and vegetables, providing a balance between strength and sustainability.

In the context of increasing environmental awareness, the eco-friendliness of paper, paperboards, and fiberboards has become a topic of discussion. The renewable and biodegradable nature of the paper makes it a popular choice for sustainable packaging initiatives. However, the production of paper can have a significant environmental impact, including deforestation, water pollution, and greenhouse gas emissions [27].

14.3.2 Flexible Films

The utilization of flexible films in food packaging represents a dynamic and evolving domain within the packaging industry, characterized by its adaptability, versatility, and technological advancements. The use of flexible films in food packaging is a rapidly growing and evolving field within the packaging industry. They are made of materials that can yield when filled or closed and can be easily modified in terms of shape, as and when required. This kind of packaging is used for consumer products, as well as in industrial applications. Common examples include bags, pouches, shrink films, tubes, sleeves, and carded packaging [28].

Flexible packaging films are available in a wide range of options, each serving a specific purpose when it comes to protecting or enhancing the appeal of different products. The most common types of flexible packaging films are:

1. **Linear Low-Density Polyethylene (LLDPE):** The most flexible of plastic sheeting films, LLDPE has an impressive potential to absorb impact without puncturing or tearing. Apart from flexibility, these films are also best known for their durability and chemical resistance properties. The most common examples are plastic bags, plastic wraps, pouches, toys, covers, lids, pipes, buckets, and containers, mainly flexible tubing [29].
2. **Low-Density Polythene (LDPE):** An incredibly flexible and lightweight plastic material, LDPE is best known for its low-temperature flexibility, durability, corrosion resistance, and chemical and impact resistance. It is the most sought-after material in the food packaging industry because it ensures compliance with food safety standards, and is recyclable [29, 30].
3. **Bi-Axially-Orientated Polypropylene (BOPP):** A popular choice for food packaging, BOPP films are known for their high tensile strength, clarity, and resistance to moisture and oils. They are commonly used for packaging snacks, confectionery, and bakery products [29].
4. **Polyethylene Terephthalate films (PET):** PET films are known for their excellent barrier properties, making them ideal for packaging products that require protection from oxygen, moisture, and other gases. They are commonly used for packaging carbonated beverages, water, and other liquids [29].
5. **Polyvinyl Chloride (PVC):** PVC films are known for their excellent clarity, flexibility, and resistance to oils and chemicals. They are commonly used for packaging meat, cheese, and other perishable food items [29].

Flexible packaging films offer several benefits, including increased shelf life of perishable items, concealment of smells, and ease of transportation. Stand-up pouches, developed using flexible film, have become a popular option for food products, especially those with high-liquid contents [31].

14.3.3 Metallized Films

Metallized films are engineered by depositing a thin layer of metal, typically aluminum, onto a flexible polymer substrate. This metallization process imparts a range of advantageous properties to the films, including enhanced barrier properties against moisture, gases, and light. They offer the lustrous metallic appearance of aluminum foil at a diminished weight and cost. This makes it pop out from the rest of the monotonous packets on the aisle. Metallized films are broadly utilized for beautiful purposes and food bundling and for specialty applications including protection and hardware. Apart from that, they act as an additional barrier, preventing the contents inside from interacting with the environment [32].

Metallization is performed utilizing an actual fume statement process. Aluminum is the most widely recognized metal utilized for deposition, yet different metals, for example, nickel or chromium are likewise utilized. The metal is warmed and dissipated under vacuum. This consolidates on the cool polymer film, which is loosened up close to the metal fume source. This covering is a lot slenderer than a metal foil could be made, in the scope of 0.5 microns. This covering will not blur or stain over the long run. While saturated PP (Polypropylene) and PET are the most widely recognized films utilized for metallization, nylon, polyethylene, and cast PP are likewise utilized [33].

However, ensuring food safety is a primary concern, and related issues associated with metalized films should be addressed. The potential for metal migration into the food matrix is scrutinized, emphasizing the need for appropriate coatings and adherence to regulatory standards. The compatibility of metalized films with various food types ensures that the unique properties of these films align with the safety requirements of diverse packaged products.

14.3.4 Packaging in Flexible Films and Laminates

Flexible films and laminates encompass a diverse range of materials, including polymers, paper, and sometimes metals, strategically combined to harness specific properties. It explores how the combination of layers in laminates achieves a balance between protection and presentation, making them suitable for a broad spectrum of food products with intricate properties such as flexibility, sealability, and barrier capabilities. Two or more layers of materials, such as polyethylene, polyester, nylon, film foil laminations, and more are used to make them. Flavored beverages, food items such as sauces, and preserved foods and powders are often found in such laminated packaging [34]. In the case of lamination films, there is an impressive range of options that you can choose from, in accordance with the specifications of your products, some types of lamination films are as follows (Figure 14.4).

1. Cold Lamination Films
2. Thermal Lamination Films
3. BOPP Lamination Films
4. PET Lamination Films

Flexible packaging in the food packaging market commonly utilizes laminates consisting of two primary webs. One of these webs is a co-extrusion film predominantly derived from polyethylene, featuring multiple layers. Within this film, a tie layer and a barrier layer are incorporated to guard against humidity and gases like oxygen. The second web typically comprises stretched BOPP or BOPET film, contributing essential mechanical properties, strength, stiffness, and, at times, additional blocking and UV protection.

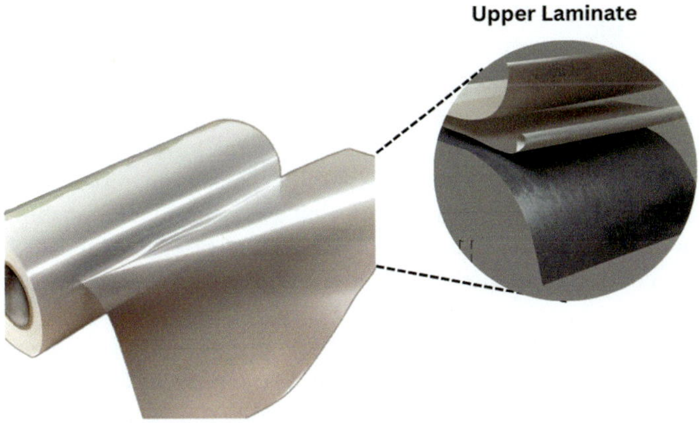

FIGURE 14.4 Layers in laminated packaging.

In certain instances, the final packaging may involve more than two layers, introducing elements such as a paper layer to impart a rustic and natural aesthetic or aluminum to prevent gas penetration and provide comprehensive protection from light. The specific layer composition depends on the intended application.

The structure of the laminate film offers an additional advantage when printing is involved. The printed content is securely enclosed between the two webs, shielding it from potential external damage throughout subsequent processing, transportation, and shelf display [35].

14.4 IMPACT OF PACKAGING ON FOOD QUALITY, SHELF LIFE, AND SAFETY

The safety, shelf life, and quality of food are all greatly impacted by packaging. Food products may become contaminated, spoil, or experience other negative consequences because of inadequate or improper packaging. With the identification of pathogens that can cause foodborne illness, the food industry's concern over packaging as a potential source of contamination has grown. Manufacturers need to make sure the materials they use in their packaging are appropriate for the food they are packaging to increase food safety. The microbiological safety of food can be impacted by packaging materials made of paper, plastic, and other materials. Pathogens in the packing material have immense potential to contaminate the food and cause cross-contamination. Packaging materials should be designed to maintain the freshness and safety of the food during storage, transportation, and distribution. Materials such as metal, glass, plastics, and paper should be carefully selected to ensure compatibility with the food product and appropriate barriers against moisture, oxygen, and other potential contaminants.

For example, food packaged in a vacuum-sealed plastic bag or can has a significantly longer shelf life compared to a similar product packaged in a non-barrier material. This is because the barrier formed by the plastic or metal allows the food to retain its freshness longer, reducing the risk of spoilage. To maintain food safety and freshness, manufacturers must also ensure that packaging materials are handled correctly. Food contamination, deterioration, or spoiling can be caused by improper handling or storage conditions. To maintain product integrity and reduce the risk of foodborne illness, manufacturers must establish and follow appropriate handling protocols.

Furthermore, developments in materials science and food packaging technology could result in the creation of even more efficient and eco-friendly packaging options. For instance, the food industry can promote a circular economy and reduce waste by developing more robust, flexible, and biodegradable packaging materials.

14.5 EMERGING APPROACHES IN FOOD PACKAGING SYSTEMS

14.5.1 Edible Packaging

Edible packaging, also known as food-grade packaging or edible wrapping, is a sustainable packaging alternative that utilizes food ingredients such as wheat, corn, potato, and vegetable oil as packaging materials. These biodegradable packaging systems have gained popularity due to their potential environmental benefits and versatility. Edible packaging represents a groundbreaking departure from traditional packaging methods, offering a symbiotic relationship between the packaging and the product it encases. As consumers increasingly prioritize sustainability and minimal environmental impact, edible packaging has gained traction as a novel solution that aligns with these values.

Edible packaging finds resonance in preserving fresh produce. Fruits and vegetables enveloped in edible coatings not only extend their shelf life but also eliminate the need for additional packaging materials, presenting a sustainable alternative to traditional wrapping [36]. Single-use items, such as condiment sachets or individual portions, benefit from edible packaging. These wrappers dissolve upon contact with moisture, reducing packaging waste and contributing to an eco-friendlier disposal method [37]. One of the foremost advantages of edible packaging lies in its potential to significantly reduce environmental impact. Traditional packaging materials, often non-biodegradable, are replaced with edible alternatives, aligning with the global call for sustainable practices. Edible packaging enhances the overall consumer experience by eliminating the need to unwrap or dispose of traditional packaging. It offers a seamless and enjoyable interaction with the product, promoting a positive brand image [38].

The future of edible packaging holds promises with ongoing technological advancements. Incorporating nanotechnology, active packaging elements, and bio-based materials are areas of exploration that may further enhance the functionality and applicability of edible packaging. However, Ensuring the structural integrity of edible packaging while maintaining an adequate shelf life poses a significant challenge. Balancing the desire for a sustainable solution with the practicalities of packaging performance remains an ongoing area of research and development.

There are several types of edible packaging materials:

1. **Gelatins:** Gelatins are obtained from collagen and are used in the production of transparent and translucent packaging materials. They can be mixed with natural and artificial food colorings to enhance the packaging aesthetics [39].
2. **Polysaccharides:** Polysaccharides are long chains of sugar molecules and are derived from plants like corn, wheat, and potato. These polysaccharides can be used to form various edible packaging materials such as wraps, bags, and films. Polysaccharides commonly applied in edible packaging include cellulose, hemicellulose, starch, chitosan, and polysaccharide gums. These polysaccharides are used as the main matrix of packaging materials and processed into polysaccharide-based edible films or layers by casting, coating, electrospinning, or extrusion technologies [40].
3. **Carrageenan:** Carrageenan is a naturally occurring polysaccharide that is harvested from red algae. It can be used to form transparent, opaque, and breathable packaging materials. It is widely used in the food industry for its gelling, thickening, and stabilizing properties [41].
4. **Whey Protein Concentrate (WPC):** WPC is a byproduct of the dairy industry and can be used to form flexible, waterproof packaging materials. According to one study, films/coatings made from whey proteins are colorless, odorless, flexible, and transparent with outstanding mechanical and barrier properties compared with polysaccharide and other-protein polymers. They have high water vapor permeability, low tensile strength, and excellent oxygen permeability compared with other protein films [42].

14.5.2 Active Packaging

Unlike conventional passive packaging, active packaging interacts dynamically with the packaged food to extend shelf life, enhance safety, and preserve quality. The objective is not merely containment but to impart functional attributes that positively influence the product throughout its shelf life. This section navigates through the innovative landscape of active packaging, elucidating its applications, benefits, challenges, and the ongoing research shaping its trajectory. Active packaging systems include oxygen scavengers, ethylene scavengers, flavor and odor absorbers/releasers, and antimicrobial and antioxidant packaging technologies [43].

Oxygen Scavenging: Packaged foods include a certain amount of headspace gases and entrained oxygen. Permeation of oxygen into plastic containers is also of concern. Molecular oxygen (O_2) can be reduced to a variety of intermediate species by the addition of between one and four electrons, thereby forming superoxide, hydroxy radical, hydrogen peroxide, and water. Except for water, the remaining three intermediate species are very reactive. These reactive oxygen species are free radicals in nature so the oxidative reactions in which they participate are therefore autocatalytic [44]. Active packaging utilizes oxygen scavengers to mitigate the detrimental effects of oxygen on food quality. This is particularly crucial for products prone to oxidation, such as fats and oils as they prevent oxidation phenomena: rancidification of fats and oils and consequent emergence of off-odors and off-flavors, loss or change of colors characteristic of food, loss of oxygen-sensitive nutrients (vitamins A, C, E, unsaturated fatty acids, etc.) [45].

Ethylene Absorption: Ethylene, a natural plant hormone, accelerates the ripening of fruits and vegetables. Active packaging incorporates ethylene absorbers to regulate fruit ripening, extending the freshness of produce [28]. One of the main mechanisms of action of ethylene scavengers is based on the use of potassium permanganate, which oxidizes ethylene to carbon dioxide and water. The typical permanganate content is between 4% and 6% [46]. Potassium permanganate oxidizes ethylene and changes color from purple to brown, and thus, a color change indicates its residual ethylene absorbing capacity, but because of its toxicity potassium permanganate cannot be used in direct contact with food.

Release of Antimicrobial Agents: Active packaging systems can release antimicrobial agents, inhibiting the growth of spoilage microorganisms and pathogens. This enhances food safety and extends the shelf life of perishable items.

Moisture Control: For products sensitive to moisture, such as certain baked goods and snacks, active packaging incorporates moisture-absorbing or desorbing materials to maintain the desired level of humidity.

Nanotechnology in Active Packaging: Nanotechnology has ushered in a new era in active packaging, enabling the development of nanocomposites with enhanced barrier properties and controlled release functionalities [47].

Intelligent Packaging: The headspace of food packages undergoes changes in their composition over time. Devices capable of identifying, quantifying, and/or reporting changes in the atmosphere within the package, the temperatures during transfer and storage, and the microbiological quality of food are the basis of intelligent packaging. The indicators should be easily activated and exhibit a change (or show an indication) that is easily measurable and irreversible, time- and temperature-dependent changes must be reproducible and ideally matched or readily correlated with the food quality, and provide information regarding the status of the package [48].

Apart from aspects of quality, safety, and distribution already outlined, intelligent packaging offers considerable potential as a marketing tool. The correct use of one or more types of active or intelligent packaging will increase shelf life and food safety. However, the development and implementation of this type of packaging will depend on the acceptance and cost-effectiveness for the industry and consumers [49].

14.5.3 Intelligent Food Packaging

The advent of intelligent food packaging stands as a pioneering approach that goes beyond traditional containment. Intelligent packaging systems actively interact with the environment and the packaged food, providing real-time information, thereby enhancing safety, and ensuring product quality. Intelligent food packaging represents a paradigm shift from passive containment to an interactive and responsive system. These packaging solutions are equipped with sensors, indicators, and data communication technologies that enable them to monitor, analyze, and communicate information about the status and conditions of the packaged food. Intelligent packaging systems employ sensors to monitor factors like temperature, humidity, and gas composition within the package. This real-time monitoring ensures the freshness of perishable goods by providing data on environmental conditions [50].

Sensors embedded in intelligent packaging can assess the quality attributes of the food, such as texture, color, and aroma. This data aids in maintaining and assuring the desired quality throughout the product's shelf life [28].

Intelligent food packaging integrates various sensor technologies, including temperature sensors, gas sensors, and radio-frequency identification (RFID) tags. These sensors provide real-time data on the conditions inside the package [48]. RFID and other communication systems enable the transfer of data from the packaging to external devices. This connectivity facilitates efficient supply chain management and allows consumers to access information about the product. This information aids in quality assurance, thereby extending the shelf life of products, and preventing premature spoilage. The benefits of intelligent food packaging extend to enhanced food safety, reduced food waste, and efficient supply chain management. However, challenges such as cost, complexity, and the need for consumer acceptance necessitate careful consideration. The integration of intelligent food packaging with sustainability practices and ongoing advancements in nanotechnology presents exciting prospects for this transformative approach to food packaging.

Looking ahead, the future of intelligent food packaging aligns with sustainable practices, exploring eco-friendly materials and energy-efficient sensor technologies. Ongoing research in nanotechnology holds the promise of enhancing the sensitivity and efficiency of sensors in these intelligent packaging systems. As technology continues to advance, intelligent food packaging is poised to play a pivotal role in ensuring a smarter, safer, and more sustainable future for the food industry. In conclusion, the exploration of intelligent food packaging in this chapter provides a comprehensive understanding of its intricacies, laying the foundation for informed discussions and advancements in the field.

14.5.4 Eco-Friendly Packaging

In the ever-changing world of food packaging, the attention is increasingly on eco-friendly packaging, signifying a considerable shift from past methods. This strategy is more than a transitory fad; it is a concerted effort to ease environmental concerns.

Eco-friendly packaging is not just a buzzword; it's a commitment to reduce our ecological footprint. It champions the use of materials like cornstarch, sugarcane, and bamboo-biodegradable wonders that gracefully return to nature. Eco-friendly packaging embraces the use of biodegradable materials derived from renewable sources; these materials break down naturally, minimizing the environmental burden associated with traditional packaging disposal [51].

Compostable packaging materials, designed to decompose in industrial composting facilities, contribute to waste reduction. Polylactic acid (PLA) and compostable films are notable examples, aligning with circular economy principles [52]. Eco-friendly packaging integrates life cycle assessment principles, considering the environmental impact at each stage of the packaging lifecycle. This comprehensive approach informs decision-making, ensuring a holistic understanding of sustainability. The selection of eco-friendly materials and streamlined production processes contribute to a reduced carbon footprint. Sustainable sourcing and manufacturing practices align with the global imperative to mitigate climate change.

Yet, the journey towards eco-friendly packaging is not without its hurdles. It is difficult to strike the appropriate balance between performance and sustainability, and current research strives to improve the strength and functionality of these materials [53]. Educating consumers about the benefits and proper disposal practices of eco-friendly packaging is crucial for widespread adoption. Consumer perception plays a pivotal role in shaping the success of sustainable packaging initiatives.

As we gaze into the future, innovations in materials steal the spotlight, with researchers delving into alternatives that promise a greener tomorrow. The integration of eco-friendly packaging into the circular economy narrative brings us closer to a world where waste is minimized, and materials find new life.

14.6 LABELING AND CODING

14.6.1 Glued-On Labels

Labeling and coding are integral aspects of packaging, providing crucial information about products and facilitating traceability. Glued-on labels are a widely used labeling technique where a label, typically made of paper or plastic, is affixed to the packaging using adhesive. This method allows for versatile labeling options, ranging from providing essential product information such as ingredients, nutritional facts, and usage instructions to serving as a branding tool for product recognition. The labels can be applied to various packaging materials, including glass, plastic, and cardboard, offering flexibility in design and application.

The process of applying glued-on labels involves precision and attention to detail. Labeling machines are often employed in industrial settings for efficient and accurate application. Factors such as the type of adhesive used, label material, and surface characteristics of the packaging play a crucial role in ensuring adhesion and longevity. Considerations for label placement, orientation, and readability are essential to create a visually appealing and informative packaging presentation [54].

While glued-on labels offer versatility, challenges such as label peeling, environmental impact, and production speed may arise. Innovations in adhesive technology, including the development of eco-friendly adhesives, aim to address environmental concerns. Additionally, advancements in labeling machinery contribute to increased efficiency and accuracy in the application process.

Looking ahead, the future of glued-on labels aligns with the broader trends in smart packaging and sustainability. Integration with technologies like RFID for enhanced traceability and interactive features may shape the evolution of glued-on labels.

14.6.2 Self-Adhesive Labels

Self-adhesive labels, also known as pressure-sensitive labels, represent a modern and efficient labeling solution. These labels come equipped with an adhesive backing that, when pressure is applied, adheres to various packaging surfaces. The simplicity of application, coupled with a wide range of design possibilities, makes self-adhesive labels a popular choice for product identification and branding. The genius of these labels lies in their simplicity – no water, heat, or additional solvents are needed for activation. This ease of use makes self-adhesive labels a preferred choice across diverse industries for packaging everything from consumer goods to pharmaceuticals. What sets self-adhesive labels apart is their adaptability to different packaging materials and shapes. From cylindrical bottles to irregularly shaped containers, self-adhesive labels conform with grace. This adaptability extends to the label's design, offering brands unparalleled freedom for creativity. With vibrant graphics, intricate details, and even tactile elements, self-adhesive labels become a canvas for brand storytelling and product differentiation [55].

The application process of self-adhesive labels is streamlined through labeling machines, ensuring precision and efficiency in high-volume production settings. These machines automate the label dispensing and application process, contributing to consistent placement and minimizing waste. The integration of self-adhesive labeling technology with production lines exemplifies the commitment to both aesthetics and operational efficiency. While self-adhesive labels bring unparalleled advantages, environmental considerations are gaining prominence. Innovations in label materials, including recyclable and compostable options, align with the global push toward sustainability.

The future trends in self-adhesive labels involve a harmonious blend of technological advancements, such as smart labels with near field communication (NFC) capabilities, and eco-conscious materials to meet the evolving demands of consumers and the planet.

14.6.3 In-Mold Labels

In-mold labels represent a fusion of label and packaging material during the manufacturing process. Unlike traditional labels applied to the surface, in-mold labels are placed inside the mold before the packaging material is formed, becoming an intrinsic part of the final product. This method is particularly prevalent in the production of plastic containers, creating a harmonious and durable label-packaging integration.

The production process of in-mold labels involves placing pre-printed labels into the mold cavity, followed by injecting or blow-molding the packaging material. The result is a label that becomes an inherent part of the packaging, eliminating concerns about label detachment, wear, or environmental exposure. This process not only enhances the visual appeal of the packaging but also contributes to its overall resilience.

In-mold labels find applications across a spectrum of industries, from food and beverage to personal care and beyond. Their versatility shines through various packaging forms, including containers, lids, and even closures. The labels seamlessly conform to the contours of the packaging, allowing for intricate and vibrant designs that enhance brand visibility.

While in-mold labeling presents numerous advantages, challenges such as the need for precision in label placement and potential production line adjustments may arise. Innovations in printing technologies, materials, and label application processes continuously address these challenges, ensuring that in-mold labels maintain their efficiency and aesthetic appeal.

Looking ahead, the future of in-mold labels aligns with the broader trends in sustainable packaging and smart labeling technologies. Integrating eco-friendly materials into the in-mold labeling process and exploring smart label features for enhanced consumer engagement represent ongoing areas of innovation.

14.6.4 Holographic Labels

Holographic labels utilize holography, a three-dimensional (3D) imaging technique, to create visually stunning and dynamic effects on packaging. Unlike traditional labels, holographic labels produce an illusion of depth and motion, adding a layer of sophistication and intrigue to product presentation. Beyond their aesthetic allure, holographic labels often serve as anti-counterfeiting measures due to their intricate and difficult-to-replicate designs [56].

The manufacturing of holographic labels involves the use of laser technology to record interference patterns on a photosensitive surface. This process results in a holographic image that, when illuminated, appears to be 3D. The complexity of these designs, combined with the precision required in manufacturing, contributes to the uniqueness and security features of holographic labels.

Holographic labels find applications across a wide array of industries, including cosmetics, pharmaceuticals, and consumer electronics. The ability to integrate vibrant colors, dynamic movement, and even micro text within the holographic design allows brands to create visually striking packaging that enhances product visibility and brand recognition [57].

One of the standout features of holographic labels is their role in enhancing product security. The intricate designs and optical effects make them challenging to replicate, serving as a deterrent to counterfeiters. Authentication features embedded within holographic labels, such as microtext or hidden images, provide a level of security that goes beyond the visual appeal.

Looking forward, holographic labels are anticipated to continue evolving with advancements in holographic technology and the integration of smart features. The marriage of holography with augmented reality (AR) and NFC technologies holds promise for creating immersive and interactive packaging experiences.

14.6.5 Bar Codes

The genesis of bar codes traces back to the late 1940s, gaining prominence in the 1970s as the Universal Product Code (UPC) made its debut. This linear symbology, consisting of parallel lines of varying widths, became a ubiquitous feature on consumer products, revolutionizing the retail landscape. Since then, bar codes have evolved to encompass diverse symbologies, including 2D codes like Quick Response (QR) codes, enhancing their versatility.

Bar codes operate on a fundamental principle – encoding information in a visual pattern that can be swiftly scanned by a reader device. The width and spacing of the bars represent alphanumeric characters or other data elements. Different symbologies cater to varied data requirements, with some encoding numerical data, while others accommodate alphanumeric and special characters, offering flexibility across industries.

The primary domain of bar codes remains retail, where they streamline point-of-sale transactions and automate inventory management. However, their reach extends far beyond, permeating industries such as healthcare, logistics, and manufacturing. In healthcare, bar codes enhance patient safety through medication verification, while in logistics, they facilitate real-time tracking and traceability of products across the supply chain.

Despite their widespread adoption, challenges such as limited data capacity in traditional linear bar codes have prompted innovations like 2D bar codes. QR codes, for instance, accommodate more information and enable linkages to multimedia content, providing a richer user experience. Advances in barcode scanning technology, including mobile scanning applications, contribute to enhanced accessibility and ease of use.

Looking ahead, the future of bar codes aligns with the emerging trends in Internet of Things (IoT) and the integration of smart technologies. Bar codes, complemented by RFID and sensor technologies, are poised to play a pivotal role in creating interconnected and data-driven supply chains [58, 59].

14.6.6 Radio Frequency Identification

RFID is a technology that allows for the tracking and identification of objects without physical contact. At its core, RFID is a technology that utilizes radio-frequency waves to transmit data between an RFID tag and a reader device. The RFID system comprises two main components – the RFID tag, affixed to the packaging, and the reader device, which wirelessly communicates with the tag. The tag contains a unique identifier and, in some cases, additional data that can be accessed and updated in real-time [60].

RFID's versatility extends across a myriad of industries, from retail and logistics to healthcare and manufacturing. In retail, RFID optimizes inventory management by providing real-time visibility into stock levels. In logistics, RFID facilitates accurate tracking and tracing of goods throughout the supply chain. The healthcare sector leverages RFID for patient and asset tracking, enhancing efficiency and patient safety.

The adoption of RFID brings about a multitude of advantages. Unlike traditional bar codes, RFID tags do not require a direct line of sight for scanning, enabling rapid and bulk reading of multiple tags simultaneously. This not only expedites processes but also reduces human errors. RFID's capacity to store and update information in real-time enhances accuracy in inventory management and ensures product authenticity.

While RFID offers significant benefits, challenges such as the cost of implementation, data security concerns, and standardization hurdles have been points of consideration. Ongoing innovations in RFID technology aim to address these challenges, with developments in printable and flexible RFID tags, enhanced data encryption, and integration with IoT technologies.

Looking forward, the future of RFID in labeling and coding is intertwined with the broader trends in smart packaging and IoT. The integration of RFID with sensors, data analytics, and cloud computing is poised to create intelligent and responsive supply chain ecosystems [58].

14.7 CONCLUSION

In the fast-paced world of RTE foods, packaging, and labeling play pivotal roles in ensuring product quality, safety, and consumer appeal. This discussion has delved into various aspects of RTE packaging, from current market statuses to emerging trends, highlighting the intricate balance between innovation and sustainability. The RTE packaging landscape is diverse, reflecting a global demand for convenience. It encompasses a wide array of materials and technologies to meet the unique challenges posed by RTE products. Packaging plays a crucial role in determining the quality, shelf life, and safety of RTE foods. Proper packaging helps in preserving freshness, preventing contamination, and extending the product's shelf life. The choice of packaging material is a nuanced decision influenced by mechanical damage, permeability, grease-proof properties, temperature resistance, light exposure, and other factors. Striking the right balance is crucial to ensuring the integrity of the packaged RTE products. From traditional paper and paperboards to modern flexible films and metalized options, RTE packaging materials have evolved to address the dynamic needs of the industry. Flexible films and laminates have gained prominence for their versatility and ability to protect against external elements. Innovations in food packaging are rapidly transforming the industry. Edible packaging, active packaging, intelligent food packaging, and eco-friendly solutions are on the rise, reflecting a growing commitment to sustainability and consumer preferences. The labeling and coding of RTE packaging is evolving with advancements such as self-adhesive labels, in-mold labels, holographic labels, bar codes, and RFID technology. These not only enhances traceability but also contributes to the overall aesthetics and marketability of the product. In conclusion, the RTE food industry is at the intersection of technological innovation and environmental responsibility. Sustainable packaging solutions, coupled with intelligent labeling, are imperative for meeting consumer expectations and industry standards. As we navigate toward a more eco-conscious future, the ongoing research and development in this field will undoubtedly lead to even more sophisticated and sustainable packaging solutions for RTE foods.

REFERENCES

1. Silberbauer, A. and Schmid, M., 2017. Packaging concepts for ready-to-eat food: Recent progress. *Journal of Packaging Technology and Research*, *1*, pp.113–126. https://doi.org/10.1007/s41783-017-0019-9
2. Smithers. 2021. *Strategic forecasts for global packaging to 2024*. Retrieved July 29, 2024, from www.smithers.com/resources/2020/mar/smithers-forecasts-global-packaging-market-to-grow
3. *Packaging Industry in India – Size, Growth, Overview & Report*. Retrieved July 28, 2024, from www.mordorintelligence.com/industry-reports/packaging-industry-in-india
4. *Packaging in 2022: Sustainability Trends to Consider and More*. Retrieved July 29, 2024, from https://fresh-lock.com/blog/sustainable-packaging-trends-2022
5. Zero Waste Living Lab. *In a best-case recycling scenario plastic packaging production and plastic leakage still double by 2050*. Retrieved July 29, 2024, from https://zerowastelivinglab.enviu.org/blog/plastic-production-leakage-double/
6. Hedenqvist, M.S., 2005. Barrier packaging materials. In *Handbook of environmental degradation of materials* (pp.547–563). William Andrew Publishing.
7. Grumezescu, A.M. and Holban, A.M., 2017. *Food packaging and preservation* (Vol. 9). Academic Press.
8. Robertson, G.L., 2009. *Food packaging and shelf life: A practical guide*. CRC Press.
9. Koutsoumanis, K.P. and Gougouli, M., 2015. Use of time temperature integrators in food safety management. *Trends in Food Science & Technology*, *43*(2), pp.236–244. https://doi.org/10.1016/j.tifs.2015.02.008
10. Verma, S.K., Prasad, A. and Katiyar, V., 2024. State of art review on sustainable biodegradable polymers with a market overview for sustainability packaging. *Materials Today Sustainability*, 100776. https://doi.org/10.1016/j.mtsust.2024.100776
11. FSSAI, 2020. Food safety and standards (labelling and display) regulations.

12. *Coding and labelling of food packages – Food packaging technology*. Retrieved July 28, 2024, from https://ebooks.inflibnet.ac.in/ftp08/chapter/coding-and-labelling-of-food-packages/
13. Falua, K.J., Pokharel, A., Babaei-Ghazvini, A., Ai, Y. and Acharya, B., 2022. Valorization of starch to biobased materials: A review. *Polymers*, *14*(11), p.2215. https://doi.org/10.3390/polym14112215
14. Wróblewska-Krepsztul, J., Rydzkowski, T., Borowski, G., Szczypiński, M., Klepka, T. and Thakur, V.K., 2018. Recent progress in biodegradable polymers and nanocomposite-based packaging materials for sustainable environment. *International Journal of Polymer Analysis and Characterization*, *23*(4), pp.383–395. https://doi.org/10.1080/1023666X.2018.1455382
15. Prasad, K., Nikzad, M. and Sbarski, I., 2018. Permeability control in polymeric systems: A review. *Journal of Polymer Research*, *25*(11), p.232. https://doi.org/10.1007/s10965-018-1636-x
16. Han, J.H., 2003. Antimicrobial food packaging. *Novel Food Packaging echniques*, *8*, pp.50–70.
17. Hubbe, M.A. and Pruszynski, P., 2020. Greaseproof paper products: a review emphasizing ecofriendly approaches. *BioResources*, *15*(1), pp. 1978–2004.
18. Schaider, L.A., Balan, S.A., Blum, A., Andrews, D.Q., Strynar, M.J., Dickinson, M.E., Lunderberg, D.M., Lang, J.R. and Peaslee, G.F., 2017. Fluorinated compounds in US fast food packaging. *Environmental Science & technology Letters*, *4*(3), pp.105–111. https://doi.org/10.1021/acs.estlett.6b00435
19. *Packaging Additives Market (Functional Additives & Barrier Coatings) by Type, Application, & by Geography – 2020 | MarketsandMarkets | Last Updated on July-2024*. Retrieved July 29, 2024, from www.marketsandmarkets.com/Market-Reports/functional-additives-and-barrier-coatings-market-86258744.html
20. *Develop Strong Paper Packages with Fluorochemical-Free Grease Barriers | Where business meets science*. Retrieved July 28, 2024, from https://insights.basf.com/home/article/read/developing-a-strong-sustainable-grease-barrier-for-food-packaging
21. Singh, S., Gaikwad, K.K., Lee, M. and Lee, Y.S., 2018. Temperature-regulating materials for advanced food packaging applications: A review. *Journal of Food Measurement and Characterization*, *12*, pp.588–601. https://doi.org/10.1007/s11694-017-9672-5
22. Rooney, M. L., 1995. *Active Food Packaging*. Springer US. https://doi.org/10.1007/978-1-4615-2175-4
23. Verduin, J., Den Uijl, M.J., Peters, R.J.B. and Van Bommel, M.R., 2020. Photodegradation products and their analysis in food. *Journal of Food Science and Nutrition*, *6*(67.10), 24966. https://doi.org/10.24966/FSN-1076/100067
24. Tripathi, S., Kumar, L., Deshmukh, R.K. and Gaikwad, K.K., 2024. Ultraviolet blocking films for food packaging applications. *Food and Bioprocess Technology*, *17*(6), pp.1563–1582. https://doi.org/10.1007/s11947-023-03221-y
25. *Light Stabilizers/UV Absorbers – Selection Tips & Formulation Examples*. Retrieved July 29, 2024, from https://polymer-additives.specialchem.com/selection-guide/light-uv-stabilizers-selection-for-polymers
26. Geometries, B. and Abrasion, S., 2012. Standard test method for haze and luminous transmittance of transparent plastics. *ASTM International*, *1*, pp.1–7.
27. *Eco Friendly Paper – A Complete Guide to Sustainable Greener Paper*. Retrieved July 28, 2024, from www.trvst.world/sustainable-living/eco-friendly/eco-friendly-paper-all-you-need-to-know/
28. Kerry, J. and Butler, P., 2008. *Smart packaging technologies for fast moving consumer goods*. John Wiley & Sons.
29. Sangroniz, A., Zhu, J.B., Tang, X., Etxeberria, A., Chen, E.Y.X. and Sardon, H., 2019. Packaging materials with desired mechanical and barrier properties and full chemical recyclability. *Nature Communications*, *10*(1), p.3559. https://doi.org/10.1038/s41467-019-11525-x
30. *7 Flexible Packaging Film Types Explained | Ecoplast Ltd India*. Retrieved July 29, 2024, from www.ecoplastindia.com/flexible-packaging-film-types/
31. *5 Benefits of Flexible Packaging and Flexible Films*. Retrieved July 28, 2024, from www.industrialpackaging.com/blog/5-benefits-of-flexible-packaging-film
32. *What is Metalized Film and its Different Types*. Retrieved July 29, 2024, from www.cosmofilms.com/blog/types-of-metalized-film
33. Deshwal, G.K. and Panjagari, N.R., 2020. Review on metal packaging: Materials, forms, food applications, safety and recyclability. *Journal of food Science and technology*, *57*(7), pp.2377–2392. https://doi.org/10.1007/s13197-019-04172-z

34. *Top 4 Different Types of Lamination Films for Packaging*. Retrieved July 29, 2024, from www.ecoplastindia.com/types-of-lamination-films-for-packaging/
35. van Velzen, U.T., de Weert, L. and Molenveld, K., 2020. *Flexible laminates within the circular economy* (No. 2037). Wageningen Food & Biobased Research.
36. García, M.P.M., Gómez-Guillén, M.C., López-Caballero, M.E. and Barbosa-Cánovas, G.V., 2016. *Edible films and coatings: Fundamentals and applications*. CRC Press.
37. Huber, K. C., & Embuscado, M. E., 2009. *Edible films and coatings for food applications*, 9, pp.169–208. Springer. https://doi.org/10.1007/978-0-387-92824-1
38. Baldwin, E.A., Hagenmaier, R. and Bai, J., 2011. *Edible coatings and films to improve food quality*. CRC Press.
39. Lu, Y., Luo, Q., Chu, Y., Tao, N., Deng, S., Wang, L. and Li, L., 2022. Application of gelatin in food packaging: A review. *Polymers*, *14*(3), p.436. https://doi.org/10.3390/polym14030436
40. Zhao, Y., Li, B., Li, C., Xu, Y., Luo, Y., Liang, D. and Huang, C., 2021. Comprehensive review of polysaccharide-based materials in edible packaging: A sustainable approach. *Foods*, *10*(8), p.1845. https://doi.org/10.3390/foods10081845
41. Cheng, C., Chen, S., Su, J., Zhu, M., Zhou, M., Chen, T. and Han, Y., 2022. Recent advances in carrageenan-based films for food packaging applications. *Frontiers in Nutrition*, *9*, 1004588. https://doi.org/10.3389/fnut.2022.1004588
42. Kandasamy, S., Yoo, J., Yun, J., Kang, H.B., Seol, K.H., Kim, H.W. and Ham, J.S., 2021. Application of whey protein-based edible films and coatings in food industries: An updated overview. *Coatings*, *11*(9), p.1056. https://doi.org/10.3390/coatings11091056
43. Drago, E., Campardelli, R., Pettinato, M. and Perego, P., 2020. Innovations in smart packaging concepts for food: An extensive review. *Foods*, *9*(11), p.1628. https://doi.org/10.3390/foods9111628
44. Kordjazi, Z. and Ajji, A., 2022. Oxygen scavenging systems for food packaging applications: A review. *The Canadian Journal of Chemical Engineering*, *100*(12), pp.3444–3449. https://doi.org/10.1002/cjce.24539
45. Firouz, M.S., Mohi-Alden, K. and Omid, M., 2021. A critical review on intelligent and active packaging in the food industry: Research and development. *Food Research International*, *141*, 110113. https://doi.org/10.1016/j.foodres.2021.110113
46. Abe, K. and Watada, A.E., 1991. Ethylene absorbent to maintain quality of lightly processed fruits and vegetables. *Journal of Food Science*, *56*(6), pp.1589–1592. https://doi.org/10.1111/j.1365-2621.1991.tb08647.x
47. Mahmud, J., Sarmast, E., Shankar, S. and Lacroix, M., 2022. Advantages of nanotechnology developments in active food packaging. *Food Research International*, *154*, 111023. https://doi.org/10.1016/j.foodres.2022.111023
48. Yam, K.L., Takhistov, P.T. and Miltz, J., 2005. Intelligent packaging: Concepts and applications. *Journal of Food Science*, *70*(1), pp. R1–R10. https://doi.org/10.1111/j.1365-2621.2005.tb09052.x
49. Prasad, P. and Kochhar, A., 2014. Active packaging in food industry: A review. *Journal of Environmental Science, Toxicology and Food Technology*, *8*(5), pp.1–7.
50. Kuswandi, B., 2018. Intelligent packaging applications for fruits and vegetables. *Innovative Packaging of Fruits and Vegetables: Strategies for Safety and Quality Maintenance*, *81*, pp. 81–104.
51. Siracusa, V., Rocculi, P., Romani, S. and Dalla Rosa, M., 2008. Biodegradable polymers for food packaging: A review. *Trends in Food Science & echnology*, *19*(12), pp.634–643. https://doi.org/10.1016/j.tifs.2008.07.003
52. Ncube, L.K., Ude, A.U., Ogunmuyiwa, E.N., Zulkifli, R. and Beas, I.N., 2020. Environmental impact of food packaging materials: A review of contemporary development from conventional plastics to polylactic acid-based materials. *Materials*, *13*(21), p.4994. https://doi.org/10.3390/ma13214994
53. Kaestner, L., Scope, C., Neumann, N. and Woelfel, C., 2023. Sustainable circular packaging design: A systematic literature review on strategies and applications in the cosmetics industry. *Proceedings of the Design Society*, *3*, pp.3265–3274. https://doi.org/10.1017/pds.2023.327
54. Emblem, A. ed., 2012. *Packaging technology: Fundamentals, materials and processes*. Elsevier.
55. Pizzi, A. and Mittal, K.L. eds., 2017. *Handbook of adhesive technology*. CRC Press.
56. Jeong, T.H., 2010. Basic principles and applications of holography. Lake Forest College.

57. Žiljak, I., 2004. Holography on packaging material and an estimate on the rising trend of its application. In *The 31st International Research Conference Iarigai*. Copenhagen, Denmark.
58. Sharma, R. and Ghoshal, G., 2018. Emerging trends in food packaging. *Nutrition & Food Science*, *48*(5), pp.764–779. https://doi.org/10.1108/NFS-02-2018-0051
59. *Barcode Technology – Barcodes Inc*. (n.d.). Retrieved July 28, 2024, from www.barcodesinc.com/articles/barcode-technology.htm
60. Zuo, J., Feng, J., Gameiro, M.G., Tian, Y., Liang, J., Wang, Y., Ding, J. and He, Q., 2022. RFID-based sensing in smart packaging for food applications: A review. *Future Foods*, *6*, 100198. https://doi.org/10.1016/j.fufo.2022.100198

15 Nanotechnology and Other Novelties

Innovation among Ready-to-Eat Foods

Sivashankari M.[1*] *and Sadvatha R. H.*[2]
[1]Scientist, ICAR-National Rice Research Institute, Cuttack, Odisha, India
[2]Scientist, ICAR-Central Institute of Agricultural Engineering, Regional Station, Coimbatore, India
*Corresponding Author

15.1 INTRODUCTION

15.1.1 Significance of Nanotechnology in the Ready-to-Eat Food Industry

Nanotechnology's significance in the ready-to-eat (RTE) food industry is profound. It represents a dynamic force of innovation, reshaping the way RTE foods are produced and consumed. One of its primary advantages lies in the realm of nutrient delivery. Through the precise manipulation of materials at the nanoscale, nanotechnology facilitates the creation of nanoparticles and nanocarriers that protect sensitive nutrients during processing and optimize their delivery within the human body (Malik et al. 2023). This targeted nutrient delivery not only enhances the nutritional value of RTE foods but also aligns with consumers' growing interest in health-conscious eating.

Moreover, nanotechnology enhances sensory attributes in RTE foods, particularly through nanoemulsions that disperse flavor-enhancing compounds more uniformly, improving taste and texture. It also extends the shelf life of these products by employing antimicrobial nanomaterials and active nanocomposites in packaging, reducing food waste and contributing to food safety. Beyond these benefits, nanotechnology promotes resource efficiency, reducing the need for additives and excessive processing. However, while nanotechnology offers remarkable potential, it poses challenges, such as regulatory considerations, safety concerns, and consumer acceptance (Singh et al. 2023b). Striking a harmonious balance between realizing these advantages and addressing these challenges is pivotal to harnessing the full potential of nanotechnology in the RTE food industry, making it more sustainable, nutritious, and consumer-friendly.

15.1.2 Overview of Nanotechnology in the RTE Food Industry

Nanotechnology, operating at the scale of individual atoms and molecules, has revolutionized the RTE food industry with its myriad applications. This innovation involves the precise manipulation and engineering of materials at the nanoscale (typically 1 to 100 nanometers) (Altabbaa et al. 2023). In the context of RTE foods, nanotechnology presents itself through nanoparticles, nanoemulsions, and nanocomposites, each offering unique solutions to various challenges in the production and consumption of these foods (Sahani and Sharma 2021).

DOI: 10.1201/9781032622408-15

Nanoparticles, at the heart of nanotechnology in RTE foods, excel in enhancing delivery, taste, and shelf life. These tiny carriers can protect sensitive nutrients during processing and effectively target their release within the human body, ultimately bolstering the nutritional value of RTE foods. Nanoemulsions, with their remarkably small droplet size, disperse flavor-enhancing compounds uniformly, which leads to improved taste, texture, and overall sensory appeal (Manzoor et al. 2023). Moreover, nanocomposites integrated into packaging materials bring intelligence to packaging. They can scavenge oxygen, manage moisture, and provide real-time indicators of temperature changes, resulting in extended shelf life and reduced food waste (Drago et al. 2020). By leveraging nanotechnology in these ways, RTE food producers are not only improving the overall quality and safety of their products but also making notable strides in sustainability and consumer satisfaction (Wei 2021). However, it is essential to navigate the regulatory complexities, address safety concerns, and manage consumer perceptions effectively to ensure the responsible integration of nanotechnology into the RTE food industry.

15.1.3 Objectives of the Chapter

This chapter aims to explore the potential of nanotechnology in RTE foods, elucidating the principles and applications of nanomaterials. It delves into the benefits and challenges associated with their use and showcases case studies of nanotechnology-driven innovations in specific RTE food categories. The chapter also examines the future trends and regulatory landscape, emphasizing the significance of further research and development.

15.2 UNDERSTANDING NANOTECHNOLOGY

15.2.1 Definition and Principles of Nanotechnology

Nanotechnology is a multidisciplinary field that operates at the nanoscale, dealing with structures and materials at dimensions typically ranging from 1 to 100 nanometers. It involves manipulating and engineering these minuscule structures to create new materials, devices, and systems with extraordinary properties and functions. At its core, nanotechnology is guided by the principles of understanding and controlling matter at the nanoscale. It draws from various scientific disciplines, including physics, chemistry, biology, and engineering, to explore the unique behaviors and properties exhibited by materials at such tiny dimensions (Sahu et al. 2023).

These principles of nanotechnology encompass the ability to design, synthesize, and manipulate nanoscale structures with precision. This precise control allows for tailoring materials' properties to achieve specific outcomes. For instance, in the context of RTE foods, nanotechnology is employed to engineer nanoparticles and nanocarriers that enhance nutrient delivery and stability, thereby improving the nutritional content and safety of food products (Khan et al. 2023). These principles of precise manipulation at the nanoscale are central to realizing the potential of nanotechnology across a wide range of industries, from electronics and medicine to food and materials science.

15.2.2 Nanomaterials Used in RTE Food Innovation

15.2.2.1 Nanoparticles for Enhanced Nutrient Delivery

Nanoparticles have emerged as pioneering vehicles for enhancing nutrient delivery in the realm of RTE foods. Notably, liposomes and solid lipid nanoparticles play a pivotal role in this innovation (Dwivedi et al. 2023). By encapsulating sensitive nutrients within these nanoscale carriers, such as vitamins, antioxidants, or bioactive compounds, these particles shield them from degradation during processing and storage (Aguilar-Pérez et al. 2023; Singh et al. 2023b). This protection not only ensures the preservation of nutrient integrity but also enables their targeted release within the human body. This means that sensitive nutrients can reach their intended destination, such as the bloodstream or specific organs, with greater efficiency. As a result, consumers of RTE foods can benefit from improved bioavailability and heightened nutritional value. The application of nanoparticles in nutrient delivery represents a significant stride in making RTE foods not only convenient but also a reliable source of essential nutrients for a health-conscious and fast-paced world (Hussein et al. 2023).

15.2.2.2 Nanoemulsions for Improved Sensory Attributes

Nanoemulsions, a hallmark of nanotechnology, are transforming the taste and texture of RTE foods. These emulsions, with their extraordinarily small droplets at the nanoscale, possess the unique ability to disperse flavor-enhancing compounds and ingredients with exceptional uniformity. This uniform dispersion translates to an even distribution of flavors and textures throughout the food matrix. For consumers, this means a more consistent and pleasurable eating experience. Whether it is the uniform spread of a creamy sauce in a microwavable meal or the even distribution of flavor-enhancing compounds in a snack, nanoemulsions ensure that each bite or spoonful is as delightful as the last. Thus, in the fast-paced world of RTE foods, where convenience meets flavor, nanoemulsions play a pivotal role in elevating sensory attributes, making these foods not just quick and easy but also a gratifying culinary experience (Luciano et al. 2023).

15.2.2.3 Nanocomposites for Enhanced Packaging and Shelf Life

Nanocomposites represent a groundbreaking approach to packaging in the RTE food industry, where the preservation of quality and safety is paramount. These innovative materials seamlessly integrate nanoscale elements into conventional packaging, conferring a host of benefits. They enhance barrier properties, preventing the ingress of oxygen and moisture, which are common culprits in food spoilage. Additionally, nanocomposites can exert antimicrobial effects, inhibiting the growth of harmful microorganisms and pathogens (Jiang et al. 2023). This dual-action protection translates to extended shelf life, reducing food waste and ensuring that RTE foods maintain their freshness and nutritional value for longer periods (de Sousa et al. 2023). Furthermore, nanocomposites can incorporate intelligent functionalities, such as temperature-sensitive indicators and moisture control, ensuring that consumers receive safe and high-quality products (Versino et al. 2023). In a world where convenience and food safety are paramount, nanocomposites provide a transformative solution by effectively marrying nanotechnology with packaging, ultimately enhancing the RTE food industry's sustainability, quality, and consumer satisfaction.

15.3 APPLICATION OF NANOTECHNOLOGY IN RTE FOODS

15.3.1 Improved Nutrient Delivery

15.3.1.1 Nanocarriers for Targeted Nutrient Delivery

Nanocarriers, exemplified by nanostructured lipids and polymer-based nanoparticles, serve as precision vehicles for the targeted delivery of bioactive compounds within the realm of RTE foods (Tomić et al. 2023). These minute carriers encapsulate sensitive nutrients and bioactive, safeguarding them from degradation during processing and storage. Such protection guarantees that these nutrients reach their intended destinations within the human body with precise control. This controlled release not only enhances the stability of these bioactive compounds but also optimizes their absorption in specific areas of the body, thereby bolstering overall nutrient bioavailability (Singh et al. 2023). In the world of RTE foods, where consumers increasingly seek foods rich in functional ingredients, nanocarriers play a pivotal role in ensuring that these valuable nutrients reach their full potential.

15.3.1.2 Nanoencapsulation for Increased Stability and Bioavailability

In a parallel vein, nanoencapsulation techniques, with their ability to envelop sensitive nutrients, provide a dual advantage for RTE foods. They shield these nutrients from deterioration during processing and prolonged storage, maintaining their nutritional integrity (Gómez-Guillén et al. 2023). Simultaneously, the encapsulation process offers greater bioavailability, enabling these nutrients to be efficiently absorbed during digestion. This encapsulated release ensures a more controlled and sustained delivery of nutrients within the digestive system, thereby optimizing their effectiveness (Dwivedi et al. 2023). Consequently, in the context of RTE foods, where maintaining nutrient quality is of paramount importance, nanoencapsulation techniques represent a valuable tool in enhancing both the stability and bioavailability of vital nutrients, aligning with the evolving preferences of health-conscious consumers.

15.3.2 Enhanced Sensory Attributes

15.3.2.1 Nanosensors for Real-Time Quality Monitoring

Nanosensors represent a technological marvel in the RTE food industry by enabling real-time, precise monitoring of critical quality parameters. These nanoscale sensors can detect changes in freshness, spoilage, and contamination with exceptional sensitivity and specificity (Muthukumar et al. 2023). When integrated into food packaging or directly into food products, nanosensors provide a continuous stream of data, allowing for immediate interventions to maintain product integrity and safety. This real-time monitoring capability is especially valuable in the RTE food sector, where rapid response to quality issues is crucial (Mahesha et al. 2023). Nanosensors empower manufacturers to reduce food waste, enhance food safety, and guarantee that consumers enjoy RTE foods at their peak quality, aligning with the industry's commitment to efficiency and excellence (Shruti et al. 2023; Singh and Kumar 2023a).

15.3.2.2 Nanoemulsions for Improved Taste and Texture

Nanoemulsions, a hallmark of nanotechnology, are revolutionizing the taste and texture of RTE foods. With their exceptionally tiny droplets at the nanoscale, they possess a unique ability to create homogeneity and ensure the uniform distribution of flavor-enhancing compounds and ingredients throughout RTE food products (Manzoor et al. 2023). This uniform dispersion is a game-changer, as it guarantees that every bite or spoonful delivers an identical and delightful sensory experience (Luciano et al. 2023). Whether it is the even spread of a creamy sauce in a microwavable meal or the consistent distribution of flavor-enhancing compounds in a snack, nanoemulsions play a pivotal role in ensuring that each bite is as delicious as the last. In the fast-paced world of RTE foods, where convenience meets flavor, nanoemulsions are pivotal in elevating sensory attributes, making these foods not just quick and easy but also a gratifying culinary experience.

15.3.3 Extended Shelf Life and Packaging

15.3.3.1 Antimicrobial Nanomaterials for Food Preservation

In the realm of RTE foods, the utilization of antimicrobial nanomaterials marks a groundbreaking approach to food preservation. Materials like silver nanoparticles and nano-clay have demonstrated remarkable effectiveness in inhibiting the growth of microorganisms that cause food spoilage and contamination (Manna 2023). Their nanoscale dimensions provide a higher surface area for interaction with microbes, making them potent agents against bacterial, fungal, and viral pathogens. By integrating these antimicrobial nanomaterials into packaging materials or directly into food products, the RTE food industry can significantly extend shelf life, reduce food waste, and enhance food safety (Tomic et al. 2023). This innovation ensures that consumers receive RTE foods with a longer storage life while enjoying the peace of mind that their meals are free from harmful microorganisms. As the industry continues to prioritize convenience, quality, and safety, antimicrobial nanomaterials emerge as a transformative solution, aligning perfectly with the needs and expectations of modern consumers.

15.3.3.2 Active Nanocomposites for Intelligent Packaging

Active nanocomposites represent a pioneering advancement in intelligent packaging for the RTE food industry. These innovative materials, seamlessly integrate nanoscale elements into conventional packaging, some equipped with a range of intelligent functionalities. They can scavenge oxygen, which is often a primary cause of food degradation, and control moisture levels, maintaining the optimal environment within the package (Huang et al. 2023). Furthermore, some active nanocomposites include temperature-sensitive indicators that provide real-time information about the freshness and quality of the RTE food (Versino et al. 2023). This dynamic packaging responds to environmental cues and adjusts to ensure that the product inside remains at its peak quality. The result is extended shelf life, reduced food waste, and a superior eating experience for consumers, aligning perfectly with the industry's dedication to delivering not just convenience but also freshness and reliability.

15.4 BENEFITS AND CHALLENGES OF NANOTECHNOLOGY IN RTE FOODS

15.4.1 Benefits

15.4.1.1 Improved Product Quality and Nutritional Value

Nanotechnology is a game-changer in the RTE food industry, offering the means to significantly enhance the nutritional value and overall quality of these products. At the heart of this innovation is the precise delivery of sensitive nutrients, bioactive compounds, and functional ingredients. By encapsulating these vital components within nanocarriers and nanoencapsulation techniques, nanotechnology safeguards them from degradation during processing and storage (Khan et al. 2023). This preservation not only ensures the retention of nutritional integrity but also optimizes their absorption within the human body. As a result, consumers of RTE foods benefit from improved bioavailability, heightened nutritional content, and an array of potential health benefits, aligning seamlessly with the ever-growing demand for nutritious, functional, and convenient food options. In the RTE food landscape, where quality and health-conscious choices are paramount, nanotechnology proves to be a groundbreaking tool for producers striving to meet and exceed consumer expectations.

15.4.1.2 Enhanced Safety and Extended Shelf Life

In the dynamic world of RTE foods, nanotechnology brings forth a dual advantage that significantly elevates both food safety and sustainability. The antimicrobial properties of nanomaterials, such as silver nanoparticles and nano-clay, play a pivotal role in preventing the growth of microorganisms, including bacterial, fungal, and viral pathogens (Manikandan and Min 2023). This potent protection assures consumers of safer, contamination-free RTE foods. Simultaneously, the integration of nanocomposites into packaging materials empowers the extension of shelf life by creating a protective barrier against the two common culprits of food degradation: oxygen and moisture. As a result, RTE food products not only maintain their quality and safety for an extended period but also contribute to the reduction of food waste, a significant environmental and economic concern. In a world where food safety and sustainability are paramount, the amalgamation of nanotechnology with the RTE food industry proves to be a remarkable solution that aligns perfectly with the expectations of modern consumers and the industry's commitment that aligns perfectly with the expectations of modern consumers and the industry's commitment to excellence.

15.4.1.3 Efficient Resource Utilization

Nanotechnology is at the forefront of revolutionizing resource utilization in the RTE food industry, offering a pathway to more sustainable and resource-efficient production. The precise manipulation of materials at the nanoscale, particularly through nanoparticles and nanocarriers, ensures the enhanced stability and efficient delivery of nutrients and bioactive compounds. This enhancement significantly reduces the need for additives and preservatives, contributing to cleaner and more natural food formulations. By minimizing the use of these additional ingredients, RTE food producers can create products that are not only healthier and more nutritionally robust but also less resource-intensive (Peker and Günal 2023). This aligns with the growing consumer preference for clean-label and minimally processed foods while reducing the overall environmental footprint of RTE food production. In essence, nanotechnology paves the way for a more sustainable RTE food industry by enhancing resource utilization and advancing the industry's commitment to quality, health, and eco-consciousness.

15.4.2 Challenges

15.4.2.1 Regulatory Considerations and Safety Concerns

The integration of nanomaterials in food, although offering remarkable benefits, introduces a complex landscape of regulatory considerations and safety concerns (Manful et al. 2023). One major challenge revolves around clear and accurate labeling, as consumers have the right to know when nanotechnology is employed in their food products. Ensuring transparent and comprehensive labeling practices is vital to maintaining consumer trust (Khan et al. 2023). Additionally, rigorous safety assessments and risk evaluations are essential, as the unique properties of nanomaterials require a thorough examination of their potential impacts on human health and the environment. Striking a balance between innovation and safety is crucial, as the regulatory framework must evolve to keep pace with the rapid advancements in nanotechnology. This dynamic regulatory environment seeks to protect consumers while still encouraging the responsible application of nanotechnology to enhance the quality and safety of RTE foods (Singh and Kumar 2023b).

15.4.2.2 Consumer Acceptance and Perception

Consumer acceptance and perception of nanotechnology in food products represent a vital dimension of its successful integration into the RTE food industry. Many consumers may be unaware of or uncertain about the use of nanotechnology in their food, which can lead to skepticism or concerns regarding safety and long-term health effects (Singh and Kumar 2023b). Transparent communication and education are pivotal in addressing these challenges. RTE food producers need to convey the benefits and safety of nanotechnology clearly, providing consumers with information to make informed choices. Building trust and confidence in the technology involves highlighting its role in improving product quality, safety, and sustainability while addressing any potential risks. By fostering open dialogue and maintaining transparency, the industry can bridge the gap between the potential of nanotechnology and consumer acceptance, ensuring that RTE foods continue to meet consumer expectations and preferences (Thanakkasaranee et al. 2023).

15.5 FUTURE TRENDS AND OUTLOOK

15.5.1 Emerging Nanotechnologies in RTE Foods

In the ever-evolving landscape of RTE foods, this chapter delves into the realm of emerging nanotechnologies, heralding a new era of innovation and excellence. The exploration encompasses cutting-edge developments like nanostructured delivery systems that promise to further enhance nutrient bioavailability and controlled release within the human body. Moreover, the chapter casts a spotlight on nanosensors designed for real-time quality assurance, elevating the precision and efficiency of monitoring freshness and safety in RTE food products (Kaushani et al. 2022). These emerging nanotechnologies represent the future of the RTE food industry, thereby offering exciting prospects for even higher product quality, enhanced safety standards, and improved consumer experiences. As the industry continues to embrace innovation, these emerging technologies are positioned to play a pivotal role in shaping the RTE food products of tomorrow, aligning seamlessly technologies are positioned to play a pivotal role in shaping the RTE food products of tomorrow, aligning seamlessly with the industry's commitment to excellence and consumer satisfaction.

15.5.2 Potential Impact on food Industry and Consumer Experience

The adoption of nanotechnology in the realm of RTE foods is poised to be a game-changer, heralding a transformation in the food industry and elevating the consumer experience. This innovation promises to deliver a wave of novel products with superior nutritional profiles, enriched sensory attributes, and significantly prolonged shelf life. RTE foods infused with nanotechnology will offer consumers not only convenience but also heightened quality and safety, aligning seamlessly with the evolving demands of modern lifestyles. This technology's potential to ensure better nutrient delivery, enhance taste and texture, and extend product freshness positions it as a key driver of innovation. As the industry continues to evolve and cater to the expectations of health-conscious and convenience-seeking consumers, the integration of nanotechnology offers a gateway to an exciting future of RTE foods that are not quick and easy but also superior in quality and sensory delight (Chowdhury et al. 2023).

15.5.3 Regulatory Landscape and Future Challenges

The evolving regulatory landscape related to nanotechnology in food brings forth a series of dynamic challenges and opportunities. As nanotechnology continues to play an increasingly vital role in RTE foods, there is a pressing need for comprehensive, science-based regulations. These should encompass safety assessments, clear labeling practices, and transparent communication with the public to allay concerns and ensure informed decision-making. Simultaneously, the RTE food industry must adhere to ethical and responsible practices, maintaining a delicate balance between innovation and consumer well-being (Kaur et al. 2023). This evolving regulatory environment, while posing challenges, also opens doors for collaboration, innovation, and responsibility. It calls for proactive strategies and cooperation between industries, governments, and consumers to harness the full potential of nanotechnology while uploading safety, ethics, and consumer confidence in the RTE food sector.

15.6 CONCLUSION

15.6.1 Summary of Key Points Covered in the Chapter

Nanotechnology has emerged as a driving force in the world of RTE foods, ushering in a new era of innovation and solutions. At the heart of this transformation are nanomaterials, encompassing nanoparticles, nanoemulsions, and nanocomposites, which represent the vanguard of nanotechnology applications within the RTE food sector. These materials serve as powerful tools, facilitating advancements in nutrient delivery and sensory enhancement. Nanoparticles, in particular, are harnessed to enhance nutrient delivery, protecting fragile compounds and enhancing their bioavailability. Simultaneously, nanoemulsions ensure an even distribution of flavors and textures, elevating the overall eating experience.

Nanotechnology's influence extends beyond the culinary realm, impacting packaging solutions that go beyond conventional boundaries. Nanocomposites with barrier properties, antimicrobial effects, and freshness indicators offer intelligent packaging, safeguarding product quality and extending shelf life. Moreover, the antimicrobial properties of nanomaterials contribute significantly to enhanced food safety, inhibiting the growth of microorganisms and reducing food waste.

Yet, as nanotechnology evolves, it also introduces a series of regulatory challenges, from clear labeling to safety assessments and risk evaluations. A key aspect of this evolution is the need for transparent communication to bridge the gap between consumer perception and acceptance of nanotechnology in food. This communication is vital as the industry explores emerging nanotechnologies like nanostructured delivery systems and nanosensors for quality assurance (Maishu and Atemenkeh, 2022). The potential impact of nanotechnology on the food industry is profound, offering the promise of enriched product quality, safety, and superior consumer experiences. As this transformation unfolds, the RTE food industry must uphold ethical and responsible practices and employ proactive regulatory strategies, all underpinned by collaboration among industries, governments, and consumers. This multifaceted approach is essential to harness the full potential of nanotechnology while maintaining safety, ethics, and consumer confidence in the RTE food sector.

15.6.2 Importance of Nanotechnology in Driving Innovation in RTE Foods

The significance of nanotechnology in the RTE food industry is paramount, as it serves as a catalyst for innovation and a solution to crucial challenges. Nanotechnology's ability to precisely manipulate materials at the nanoscale introduces a new era of food production, where nutrient delivery is optimized, sensory attributes are elevated, and shelf life is significantly extended. This transformative technology has the potential to reshape the way consumers interact with RTE foods, offering them not just convenience but also enhanced nutritional profiles, superior taste and texture, and prolonged freshness.

Moreover, nanotechnology introduces a shift toward more sustainable and resource-efficient practices in food production, reducing the need for additives and preservatives. While the regulatory landscape evolves to keep pace with these advancements, the industry is presented with both opportunities and challenges. Through responsible practices, clear communication, and proactive strategies, nanotechnology can uphold safety and ethics, thereby ensuring consumers embrace it. As the RTE food industry strives to meet the demands of modern consumers who seek both convenience and quality, the role of nanotechnology in driving innovation is irrefutable, promising a future where RTE foods are not only quick and easy but also a pinnacle of culinary delight and sustainability.

15.6.3 Potential for Further Research and Development

The chapter's conclusion underscores the vast potential for further research and development in nanotechnology applications within the RTE food industry. As nanotechnology continues to unveil new possibilities, researchers and industry experts have an exciting frontier of exploration ahead. There is an opportunity to delve deeper into the design and engineering of nanoscale delivery systems to optimize nutrient bioavailability, enhance sensory attributes, and create intelligent packaging that adapts to environmental conditions. Such research promises to further elevate the quality, safety, and sustainability of RTE foods.

Sustainability is a key theme in this chapter, and it is an area ripe for exploration. Researchers can focus on developing nanomaterials and applications that reduce waste, minimize the use of additives, and enhance resource utilization. This aligns perfectly with the broader trend toward eco-conscious food production. Additionally, as nanotechnology in food advances, continued investigation into safety, regulation, and consumer acceptance is essential. This research should be guided by ethical considerations, striving for a balance between innovation and responsibility.

In essence, the future of nanotechnology in RTE foods holds great promise, and research and development efforts will be instrumental in unleashing its full potential, ensuring that the industry evolves in a sustainable and consumer-centric manner. This chapter encourages a forward-thinking, research-driven approach to nanotechnology's integration in the RTE food sector, ultimately benefiting both producers and consumers.

REFERENCES

Aguilar-Pérez, Katya M., Gustavo Ruiz-Pulido, Dora I. Medina, Roberto Parra-Saldivar, and Hafiz MN Iqbal. "Insight of nanotechnological processing for nano-fortified functional foods and nutraceutical—opportunities, challenges, and future scope in food for better health." *Critical Reviews in Food Science and Nutrition* 63, no. 20 (2023): 4618–4635.

Altabbaa, Sana, Neharika Ann Mann, Neelam Chauhan, Kumar Utkarsh, Nitika Thakur, and Ghada Abd-Elmonsef Mahmoud. "Era connecting nanotechnology with agricultural sustainability: issues and challenges." *Nanotechnology for Environmental Engineering* 8, no. 2 (2023): 481–498.

Chowdhury, Md Anamul Hasan, Md Ashrafudoulla, Senakpon Isaïe Ulrich Mevo, Md Furkanur Rahaman Mizan, Si Hong Park, and Sang-Do Ha. "Current and future interventions for improving poultry health and poultry food safety and security: A comprehensive review." *Comprehensive Reviews in Food Science and Food Safety* 22, no. 3 (2023): 1555–1596.

de Sousa, Marcos Silva, Andersen Escobar Schlogl, Felipe Ramalho Estanislau, Victor Gomes Lauriano Souza, Jane Sélia dos Reis Coimbra, and Igor José Boggione Santos. "Nanotechnology in packaging for food industry: Past, present, and future." *Coatings* 13, no. 8 (2023): 1411.

Drago, Emanuela, Roberta Campardelli, Margherita Pettinato, and Patrizia Perego. "Innovations in smart packaging concepts for food: An extensive review." *Foods* 9, no. 11 (2020): 1628.

Dwivedi, Khusbu, Ashok Kumar Mandal, Obaid Afzal, Abdulmalik Saleh Alfawaz Altamimi, Ankit Sahoo, Manal A. Alossaimi, Waleed H. Almalki et al. "Emergence of nano-based formulations for effective delivery of flavonoids against topical infectious disorders." *Gels* 9, no. 8 (2023): 671.

Gómez-Guillén, María Carmen, Selene Pérez-García, Ailén Alemán, María Elvira López-Caballero, Carmen G. Sotelo, and María Pilar Montero. "Development of a ready-to-eat fish product enriched with fish oil entrapped in a κ-carrageenan egg white fish protein hydrolysate dry powder." *Foods* 12, no. 11 (2023): 2272.

Huang, Shuting, Varee Tyagi, Bhaswati Bhattacharya, and Yixiang Wang. "Nanotechnology: A new approach to advanced food packaging." In *Nanotechnology Applications for Food Safety and Quality Monitoring*, pp. 241–264. Academic Press, 2023.

Hussein, Abdallah Omar, Tong Woei Yenn, Leong Chean Ring, and Syarifah Ab Rashid. "Potential use of nanotechnology to reduce postharvest spoilage of fruits and vegetables." In *Materials Innovations and Solutions in Science and Technology: With a Focus on Tropical Plant Biomaterials*, pp. 13–23. Springer, 2023.

Jiang, Guihun, Karna Ramachandraiah, and Chen Zhao. "Advances in the development and applications of nanofibers in meat products." *Food Hydrocolloids* (2023): 109210.

Kaur, Kirandeep, Arun B. Prasad, Chou-Yi Hsu, Joseph Oduor Odongo, Satyam Sharma, Yathrib Ajaj, Irfan Rashid Sofi, and Zahid Nabi. "Nanotechnology at workplace: Risks, ethics, precautions and regulatory considerations." In *Modern Nanotechnology: Volume 2: Green Synthesis, Sustainable Energy and Impacts*, pp. 429–453. Cham: Springer Nature Switzerland, 2023.

Kaushani, K. G., N. L. Rathnasinghe, N. Katuwawila, R. A. Jayasinghe, A. H. L. R. Nilmini, and G. Priyadarshana. "Trends in smart packaging technologies for sustainable monitoring of food quality and safety." *International Journal of Research and Innovation in Applied Science*, 7, no.7, 2022: 7–30.

Khan, Shadman, Jonathan K. Monteiro, Akansha Prasad, Carlos D. M. Filipe, Yingfu Li, and Tohid F. Didar. "Material breakthroughs in smart food monitoring: Intelligent packaging and on-site testing technologies for spoilage and contamination detection." *Advanced Materials* (2023): 2300875.

Luciano, Winnie Alencar, Tatiana Colombo Pimentel, Fabrícia França Bezerril, Carlos Eduardo Barão, Vanessa Aparecida Marcolino, Rafaela de Siqueira Ferraz Carvalho, Marcos dos Santos Lima, Olga Martín-Belloso, and Marciane Magnani. "Effect of citral nanoemulsion on the inactivation of *Listeria monocytogenes* and sensory properties of fresh-cut melon and papaya during storage." *International Journal of Food Microbiology* 384 (2023): 109959.

Mahesha, K. N., Apoorva Guddaraddi, Harish Reddy, and Sumanth Kumar G. V. "Unleashing the potential of nanotechnology in agriculture." *Advanced Innovative Technologies in Agricultural Engineering for Sustainable Agriculture* (2023): 97.

Maishu, Samka Peregrine, and Ngwa Celestine Atemenkeh. "Applications of nanosensors as pathogen detectors in packaged food." *Nanotechnology in Intelligent Food Packaging* (2022): 291–312.

Malik, Shiza, Khalid Muhammad, and Yasir Waheed. "Emerging applications of nanotechnology in healthcare and medicine." *Molecules* 28, no. 18 (2023): 6624.

Manful, Maame Ekua, Lubna Ahmed, and Catherine Barry-Ryan. "New approach methodologies (NAMs) for safety testing of complex food matrices: A review of status, considerations, and regulatory adoption." *Trends in Food Science & Technology* (2023): 104191.

Manikandan, Velu, and Sea C. Min. "Roles of polysaccharides-based nanomaterials in food preservation and extension of shelf-life of food products: A review." *International Journal of Biological Macromolecules* (2023): 126381.

Manna, Atanu. "Bacteriocin-mediated food preservation in conjugation with silver nanoparticles: A green approach." *Food Chemistry Advances* (2023): 100464.

Manzoor, Mehnaza, Priyanshu Sharma, Mohd Murtaza, Amit K. Jaiswal, and Sundeep Jaglan. "Fabrication, characterization, and interventions of protein, polysaccharide and lipid-based nanoemulsions in food and nutraceutical delivery applications: A review." *International Journal of Biological Macromolecules* (2023): 124485.

Muthukumar, Janani, Sanjusree Kumar, Nisha Arunachalam, Lokesh Kumar, and Ramalingam Chidambaram. "Quality assurance of packaged food using nanotechnology." In *Nanotechnology Applications for Food Safety and Quality Monitoring*, pp. 341–372. Academic Press, 2023.

Peker, Hatice, and Ahmet Murat Günal. "Sustainable nutrition." *Sustainable Social Development* 1, no. 2 (2023).

Sahani, Shalini, and Yogesh Chandra Sharma. "Advancements in applications of nanotechnology in global food industry." *Food Chemistry* 342 (2021): 128318.

Sahu, M. K., R. Yadav, and S. P. Tiwari. "Recent advances in nanotechnology." *International Journal of Nanotechnology and Nanomedicine* 9, no. 2 (2023): 015–023.

Shruti, Asparshika, Nirgaman Bage, and Pradip Kar. "Nanomaterials based sensors for analysis of food safety." *Food Chemistry* (2023): 137284.

Singh, Akhilesh Kumar, Priti Pal, Brijesh Pandey, Gulden Goksen, Uttam Kumar Sahoo, Jose M. Lorenzo, and Prakash Kumar Sarangi. "Development of "Smart Foods" for health by nanoencapsulation: Novel technologies and challenges." *Food Chemistry*: X 20 (2023): 100910.

Singh, Ragini, and Santosh Kumar. "Intelligent nano-based sensor for quality detection of food products." In *Nanotechnology Advancement in Agro-Food Industry*, pp. 119–156. Singapore: Springer Nature Singapore, 2023a.

Singh, Ragini, and Santosh Kumar. "Regulatory and safety concerns regarding the use of active nanomaterials in food industry." In *Nanotechnology Advancement in Agro-Food Industry*, pp. 269–306. Singapore: Springer Nature Singapore, 2023b.

Singh, Rajesh, Shradha Dutt, Priyanka Sharma, Ashok K. Sundramoorthy, Aman Dubey, Anoop Singh, and Sandeep Arya. "Future of nanotechnology in food industry: Challenges in processing, packaging, and food safety." *Global Challenges* 7, no. 4 (2023b): 2200209.

Thanakkasaranee, Sarinthip, Kambiz Sadeghi, and Jongchul Seo. "Packaging materials and technologies for microwave applications: A review." *Critical Reviews in Food Science and Nutrition* 63, no. 23 (2023): 6464–6483.

Tomić, Ana, Olja Šovljanski, and Tamara Erceg. "Insight on incorporation of essential oils as antimicrobial substances in biopolymer-based active packaging." *Antibiotics* 12, no. 9 (2023): 1473.

Versino, Florencia, Florencia Ortega, Yuliana Monroy, Sandra Rivero, Olivia Valeria López, and María Alejandra García. "Sustainable and bio-based food packaging: A review on past and current design innovations." *Foods* 12, no. 5 (2023): 1057.

Wei, Yu-Ping. "The effect of food safety-related attributes on customer satisfaction of ready-to-eat foods at hypermarkets." *Sustainability* 13, no. 19 (2021): 10554.

16 Ready-to-Eat Recipes and Research Formulations (Functional, Blended, and Herbal) for Special Needy Groups (Children, Elderly, and Athletes) under Innovative Practices in Terms of Fortification, Supplementation, and Enrichment Means

Prajya Arya[1] and Pradyuman Kumar[1]*
[1]Department of Food Engineering and Technology, Sant Longowal Institute of Engineering and Technology (Deemed-to-be University), Longowal, Punjab, India
*Corresponding Author

16.1 INTRODUCTION

The conception of such a food product that extinguishes sudden hunger and provides a satiety feeling gave a term recognized as convenience food. Convenience foods aid in saving time and energy, alter diet patterns, assist in single households, help elderly people from malnutrition, and maintain social, economic, and cultural acceptance (Nakano and Washizu, 2020). Convenience foods contain a wide variety including various processed and semi-processed foods that primarily include ready-to-eat (RTE). The concept of RTE has emerged for overcoming the vast time management issues of today's working generation. RTE food is commonly referred to as precooked or pre-processed meals that aid in saving time and energy, are easy to cook, convenient, and maintain taste and texture of the food. RTE foods cover the vast FMCG sector including different foods such as noodles, rice, cooked meat, poultry, fruits, and vegetables with other processed ingredients (Yang et al., 2022). The food products that can be consumed directly without further processing prior to serving are recognized as

DOI: 10.1201/9781032622408-16

RTE, which commonly includes salads, cooked meats, desserts, sandwiches, meats, and cheese that can be served cold (Taylor and Emmambux, 2008).

RTE foods aid in the present lifestyle drastically due to immense popularity and aid in quick cooking and processing. The increasing trend of RTE foods is more toward working professionals and nuclear families where time management is a huge concern. The alarming situation of reducing the cooking space in households is alternatively replaced with convenient foods. The large concept of RTE foods includes different processing techniques such as ready-to-eat, shelf-stable products, refrigerated, frozen foods, and food that requires minimal heat processing. RTE foods widely encompass the food market in the form of fried snacks, pulsed-based extruded products, processed cheeses, retort technology-based processed foods, frozen foods, relish, bakery products, and other edibles that are made to focus on quenching hunger in no time (Venkatachalapathy et al., 2020). The renowned market and thrust of RTE are commonly popular among college-going students. It is being replaced in the name of quick, easy, and fulfilling hunger solutions in the form of unique food consumption patterns. The focus lies on the concept of developing RTE that tastes good regardless of food consumption habits. RTE-based foods can be an alternative solution to the home cooked food but not all the time (Choi, 2022). The relation of RTE foods varies with the age groups undoubtedly more common in children and teenagers. The popularity of RTE foods in young children is aiding in developing new ways to develop and deliver more nutritious foods to consumers (Papanikolaou and Fulgoni, 2017). RTE food is commonly included in some parts of breakfast as a meal replacer. Cereal-based RTE snacks are rich in nutrients like calcium, fiber, potassium, and different vitamins, but to fulfill the nutrient and energy requirement index of the body there is more incorporation of high-fiber grains in RTE foods (Fulgoni and Buckley, 2015). The consumption of RTE foods may lead to some adverse health effects such as obesity, type 2 diabetes, hypertension, and cardiovascular diseases that aid in the rise in the mortality rate (Priebe and McMonagle, 2016). The balance between RTE foods and alarming health concerns gave rise to the concept of recipe modification and formulation in such a manner that can control bodily disorders while supplying fulfilling nutritional foods (Xu et al., 2016).

This chapter focuses on the various aspects of RTE food and its related aspects such as RTE recipes, formulations, health benefits, requirements of different age groups, and innovative practices. The various types of RTE-based food available in the market are represented in Figure 16.1.

16.2 RTE TYPES

The increasing demand for RTE foods forced many scientists in the food community to look for various available food source matrixes. The wide availability of choices for selecting RTE food sources acts and aids in the formulations of different varying RTE recipes. RTE types of foods are based on consumer preferences that are more conveniently handled, served, and consumed along with taste, health, and price-related aspects that help in maintaining market fulfillment (Anusha et al., 2020). The wide varieties of RTE foods available in the market are due to certain changes in lifestyle like less time for preparing meals, and hectic and overloaded work schedules. The increasing factors such as employment status, income level, work–life balance, and convenience orientation merged the mindset more toward the consumption of RTE foods. These changes increase the demand for RTE foods along with different types of options seeking for packaged foods (Vijayeta, 2015). The vast area for RTE foods is primarily looking for traditional recipes along with their blend of new technologies to make RTE more available, nutritious, and shelf stable. RTE foods fall under a wide variety of available options such as bottled rice-fish (Bigueja et al., 2022), bakery items, hard-boiled sugar confectionery, savories (Jonnalagadda et al., 2004), extruded products (Reddy et al., 2014), breakfast cereals (Kumari et al., 2019), and many more.

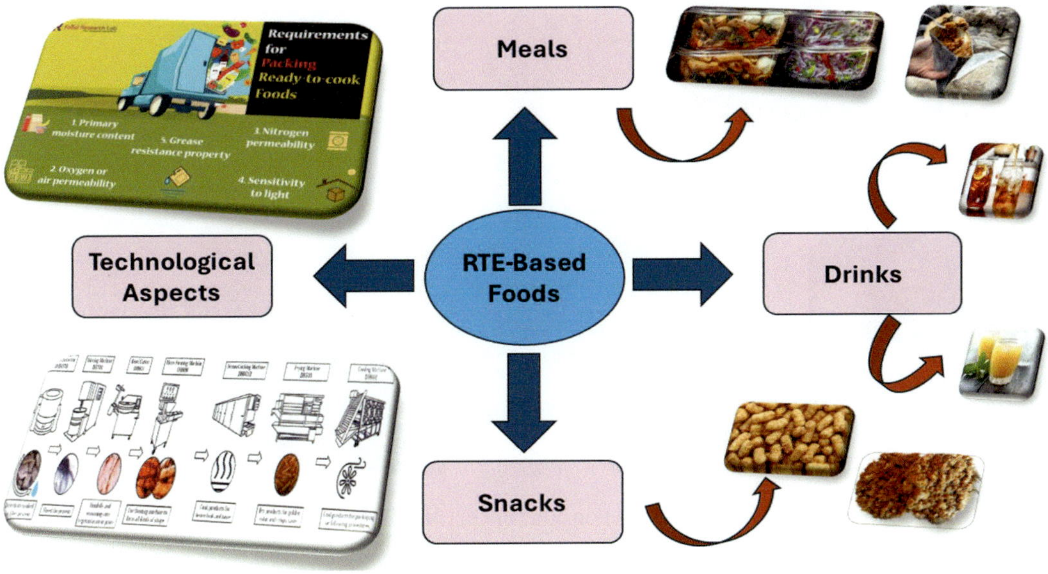

FIGURE 16.1 Pictorial representation of the various types of RTE foods in market.

16.3 RTE RECIPES

The requirement of various RTE foods on the market shelves can be fulfilled by different varieties that include formulations, standardization, process development, and process execution. The RTE food recipes depend on the type of food, targeted population (age group that consumes RTE food), shelf life, categories, and raw materials. RTE food recipes are based on a few important criteria such as health, mood, convivence, sensory appeal, natural content, price, weight control, familiarity, and ethical concerns (Chaudhury, 2010). RTE food recipes are primarily affected by the type of raw material selected and the technique adopted for the RTE food formulations. The extruded type RTE food recipe is mainly concerned with the type of different cereal grains such as wheat, millet, cereals, pulses, and a mixture of different food grains (Reddy et al., 2014). The millet-based RTE foods recipe is based on the interaction of the millet starch component with the technology used for desired food types. Millet-based RTE recipes impose difficulties in structural properties that cause the intended usage of the other ingredients in recipes (Balasasirekha and Santhoshini, 2016). Another formulated RTE-based parboiled rice that was enriched with iron and folic acid under the effect of milling. The better bioaccessibility and bioavailability of such micronutrients were observed with *in vitro* studies revealing about iron 14.7–32.1% and folic acid 55.1–91.9% with glycemic index 56–69 (Wahengbam et al., 2019). The study of cranberry juice concentrates (CJC) was observed against three pathogenic bacteria: *Escherichia coli* O157:H7, *Listeria monocytogenes*, and *Salmonella typhimurium* with pH level 2.4 in the presence of organic acids. CJC showed a reduction in the log cycle of *E. coli* O157:H7, *L. monocytogenes,* and *S. typhimurium* with 2.5, 1.8, and 5 log reduction, respectively, without altering sensory properties of RTE juice (Harich et al., 2017).

16.4 DIFFERENT RTE FORMULATIONS

The concept of food formulation falls under the wide range umbrella of different aspects of food engineering, food processing, food technology, and their combinations. The word "formulation" in

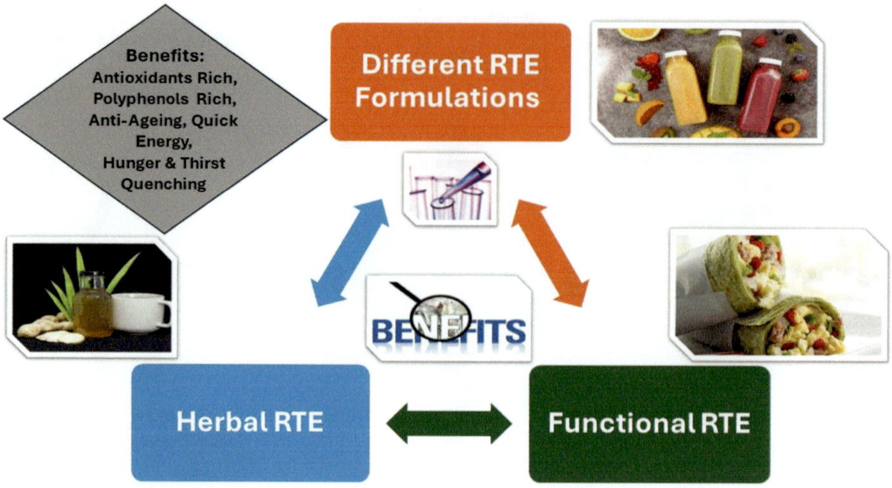

FIGURE 16.2 Depiction of different formulation and benefits of RTE foods.

food sectors implies various activities of creating, designing, developing, and implementing food products with some functional and financial motives to fulfill desirable ends (Pathania et al., 2021). The continuous addition in consumer demand for healthy and convenient food has mended the ways of research to solve the query. This has led to increased production, formulation, and nutritional fulfillment of RTE-based foods in common man life by up to 30% in the past 10 years (de Oliveira et al., 2015). There are diverse types of formulations being formulated and tested across the world that provide a review of the requirements of RTE-based foods and their market values. Such a combination of maize, pea, and anchote flours was formulated in six different types of diets for fulfilling the nutritional supplements under the name of complementary food. Among the six types of diet, five types contain the most suited combination of macronutrients and micronutrients as per CODEX standards that predict its better market stability in RTE-formulated foods (Gemede, 2020). Another recent study revealed the food-to-food fortification of RTE-based cereals-based porridges that aids elderly people in better uptake and retention of iron and zing from a plant-based meal. The meal produced from the extrusion cooking reduced the phytate content of porridge by 30–35% which enhances bioaccessibility of micronutrients (Adetola et al., 2022). Among the different formulations of RTE-based foods, it is commonly fortified with functional bioactive compounds, their blend or herbal mixture of some nutritionally rich ingredients as discussed in below-mentioned sections and shown in Figure 16.2.

16.4.1 Functional RTE

The upcoming trend in the food market loaded with various ingredients in RTE foods falls under the category of "healthier food" as packaged. This drift gains a good commercial interest in the retail marketplace by including and identifying a part of "clean label." This includes food with the least number of artificial agents and a small list of mostly natural ingredients used in the preparation of food (Koyratty et al., 2014). Another vogue represents the modified RTE foods with less amount of sugar, fat, and salt that normally trigger several disorders in human metabolic systems. These methods and processing factors of preparation of RTE foods aid customers to become more aware of the food consumption trend and health aspects of such foods. Institute of Food Technologists (IFT) has proposed a clear meaning of clean food which implies that preparing a product with fewer ingredients and these ingredients must be recognized by customers and foreseen as wholesome (IFT, 2021). There are some of the functional ingredients that are commonly used under the label of clean food but also provide shelf life and could act as taste enhancements. These ingredients include carrot, leafy vegetables, spinach, cole, wax gourd, eggplant, carrot, root vegetable, and mustard leaf. The presence of various bioactive functional compounds makes the food matrices more nutritious and healthier in view of RTE base food products (Bodie et al., 2024). Another study imposed the usage of functional bioactive compounds from black rice anthocyanin content, and antioxidant content as potential sources in baked RTE foods that enhance customer acceptance and nutritional benefits (Das et al., 2023).

16.4.2 Blended RTE

The mixture of a few nutritionally rich ingredients aids in the betterment of health via RTE foods. A similar study was conducted for spray-dried encapsulated ascorbic acid powder and iron gluconate powder pasta modification with apple pomace finger millet-based pasta. This research work focused on the incorporation of spray-dried encapsulated ascorbic acid and iron gluconate powder in modified pasta. The fortified pasta showed that retention of iron content was 23.47 ± 0.08 mg/100 g and ascorbic acid was 87.51 ± 0.11 mg/100 g with reduced cooking loss (Kumari et al., 2023). Another blend of RTE-based iron-rich extruded snacks was prepared from barnyard millet, defatted soy flour, amla, and rice flour. The retention of nutritional components in extruded snacks such as 15.71 mg/100 g of iron and 18.91 g/100 g of protein shows better consumer acceptance and sensory attributes in regard to RTE foods (Suri et al., 2020). A blend of green gram flour and bhimkol flour has been utilized for RTE-based breakfast cereal production. This study comprises various mixture proportions and processing parameters. The nutritional content of extrudate for vitamin B1 was 0.847 μg/g and dietary fiber was 5.16%, which acts as RTE-based food as well as weaning foods (Borah et al., 2023). The formulation of RTE-based high-energy paste prepared from a homogeneous mixture of peanut paste, edible oil, icing sugar, cereals, legume flour, milk powder, and vitamin-mineral premix for severely acute malnutrition. The prepared energy paste was quite stable for up to 60 days in 40 μm metallic foil laminated low-density polyethylene among varying packages, their thickness, and material that aids in enhancing the shelf life of energy paste (Raigar and Mishra, 2022).

16.4.3 Herbal RTE

Production of the desirable and acceptable herbal ingredients-based RTE food is an unfeigned query that needs to be addressed properly. As the word “herbal” shows on a food package, it gives reliable information and trust to the consumer that food is prepared from some of the knowledge, skills, and practices that were based on some relevant theories, beliefs, and experiences indigenous to different culture that aids in control, prevention of some metabolic disorder via foods (Bozzetti, 2003). The usage of natural metabolite from plant extracts exerts effects on some specific cellular mechanism that impacts certain changes in metabolic pathways by modifying food matrices in which herbal extracts have been added (Akkol et al., 2022). The usage of different herbal ingredients in RTE food enhanced the sale of the food products. These ingredients include extracts or oils of lavender, rose, tea tree, sandalwood, bilberry, cranberry, elderberry, ginkgo leaf, grape seed, turmeric, and saw palmetto. These ingredients are sometimes diluted with low-quality extracts or other chemical preservatives that might hamper the RTE foods' quality under the tag of herbal ingredients (Gafner et al., 2023). The consumption rate of herbal ingredients-based RTE foods has significantly increased after the COVID-19 pandemic creating awareness in people more toward their health via foods (Pin and Daniel, 2023).

16.5 HEALTH BENEFITS OF RTE-BASED FOODS

There are certain checkpoints where crucial statement plays a vital role related to the quality and health aspects of such foods. Among the various types of RTE foods, there are certain types of nutritional benefits that ascertain human health from farm to fork. Among the various ingredients, antioxidants, phytochemicals, anthocyanins, and bioactive compounds aid in seeking nutrition and prioritizing the taste, odor, flavor, and color of the food (Calligaris et al., 2022). Among such studies, one research focused on enhancing the diet pattern of elderly people by modifying the soup matrices and enriching them with certain agricultural commodities such as brown rice, pumpkin, sweetcorn, and red tilapia followed by oil and water. There were many formulations that were tried and analyzed and, among them, the seventh formulation provided the most suitable results in view of nutritional uptake and sensory perception. The research also predicted good keeping quality of the soup for up to 6 months at 25°C (Sinchaipanit et al., 2023). Another study revealed that tomato and carrot puree contain antioxidants, natural color, and bioactive compounds when processed with high-pressure treatment. The retention of more than 90% of ascorbic acid was reported in one study as compared to thermally processed puree in RTE category food (Patras et al., 2010).

16.6 RTE REQUIREMENT FOR VARIOUS AGE GROUPS

The choice and preferred taste of food vary as per age group, location, gender, and type of food product. Today's generation, regardless of age group, is generally dependent on RTE foods that impart convenience in preparing quick meals due to hectic lifestyles. The authenticity of the subject, i.e. which age group requires which type of RTE foods, could be resolved under the umbrella of age categories such as neonates, children, young people, athletes, and elderly, location wise population such as rural or urban (Choi, 2022) as shown in Figure 16.3. The preference of RTE-based foods was primarily observed in the younger age group, as they tend to purchase more of the packaged food items as compared to other ones. This shift in behavior is triggered by daily lifestyle, eating out behavior, eating habits, food choices, and location and demographic changes. A study conducted by

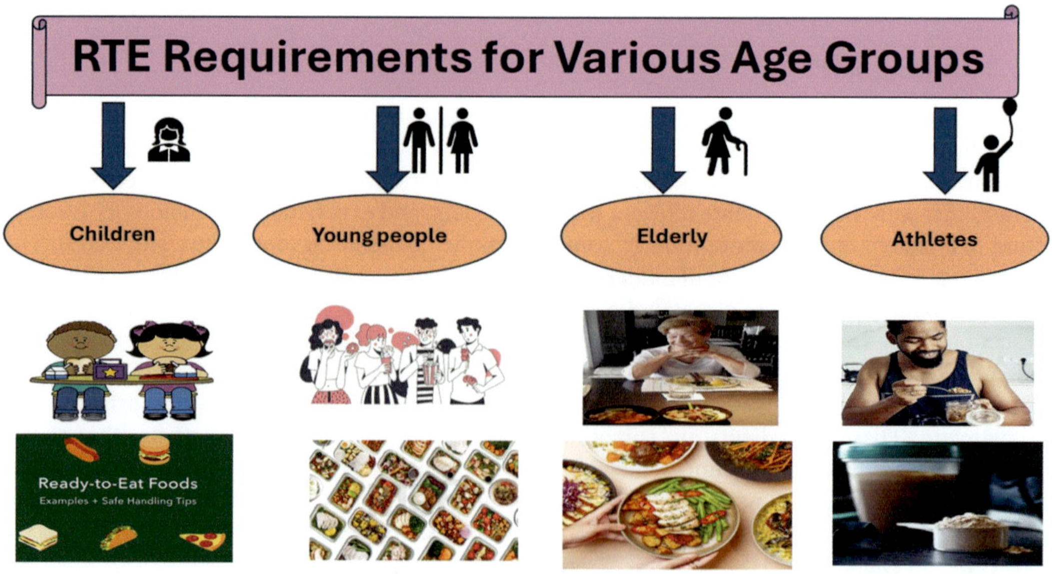

FIGURE 16.3 Representation of RTE-based foods for people of various ages.

Choi (2022) observed the purchase of RTE-based food items in college-age groups in an urbanized area. This research depicted the higher intake of RTE-based foods that may cause some serious health problems if the RTE food was not up to the quality criteria. Another age group ranging from 0.5 to 17 years of children was studied for the preference of RTE foods based on cereals revealing about much higher intake of carbohydrates, and sugar followed by some of the micronutrients varying from food types. RTE cereals-based foods contain a good amount of thiamine, niacin, folate, zinc, and potassium which enhances the overall quality of foods and aids in consumer health (Smith et al., 2019).

While looking at the requirements of athletes in RTE-based foods or drinks, there is a wide range of nutritional aspects that need to be looked at. Sprout nutrition is always related to quick energy, maintaining performance, reducing fatigue, suppressing the risk of disease and injury, maintaining an ionic balance of the body, and proper muscle mass (Purcell and Section, 2013). Energy drinks also fall under the category of RTE foods that alter consumer energy very quickly depending on male and female. A recent study revealed the consumption of energy drinks by athletes that predicted a consumption rate of 41.7% by aiming to control body weight and energy fulfillment simultaneously. Generally, energy drinks contain caffeine for ergogenic benefits with caffeine dosage ranging from 3 to 9 mg/kg which provides a quick boost of energy (Hoffman, 2010). Sometimes energy drinks aid in reducing body weight by increasing lipolysis, fat oxidation, and reduction of glycogen breakdown when occurring with thermogenic effects, and enhanced metabolic rate aids in fat loss (Haller et al., 2004).

16.7 RTE WITH INNOVATIVE PRACTICES

The concept of using various bioactive ingredients in different food formulations aims to enhance their nutritional value. The addition of bioactive ingredients is carried out under different names: fortification, supplementation, and enrichment. Greater consumption of RTE foods has also raised awareness toward the better uptake of nutrients from foods. This trend made the consumers more aware and alert toward better food choices and perceived quality that aids in the upliftment of the RTE-based food market up to great value. The addition of bioactive compounds in RTE-based foods must be checked against their stability in packaged condition that is stable at room temperature or refrigeration and must not degrade during minimal processing (Chandrasekara and Shahidi, 2015). Fortified foods are widely consumed in simplest and practical forms such as the addition of vitamin A in margarine in early 1920 in Denmark (Cowgill, 1964), vitamin D in milk in the US in 1930 for prevention of rickets in children (Dary and Mora, 2002). There are many other trial foods and marketed foods that initiate the use of fortification in food matrices that aid in the nutritional quality of food. Along with that, there are some genuine concerns related to RTE foods and fortification such as whether they are bioavailable, higher cost, self-perception, self-prescription, enhanced risk of toxicity, and degradation during storage (Gupta, 2014). The study was conducted on various age groups including toddlers, children, adolescents, and adults to study consumption of fortified RTE cereal-based food at different levels. This research revealed about more better energy level and utilization of nutrients in estimated average requirement (EAR) when executed with optimized fortification (Smith et al., 2020).

While dealing with RTE-based foods, another important aspect of nutritional enrichment coins under the name of "supplementation" which is an effective strategy for considering children's nutritional fulfillment such as iron supplementation for iron deficiencies (Ramchandran et al., 2020). Another supplementation was achieved by RTE-based foods for multiple micronutrients (for the overall development of children), zinc supplementation (suppressed diarrhea) (Yakoob et al., 2011), and Take-Home Ration (THR) prepared from a blend of soy flour and gram flour providing good nutritional benefits (Nackers et al., 2010). This aspect of food generated a new term as Ready-to-Use Supplementary Food (RUSF) which contains a vivid range of nutrients (Ramchandran et al., 2020).

A study was conducted using population sample size from various nations including India, Kenya, Zambia, and Cambodia for severe acute malnutrition in children. It stated that supplementation of ready-to-use therapeutic food (RUTF)-based foods aids in gaining good body weight and better health (Schoonees et al., 2019).

Cereal-based RTE food was prepared with a lesser glycemic index and starch blend of chokeberry pomace powder (CPP) to enhance dietary fiber and polyphenol content of the co-extruded product with varying processing parameters. Among the varying CCP percentages, 25% revealed better techno-functional aspects with compelling sensory attributes. This predicted a good number of polyphenols available for absorption are quite high with minimal damage with slight degradation during the extrusion cooking method (Schmid et al., 2022). Another extrusion-based RTE food product was developed by utilizing polyphenol content from apple pomace to enhance antioxidant content. The retention of phenolic compounds was observed in the range of 79.9–97.1% in an unprocessed mixture while it ranged from 57 to 71% for chlorogenic acid, 56 to 70% for rutin, 38 to 51% for p-coumaric acid, to 46 to 76% phloridzin. Such retention of phenolic compounds in extruded products depicts high-temperature stability and shelf-life retention of the products and increases the nutritional supplement of the various age groups (Leyva-Corral et al., 2016). Here are some of the crucial novel RTE-based food products enlisted under Table 16.1 depicting their various aspects of nutritional profiles.

16.8 CONCERNS RELATED TO RTE FOODS

There are some of the genuine concerns related to RTE foods such as the allowable limit of ingredients, flavor, color, growth of microorganisms, and base ingredients. The type and concentration of allowed color were observed for sunset yellow which was maximum in sweet meat (18,767 ppm), non-alcoholic beverages (9450 ppm), hard boiled sugar confectionery (3811 ppm) (Jonnalagadda et al., 2004). An RTE-based food sampling conducted in UAE-based countries during "Tajrat" revealed about the presence of *Listeria* spp., *Staphylococcus aureus*, and coliform bacteria in 225, 43% and 31% samples, respectively. The fat content of the food was also found in higher limits ranging from 2.6 to 30 g/100g which impose serious threats to health when consumed frequently (Mostafa et al., 2024). The growth of *Listeria monocytogenes* was prevalent in some of the commonly occurring RTE-based foods such as starters with or without mayonnaise pasta, rice-based meals, and meat and fish foods produced by an Italian industry. The observed prevalence of *L. monocytogenes* was 17.4% with 23 samples, especially with mayonnaise showing a much higher prevalence of 56.7% (Tirloni et al., 2024). Another study showed the presence of multi-drug-resistant variants of *S. aureus* showing phenotypic enterotoxigenic virulence in RTE shrimps imposing public health risk in Delta State Nigeria causing health damage (Beshiru et al., 2021). Therefore, consumption of RTE-based foods must be checked against certain health deteriorating factors.

16.9 CONCLUSION

The concept of RTE-based foods helps many families in retaining their work-life balance. There is constant research going on in the scientific community that fulfills market demand and supply chain management. Such food inventions are wide in category for different age groups. The nutritional requirements of various people, yet less meal preparation time acted as an added advantage for incorporating bioactive compounds in such foods. The awareness of updated consumers tends toward the production of better and more convenience foods. This article focused on various RTE formulations that target different sectors, age groups, and food preferences among people. Such a wide range of RTE-based foods helps consumers to look for a better lifestyle on a day-to-day basis.

TABLE 16.1
Representation of Various Nutritional Profiles of RTE-Based Products from Different Raw Materials

S. No.	Base Material	Category	Inference	Reference
1.	Black rice bran	Cereal-based RTE pasta	Increased anthocyanin, fiber, protein in pasta. Lesser water uptake, lesser cooking time, minimized pasta firmness and pasta sticking.	Sethi et al., 2020
2.	Black rice extract powder	RTE-based bread	4% fortification of flour showed better density of loaf volume with reduced elasticity of bread with increased rate of digestion and absorption.	Sui et al., 2016
3.	Black rice protein powder	RTE-based cakes	Enhanced specific gravity of cake batter, lesser viscosity, increased chewiness darker crumb, more accepted sensory attributes with only 10–30% addition of black rice protein powder.	Mau et al., 2017
4.	Tree bean powder	RTE extruded snacks	Better phenolics, ascorbates, flavonoids, and suppressed glutathione. Presence of certain essential amino acids such as leucine, cystine, tyrosine, and methionine estimated by HPLC with 10% tree bean powder incorporation in snacks.	Sahoo et al., 2022
5.	Composite flour (maize flour [MF], finger millet [FM], defatted soy [DS], and elephant foot yam [YP])	RTE extruded product	Most suite blend of MF:FM:DS:YP in ratio of 40:30:20:10 revealed better techno-functional and sensory attributes such as expansion ratio (3.29), bulk density (0.14 g/cm), hardness (11.94 N), water absorption index (6.13 g/g), water solubility index (17.07%)	Sahu et al., 2021
6.	Beetroot powder, Chekurmanis leaf powder	RTE extruded snacks	Better antioxidant profile, high dietary fiber, better organoleptic properties	Shilpa et al., 2023
7.	Soy-fortified wheat flour	RTE extruded snacks	Better estimation of Dincer number for steamed flat cold extrudates puffed in fluidized air at temperature 220°C and 3.95 m/s velocity was determined 276,223.8. This provided a good estimation of puffing time and moisture content	Pardeshi and Chattopadhyay, 2010
8.	Mixture of rice:ragi:defatted groundnut:defatted soybean:Bengal gram (60:10:10:10:10)	RTE extruded snacks	A higher content of protein was estimated in the snacks along with expansion ratio (3.18), bulk density (0.658 g/cm), hardness (24.21 N), water absorption index (5.61 g/g), water solubility index (30.24%), fracturability (10.94 N), crispness (55.09 N).	Athira et al., 2022
9.	Folic acid fortification	RTE + ECGP (enriched cereal grain product)	Better retention of red blood cell folate concentrations that prevent pregnancy miscarriages with achievable intake of 400 μg/day and reduce neural tube damage. It could be an alternative and better source for folic acid supplementation maintaining better health and nutrition and child growth	Crider et al., 2018
10.	Fortification with liquid whey and barley flour	RTE buns	Buns were of better texture with availability of superior nutrition, lesser microbial spoilage, and higher antioxidant contents.	Dobhal et al., 2024
11.	Phosphorous, aluminum-enriched wheat flour	RTE tortillas	Helpful in preventing the cardiovascular, and renal malfunctions by providing recommended intake of aluminum and phosphorus via fortified wheat flour tortillas as compared to corn tortillas	Chiang et al., 2023

REFERENCES

Adetola, O., K. Duodu, and J. Taylor. "Food-to-food fortification of ready-to-eat cereal-based porridges: A potential sustainable food-based strategy to improve bioavailable iron and zinc in plant-based diets." *Current Developments in Nutrition* 6 (2022): 498–499. http://dx.doi.org/10.1093/cdn/nzac077.001

Akkol, K., H. B. Esra, T. H. Barak, M. Aschner, G. Ş. Karatoprak, H. Khan, and Y. Hussain. "Herbal ingredients in the prevention of breast cancer: Comprehensive review of potential molecular targets and role of natural products." *Oxidative Medicine and Cellular Longevity* 2022 (2022): 1–23. https://doi.org/10.1155/2022/6044640

Anusha, A. C. S., R. Tunung, C. Kavita, A. Ribka, and M. Chandrika. "Ready-to-eat food consumption practices, food safety knowledge and relation to gender and education level of consumers in Kuala Lumpur, Malaysia." *Food Research* 4 (2020): 1537–1544.

Athira, K., G. K. Rajesh, N. Sagarika, and M. G. Anjaly. "Development and optimization of process variables of protein enriched rice based RTE Food Products." *Indian Journal of Ecology* 49, no. 6 (2022): 2336–2342.

Balasasirekha, R., and P. Santhoshini. "Development of RTE millet mixes with dehydrated vegetable peel." *International Journal of Current Microbiology and Applied Science*, 5, no. 11 (2016): 24–37.

Beshiru, A., I. H. Igbinosa, and E. O. Igbinosa. "Characterization of enterotoxigenic *Staphylococcus aureus* from ready-to-eat seafood (RTES)." *LWT – Food Science and Technology* 135 (2021): 110042.1–10. https://doi.org/10.1016/j.lwt.2020.110042

Bigueja, M. C., G. S. Sales, I. P. Daet, and C. C. Bigueja, "Optimization of bottled rice-fish ready-to-eat (RTE) food." *United International Journal for Research and Technology* 03 (08) (2022), 1–7.

Bodie, A. R., L. A. Wythe, D. K. Dittoe, M. J. Rothrock Jr, C. A. O'Bryan, and S. C. Ricke. "Alternative additives for organic and natural ready-to-eat meats to control spoilage and maintain shelf life: Current perspectives in the United States." *Foods* 13, no. 3 (2024): 464.1–21. https://doi.org/10.3390/foods13030464

Borah, A., D. K. Das, R. Mukhopadhyay, and C. L. Mahanta. "Investigation of low amylose rice extrudates blended with germinated green gram and *bhimkol* flour." *Applied Food Research* (2023): 100315.1–8. https://doi.org/10.1016/j.afres.2023.100315

Bozzetti, F. "Nutritional issues in the care of the elderly patient." *Critical Reviews in Oncology/Hematology* 48 (2) (2003): 113–121. https://doi.org/10.1016/j.critrevonc.2003.08.001

Calligaris, S., M. Moretton, S. Melchior, A. C. Mosca, N. Pellegrini, and M. Anese. "Designing food for the elderly: The critical impact of food structure." *Food and Function* 13 (12) (2022): 6467–6483.

Chandrasekara, A., and F. Shahidi. "The use of antioxidants in ready-to-eat (RTE) and cook-chill food products." In *Handbook of Antioxidants for Food Preservation*, Woodhead Publishing Series in Food Science, Technology and Nutrition, pp. 433–446. Woodhead Publishing, 2015. http://dx.doi.org/10.1016/B978-1-78242-089-7.00018-X

Chaudhury, R. "Determinants of consumer behavior in buying RTE foods." *Journal of Business and Retail Management Research*, 5, no. 1 (2010): 1–11.

Chiang, K. C., R. A. Yokel, J. M. Unrine, K. Kalantar-Zadeh, and A. Gupta. "Dietary burden of phosphorus and aluminum concentrations of ready-to-eat wheat flour tortillas exceed that of corn tortillas: Implications for patients with renal or cardiovascular disease." *MedRxiv* (2023): 2023-09.1-12. https://doi.org/10.1101/2023.09.09.23295298

Choi, J. "A pilot study on RTE food purchasing and food-related behaviors of college students in an urbanized area." *International Journal of Environmental Research and Public Health* 19, no. 6 (2022): 3322.1–10. https://doi.org/10.3390%2Fijerph19063322

Cowgill, G. R. "Jean Baptiste Boussingault – a biographical sketch (February 2, 1802–May 11, 1887)." *Journal of Nutrition* 84, no. 1 (1964): 1–9.

Crider, K. S., Y. P. Qi, O. Devine, S. C. Tinker, and R. J. Berry. "Modeling the impact of folic acid fortification and supplementation on red blood cell folate concentrations and predicted neural tube defect risk in the United States: Have we reached optimal prevention?" *American Journal of Clinical Nutrition* 107, no. 6 (2018): 1027–1034.

Dary, O., and J. O. Mora. "Food fortification to reduce vitamin A deficiency: International Vitamin A Consultative Group recommendations." *Journal of Nutrition* 132, no. 9 (2002): 2927S-2933S.

Das, M., U. Dash, S. S. Mahanand, P. K. Nayak, and R. K. Kesavan. "Black rice: A comprehensive review on its bioactive compounds, potential health benefits and food applications." *Food Chemistry Advances* 3 (2023): 100462.1–10. https://doi.org/10.1016/j.focha.2023.100462

de Oliveira, K. Á. R., J. P. de Sousa, J. A. da Costa Medeiros, R. C. B. Q. de Figueiredo, M. Magnani, J. P. de Siqueira Júnior, and E. L. de Souza. "Synergistic inhibition of bacteria associated with minimally processed vegetables in mixed culture by carvacrol and 1,8-cineole." *Food Control* 47 (2015): 334–339.

Dobhal, A., P. Awasthi, S. Srivastava, A. Dutta, N. C. Shahi, A. Kumar, A. H. Ahmad, S. Sharma, and S. Kumar. "Impact of liquid whey and barley flour fortification on the physicochemical, microbial, and sensory characteristics of buns stored in different packaging materials." *Measurement: Food* 13 (2024): 100128.1–8. https://doi.org/10.1016/j.meafoo.2023.100128

Fulgoni III, V. L., and R. B. Buckley. "The contribution of fortified ready-to-eat cereal to vitamin and mineral intake in the US population, NHANES 2007–2010." *Nutrients* 7, no. 6 (2015): 3949–3958.

Gafner, S., M. Blumenthal, S. Foster, J. H. Cardellina, I. A. Khan, and R. Upton. "Botanical ingredient forensics: Detection of attempts to deceive commonly used analytical methods for authenticating herbal dietary and food ingredients and supplements." *Journal of Natural Products* 86, no. 2 (2023): 460–472. https://doi.org/10.1021/acs.jnatprod.2c00929

Gemede, H. F. "Nutritional and antinutritional evaluation of complementary foods formulated from maize, pea, and anchote flours." *Food Science and Nutrition* 8, no. 4 (2020): 2156–2164. https://doi.org/10.1002/fsn3.1516

Gupta, A. "Fortification of foods with vitamin D in India." *Nutrients* 6, no. 9 (2014): 3601–3623. https://doi.org/10.3390/nu6093601

Haller, C. A., P. Jacob III, and N. L. Benowitz. "Enhanced stimulant and metabolic effects of combined ephedrine and caffeine." *Clinical Pharmacology and Therapeutics* 75, no. 4 (2004): 259–273.

Harich, M., B. Maherani, S. Salmieri, and M. Lacroix. "Antibacterial activity of cranberry juice concentrate on freshness and sensory quality of ready to eat (RTE) foods." *Food Control* 75 (2017): 134–144. https://doi.org/10.1016/j.foodcont.2016.11.038

Hoffman, J. R. "Caffeine and energy drinks." *Strength and Conditioning Journal* 32, no. 1 (2010): 15–20. https://doi.org/10.1519/SSC.0b013e3181bdafa0

IFT. "What is clean label?" 2021. Available online: www.ift.org/news-and-publications/blog/2018/november/what-is-clean-label (accessed on 23 May 2023).

Jonnalagadda, P. R., P. Rao, R. V. Bhat, and A. N. Naidu. "Type, extent and use of colours in ready-to-eat (RTE) foods prepared in the non-industrial sector – a case study from Hyderabad, India." *International Journal of Food Science and Technology* 39, no. 2 (2004): 125–131.

Koyratty, N. S., B. A. Bibi, and S. A. Neeliah. "Food additive control: A survey among selected consumers and manufacturers." *British Food Journal* 116, no. 2 (2014): 353–372. https://doi.org/10.1108/BFJ-05-2012-0125

Kumari, M., P. Arya, S. K. Khatkar, and P. Kumar. "Development and characterization of apple pomace and finger millet-based pasta enriched with encapsulated micronutrient." *Food Chemistry Advances* 3 (2023): 100551.1–10. https://doi.org/10.1016/j.focha.2023.100551

Kumari, R., K. Singh, R. Singh, N. Bhatia, and M. S. Nain. "Development of healthy ready-to-eat (RTE) breakfast cereal from popped pearl millet." *Indian Journal of Agricultural Sciences* 89, no. 5 (2019): 877–881.

Leyva-Corral, J., A. Quintero-Ramos, A. Camacho-Dávila, J. de Jesús Zazueta-Morales, E. Aguilar-Palazuelos, M. G. Ruiz-Gutiérrez, C. O. Meléndez-Pizarro, and T. de Jesús Ruiz-Anchondo. "Polyphenolic compound stability and antioxidant capacity of apple pomace in an extruded cereal." *LWT – Food Science and Technology* 65 (2016): 228–236. https://doi.org/10.1016/j.lwt.2015.07.073

Mau, J.-L., C.-C. Lee, Y.-P. Chen, and S.-D. Lin. "Physicochemical, antioxidant and sensory characteristics of chiffon cake prepared with black rice as replacement for wheat flour." *LWT – Food Science and Technology* 75 (2017): 434–439.

Mostafa, H., A. S. Al Dhaheri, J. Feehan, and N. M. K. Yousif. "Assessment of consumer demographics and food safety risks associated with ready-to-eat (RTE) homemade foods purchased online in the UAE." *Heliyon* 10, no. 3 (2024): 1–13. https://doi.org/10.1016/j.heliyon.2024.e24624

Nackers, F., F. Broillet, D. Oumarou, A. Djibo, V. Gaboulaud, P. J. Guerin, B. Rusch, R. F. Grais, and V. Captier. "Effectiveness of ready-to-use therapeutic food compared to a corn/soy-blend-based pre-mix for the treatment of childhood moderate acute malnutrition in Niger." *Journal of Tropical Pediatrics* 56, no. 6 (2010): 407–413.

Nakano, S., and A. Washizu. "Aiming for better use of convenience food: An analysis based on meal production functions at home." *Journal of Health, Population and Nutrition* 39 (2020): 1–16.

Papanikolaou, Y., and V. L. Fulgoni III. "Certain grain foods can be meaningful contributors to nutrient density in the diets of US children and adolescents: Data from the national health and nutrition examination survey, 2009–2012." *Nutrients* 9, no. 2 (2017): 160.1–20.

Pardeshi, I. L., and P. K. Chattopadhyay. "Hot air puffing kinetics for soy-fortified wheat-based ready-to-eat (RTE) snacks." *Food and Bioprocess Technology* 3 (2010): 415–426. https://doi.org/10.1007/s11947-008-0100-z

Pathania, S., C. Bhatia, and B. K. Tiwari. "Food formulation and product development." *Food Formulation: Novel Ingredients and Processing Techniques* (2021): 1–4. https://doi.org/10.1002/9781119614760.ch1

Patras, A., N. P. Brunton, and F. Butler. "Effect of water immersion and sous-vide processing on antioxidant activity, phenolic, carotenoid content and color of carrot disks." *Journal of Food Processing and Preservation* 34, no. 6 (2010): 1009–1023. http://dx.doi.org/10.1111/j.1745-4549.2009.00434.x

Pin, C. H., and N. Daniel. "Processing of herbal-based natural products and functional foods: a review." *Sains Malaysiana* 52, no. 9 (2023): 2587–2598. http://dx.doi.org/10.17576/jsm-2023-5209-09

Priebe, M. G., and J. R. McMonagle. "Effects of ready-to-eat-cereals on key nutritional and health outcomes: a systematic review." *PLoS One* 11, no. 10 (2016): e0164931.1–11.

Purcell, L. K., Canadian Paediatric Society, and Paediatric Sports and Exercise Medicine Section. "Sport nutrition for young athletes." *Paediatrics and Child Health* 18, no. 4 (2013): 200–202. https://doi.org/10.1093/pch/18.4.200

Raigar, R. K., and H. N. Mishra. "Ready-to-eat high-energy food paste stability and shelf life prediction in different packaging materials." *Applied Food Research* 2, no. 2 (2022): 100230.1–6. https://doi.org/10.1016/j.afres.2022.100230

Ramchandran, P., N. Patel, H. Swaminathan, A. Jadhav, P. Shinde, P. Karnik, D. Parthasarathy, and N. G. Shah. "Efficacy of micronutrient fortified supplementary foods (ready to cook and ready to eat) and take-home ration (THR) in children with moderate acute malnutrition (MAM) using community based approach." *Pediatric Oncall Journal* 17, no. 1 (2020): 1–7. https://doi.org/10.7199/ped.oncall.2020.19

Reddy, M. K., A. Kuna, N. L. Devi, N. Krishnaiah, C. Kaur, and Y. Nagamalleswari. "Development of extruded ready-to-eat (RTE) snacks using corn, black gram, roots and tuber flour blends." *Journal of Food Science and Technology* 51 (2014): 1929–1937.

Sahoo, M. R., A. Kuna, M. P. Devi, M. Sowmya, and M. Dasgupta. "Fortification of ready-to-eat extruded snacks with tree bean powder: Nutritional, antioxidant, essential amino acids, and sensory properties." *Journal of Food Science and Technology* (2022): 1–10. https://doi.org/10.1007/s13197-021-05251-w

Sahu, C., and S. Patel. "Optimization of maize–millet based soy fortified composite flour for preparation of RTE extruded products using D-optimal mixture design." *Journal of Food Science and Technology* 58, no. 7 (2021): 2651–2660. https://doi.org/10.1007/s13197-020-04771-1

Schmid, V., E. Mayer-Miebach, D. Behsnilian, K. Briviba, H. P. Karbstein, and M. A. Emin. "Enrichment of starch-based extruded cereals with chokeberry (*Aronia melanocarpa*) pomace: Influence of processing conditions on techno-functional and sensory related properties, dietary fibre and polyphenol content as well as in vitro digestibility." *LWT – Food Science and Technology* 154 (2022): 112610.1–10. https://doi.org/10.1016/j.lwt.2021.112610

Schoonees, A., M. J. Lombard, A. Musekiwa, E. Nel, and J. Volmink. "Ready-to-use therapeutic food (RUTF) for home-based nutritional rehabilitation of severe acute malnutrition in children from six months to five years of age." *Cochrane Database of Systematic Reviews* 5 (2019), 5–51. https://doi.org/10.1002/14651858.CD009000.pub3.

Sethi, S., S. K. Nanda, and M. Bala. "Quality assessment of pasta enriched with anthocyanin-rich black rice bran." *Journal of Food Processing and Preservation* 44, no. 12 (2020): e14952.1–12. https://doi.org/10.1111/jfpp.14952

Shilpa, H. S., G. K. Sadananda, G. J. Suresha, S. Manchali, S. M. Kumar, M. B. Darshan, and S. Hongal. "Development of ready-to-eat extruded snack fortified with dehydrated beetroot powder and Chekurmanis leaf powder." *Pharma Innovation Journal* 12, no. 2 (2023): 2213–2218.

Sinchaipanit, P., A. Sangsuriyawong, P. Visetchart, and N. P. Nirmal. "Formulation of ready-to-eat soup for the elderly: Nutritional composition and storage stability study." *Foods* 12, no. 8 (2023): 1680.1–13. https://doi.org/10.3390/foods12081680

Smith, J. D., N. Jain, and R. L. Bailey. "Ready-to-eat cereal fortification: A modelling study on the impact of changing ready-to-eat cereal fortification levels on population intake of nutrients." *Public Health Nutrition* 23, no. 12 (2020): 2165–2178. https://doi.org/10.1017/S1368980019003690

Smith, J. D., Y. Zhu, V. Vanage, N. Jain, N. Holschuh, and A. H. Agler. "Association between ready-to-eat cereal consumption and nutrient intake, nutritional adequacy, and diet quality among infants, toddlers, and children in the national health and nutrition examination survey 2015–2016." *Nutrients* 11, no. 9 (2019): 1989. 1–19. http://dx.doi.org/10.3390/nu11091989

Sui, X., Y. Zhang, and W. Zhou. "Bread fortified with anthocyanin-rich extract from black rice as nutraceutical sources: Its quality attributes and in vitro digestibility." *Food Chemistry* 196 (2016): 910–916.

Suri, S., A. Dutta, N. C. Shahi, R. S. Raghuvanshi, A. Singh, and C. S. Chopra. "Numerical optimization of process parameters of ready-to-eat (RTE) iron rich extruded snacks for anemic population." *LWT – Food Science and Technology* 134 (2020): 110164.1–11. https://doi.org/10.1016/j.lwt.2020.110164

Taylor, J. R. N., and M. N. Emmambux. "Products containing other speciality grains: sorghum, the millets and pseudocereals." In *Technology of Functional Cereal Products*, pp. 281–335. Woodhead Publishing, 2008. https://doi.org/10.1533/9781845693886.2.281

Tirloni, E., G. Centorotola, F. Pomilio, M. Torresi, C. Bernardi, and S. Stella. "*Listeria monocytogenes* in ready-to-eat (RTE) delicatessen foods: Prevalence, genomic characterization of isolates and growth potential." *International Journal of Food Microbiology* 410 (2024): 110515.1–10. https://doi.org/10.1016/j.ijfoodmicro.2023.110515

Venkatachalapathy, N., Yashaswini, J. P., and Kulbushan, S. (2020). Technology used in ready to eat. Food and beverage news, FNB News – Technology used in ready to eat | FNB News.

Vijayeta, P. "Purchasing practice of the consumers towards ready to eat food products." *Asian Journal of Home Science* 10, no. 2 (2015): 290–295. https://doi.org/10.15740/HAS/AJHS/10.2/290-295

Wahengbam, E. D., A. J. Das, B. D. Green, J. Shooter, and M. K. Hazarika. "Effect of iron and folic acid fortification on in vitro bioavailability and starch hydrolysis in ready-to-eat parboiled rice." *Food Chemistry* 292 (2019): 39–46. https://doi.org/10.1016/j.foodchem.2019.04.044

Xu, M., Huang, T., Lee, A.W., Qi, L., and Cho, S. "Ready-to-eat cereal consumption with total and cause-specific mortality: Prospective analysis of 367,442 individuals." *Journal of the American College of Nutrition* 35 (2016), 217–223.

Yakoob, M. Y., E. Theodoratou, A. Jabeen, A. Imdad, T. P. Eisele, J. Ferguson, A. Jhass et al. "Preventive zinc supplementation in developing countries: impact on mortality and morbidity due to diarrhea, pneumonia and malaria." *BMC Public Health* 11, no. 3 (2011): 1–10.

Yang, X., J. Huang, Q. Wu, J. Zhang, S. Yang, J. Wang, Y. Ding et al. "Occurrence, serovars and antibiotic resistance of *Salmonella* spp. in retail ready-to-eat food products in some Chinese provinces." *LWT – Food Science and Technology* 154 (2022): 112699.1–11. https://doi.org/10.1016/j.lwt.2021.112699

Index

I

J

K

L

M

N

O

P

R

S

T

U

V

Z

For Product Safety Concerns and Information please contact our EU representative GPSR@taylorandfrancis.com Taylor & Francis Verlag GmbH, Kaufingerstraße 24, 80331 München, Germany